AIR, WATER AND SOIL POLLUTION SCIENCE AND TECHNOLOGY SERIES

WATER PURIFICATION

AIR, WATER AND SOIL POLLUTION SCIENCE AND TECHNOLOGY SERIES

Trends in Air Pollution Research
James, V. Livingston (Editor)
2005. ISBN: 1-59454-326-7

Agriculture and Soil Pollution: New Research
James, V. Livingston (Editor)
2005. ISBN: 1-59454-310-0

Water Pollution: New Research
A.R. Burk (Editor)
2008. ISBN: 1-59454-393-3

Air Pollution: New Research
James, V. Livingston (Editor)
2007. ISBN: 1-59454-569-3

Air Pollution Research Advances
Corin G. Bodine (Editor)
2007. ISBN: 1-60021-806-7

Marine Pollution: New Research
Tobias N. Hofer (Editor)
2008. ISBN: 978-1-60456-242-2

Complementary Approaches for Using Ecotoxicity Data in Soil Pollution Evaluation
M. D. Fernandez and J. V. Tarazona
2008. ISBN: 978-1-60692-105-0

Complementary Approaches for Using Ecotoxicity Data in Soil Pollution Evaluation
(Online Book)
M. D. Fernandez and J. V. Tarazona
2008. ISBN: 978-1-60876-411-2

Lake Pollution Research Progress
Franko R. Miranda and Luc M. Bernard
2008. ISBN: 978-1-60692-106-7

Lake Pollution Research Progress
(Online Book)
Franko R. Miranda and Luc M. Bernard (Editors)
2008. ISBN: 978-1-60741-905-1

River Pollution Research Progress
Mattia N. Gallo and Marco H. Ferrari (Editors)
2009. ISBN: 978-1-60456-643-7

Heavy Metal Pollution
Samuel E. Brown and William C. Welton (Editors)
2008. ISBN: 978-1-60456-899-8

Cruise Ship Pollution
Oliver G. Krenshaw (Editor)
2009. ISBN: 978-1-60692-655-0

Industrial Pollution including Oil Spills
Harry Newbury and William De Lorne (Editors)
2009. ISBN: 978-1-60456-917-9

Environmental and Regional Air Pollution
Dean Gallo and Richard Mancini
2009. ISBN: 978-1-60692-893-6

Environmental and Regional Air Pollution
(Online Book)
Dean Gallo and Richard Mancini
2009. ISBN: 978-1-60876-553-9

Traffic Related Air Pollution and Internal Combustion Engines
Sergey Demidov and Jacques Bonnet (Editors)
2009. ISBN: 978-1-60741-145-1

Sludge: Types, Treatment Processes and Disposal
Richard E. Baily (Editor)
2009. ISBN: 978-1-60741-842-9

Water Purification
Nikolaj Gertsen and Linus Sønderby (Editors)
2009. ISBN: 978-1-60741-599-2

AIR, WATER AND SOIL POLLUTION SCIENCE AND TECHNOLOGY SERIES

WATER PURIFICATION

NIKOLAJ GERTSEN
AND
LINUS SONDERBY
EDITORS

Nova Science Publishers, Inc.
New York

LIBRARY OF CONGRESS CATALOGING-IN-PUBLICATION DATA

Water purification / editors, Nikolaj Gertsen and Linus Sønderby.
    p. cm.
 Includes bibliographical references.
 ISBN 978-1-60741-599-2 (hardcover)
 1. Water--Purification. I. Gertsen, Nikolaj. II. Sønderby, Linus.
 TD430.W3646 2009
 628.1'62--dc22
                                                                    2009024619

Published by Nova Science Publishers, Inc. ✦ *New York*

CONTENTS

PREFACE

Water purification is the process of removing undesirable chemical and biological contaminants from raw water. The goal is to produce water fit for a specific purpose. Most water is purified for human consumption (drinking water) but water purification may also be designed for a variety of other purposes, including meeting the requirements of medical, pharmacology, chemical and industrial applications. In general, the methods used include physical process such as filtration and sedimentation, biological processes such as slow sand filters or activated sludge, chemical process such as flocculation and chlorination and the use of electromagnetic radiation such as ultraviolet light. The purification process of water may reduce the concentration of particulate matter including suspended particles, parasites, bacteria, algae, viruses, fungi; and a range of dissolved and particulate material derived from the minerals that water may have made contacted after falling as rain. This new book gathers the latest research from around the globe in this field.

Chapter 1 - This chapter first describes the background information about the environmental concern caused by the inadequate release of several different toxic antropic substances in different water bodies. The environmental demand for the pollution-free waste discharge to different receiving water bodies is discussed and the development of integrated treatment process using the advanced chemical oxidation technology in combination with the well-established biochemical processes is introduced as a top challenge in order to provide an adequate pollution control in residuary waters. Ozone application aiming to provide removal of pathogenic agents and degradation of refractory pollutants is presented in order to emphasize its potentialities when compared to other oxidation processes based on application of chlorine and its derivates. An overview concerning the advanced chemical oxidation technologies based on ozonation and other advanced oxidation processes is presented, as well as some guidelines for application of these alternative processes in order to perform a rational analysis for the large-scale operation in water treatment facilities. Modern ozonation systems are presented and discussed in the light of the technological advances gained, mainly, in the last three decades. Emphasis is placed on ozonation devoted to drinking water treatment and several different aspects concerning the peroxone applications are discussed. The performance presented by different advanced oxidation technologies for the wastewater treatment is discussed taking into account target industrial wastewaters containing important refractory pollutants. Removal of pharmaceuticals present in water using ozonation is discussed placing emphasis on compounds classified as endocrine disruptors.

Chapter 2 - A newly developed ecosystem model - the first model describing the ecological connectivity consisting of both benthic-pelagic and central bay-tidal flat ecosystem coupling while simultaneously describing the vertical micro-scale in the benthic ecosystem - was developed and applied to Tokyo Bay (Sohma *et al.* 2005a, 2008). The model permits the prediction/evaluation of the effects of environmental measures, such as tidal flat creation/restoration, sand capping, dredging, and nutrient load reduction from rivers, on the hypoxic estuary from the perspectives of (1) the whole estuary composed of temporal-spatial mutual linkage of benthic-pelagic or central bay-tidal flat ecosystems (holistic approach), and (2) each biochemical and physical process contributing to oxygen production/consumption (elemental approach). The model outputs demonstrated the significant ecosystem responses as follows. First, the oxygen consumption in the benthic system during summer was quite low due to the low level of dissolved oxygen (hypoxia), although reduced substances, Mn^{2+}, Fe^{2+}, and S^{2-}, were highly produced and accumulated in the pore water. This result denotes importance to use the oxygen consumption rate under the high level of dissolved oxygen as the index of hypoxia potential. Second, both the tidal flat creation and nutrient load reduction decreased the anoxic water volume and mass of detritus in Tokyo Bay. However, the creation of tidal flats led to the higher biomass of benthic fauna, while the nutrient load reduction led to the lower biomass of benthic fauna compared to the existing situation. This result clarifies the differences from a measure aimed at a "bountiful ocean; a non-hypoxic and rich production of higher level trophic biology" to a measure just aimed at a "clear ocean; a non-hypoxic and low level of particulate organic matter" and also the differences from a bountiful ecosystem to a higher water quality. Lastly, in the simulation, Tokyo Bay reproducing reclaimed tidal flats (earlier Tokyo Bay system) prevented the increase of oxygen consumption potential (hypoxia potential) and the decrease of higher trophic production to red tide, compared to the existing Tokyo Bay system with reclamation of tidal flats. This result demonstrates the higher ecosystem tolerance of the earlier Tokyo Bay to red tide, and the tidal flat's function of keeping an optimized ecological balance to environmental perturbation.

Chapter 3 - Bank filtration is a relative low cost system for raw water treatment or pre-treatment for drinking water abstraction, used in European countries since about 140 years with very good experiences, but up to now knowledge of the physical, chemical and biological processes of water purification is still insufficient, especially under consideration of the application of this technology in other countries. Focus of interest are the mechanical, physicochemical, chemical and biological processes during infiltration pathway such as the retention of particulate organic material (POM), especially of algae cells with toxic cell compounds (cyanobacteria), the turnover of natural organic matter (NOM), bacteria and viruses, and the retention of toxicology relevant micro pollutants like cyanotoxins and drugs.

The water infiltration during bank filtration is not only controlled hydraulically but determined by severe clogging processes mainly triggered by accumulation of biological components in the upper sediment layer. Clogging of the interstice is regularly observed in infiltration ponds, and up to now several mechanisms can be distinguished like physical (input of fine sediments, building of gas bubbles), chemical (precipitation mainly of carbonates) and biological processes (excretion of extracellular substances by algae and bacteria). As a consequence water permeability of the interstice will become strictly reduced. The interstice are place of an adapted biocoenosis of bacteria, fungi, algae and meiofauna, which is characterized by the occurrence of extra-cellular polymeric substances (EPS). The meiofauna counteracts the clogging process by detritivorous activity. For many contaminants like DOM, POM, pathogens (*Giardia*, *Cryptosporidium*) and cyanobacteria as well as cyanotoxins a good removal is given. The active sediment layer is the first meter of infiltration pathway with a decrease in DOC concentration of up to 50 % and high removal rates of $10^2 - 10^4$ for pathogens like bacteria and protozoa.

Chapter 4 - The present review is focused on the diverse approaches employing chitin and/or chitosan as either active or passive components which have been employed to pesticide control in environmental protection and water purification. In some cases, ideas are presented in order to contribute in the solution of particular problems associated with this topic. The first section is dedicated to introduce the most important basic elements in the work as for example classification of the pesticides (according to their toxicological effects, chemical similarities and type of plague to control), description of the chitin and chitosan and some general applications of these biomaterials on removal of pollutants. Second section briefly analyzes some strategies employed in order to minimize the effects of the indiscriminate application of pesticide, through previous actions, as controlled release and protected dosing system. Third part is dedicated to discuss the diverse action modes which chitin or chitosan can be used for pesticide removal, including their uses in coagulation/flocculation process, filtration membranes and as adsorbents. Some applications of these materials as supports of pesticide-degradating agents are presented in the section four including enzymes micro-organisms and catalysts. Finally, a section dedicated to trends in this topic is presented. In general, the review shows that these materials posses high potentialities for use in pollutants removal. Applications as adsorbent appear to be the most promising at short term, especially to either anionic or acid pesticides, considering that perhaps chitosan is the unique natural polycation. Similarly, some proposal to produce new composite materials with applications on pesticide removal, using processes involving chitin and chitosan are presented.

Chapter 5- Surfaces of Rutile TiO_2 particles have been modified with CdS particles. The TiO_2/CdS system has been used as catalyst in water purification by photo-degradation of organic contaminants such as methyl orange (a commonly encountered contaminant dye). Both UV and visible ranges have been investigated. CdS sensitization of TiO_2 to visible region has been observed, as the TiO_2/CdS system showed higher catalytic efficiency than the naked TiO_2 system in the visible region. The TiO_2/CdS system was unstable under neutral, acidic conditions and basic conditions. Leaching out, of CdS into hazardous aqueous Cd^{2+} ions, while working at pH 7 or lower occurred. This imposes limitations on future usage of CdS-sensitized TiO_2 photo-catalytic systems in water purification processes.

In an effort to solve out the leaching difficulties, and to make catalyst recovery easier, the TiO_2/CdS system has been supported onto insoluble silica particles giving Silica/TiO_2/CdS systems for the first time. The silica/TiO_2/CdS system showed lower efficiency than TiO_2 and TiO_2/CdS systems in UV regions. In the visible region, the Silica/TiO_2/CdS was less efficient than the TiO_2/CdS but more efficient than naked TiO_2. The silica support has an added application value of making catalyst recovery much easier, after reaction completion. Unfortunately the difficulty of the Cd^{2+} ion leaching out has been solved out partly only in basic media. Pre-annealing of the catalyst systems did not give significant effect on stability. Despite the numerous literature reports, on using CdS as sensitizer in degradation studies, its tendency to leach out puts a limitation on its future usage. Should this tendency not be solved out completely, replacement with other more safe dyes should be considered. Effects of catalyst concentration, catalyst recovery, contaminant concentration, temperature and pH, on catalyst efficiency, have also been studied.

Chapter 6 - ZnO is a wide band gap (3.2 eV) semiconductor, with limited photo-catalytic applications to shorter wavelengths only. However, it is suitable to use in solar light photo-degradation of different contaminants, due to a number of reasons, taking into account that the reaching-in solar radiation contains only a tail in the near UV region. The high absorptivity of ZnO makes it efficient photo-catalyst under direct solar light. Moreover, it is relatively safe, abundant and non costly. In this chapter, ZnO has been investigated as a potential catalyst for photo-degradation of methyl orange (a known dye) in aqueous solutions with direct natural solar light under different conditions. The major aim was to assess the efficiency and stability of ZnO under photo-electrochemical (PEC) conditions, and to suggest techniques to enhance such features. This will shed light on the future applicability of ZnO as a candidate for economic and friendly processes in water purification.

Recovery of ZnO particles, after reaction completion, has been facilitated by supporting ZnO onto activated carbon, to yield AC/ZnO system. The AC/ZnO was used as catalyst for contaminant photo-degradation in water solutions under direct solar light.

Both catalytic systems, naked ZnO and AC/ZnO, were highly efficient in degrading both contaminants, reaching complete removal in reasonable times. The latter system showed higher efficiency. In both systems, the reaction goes faster with higher catalyst loading, until a maximum efficiency is reached at a certain concentration, after which the catalyst concentration did not show a systematic effect.

In both catalytic systems, the rate of degradation reaction increases with higher contaminant concentrations until a certain limit is used. The contaminant degradation reaction was studied, using both catalysts, at different pH values. The pH value 8.0 gave the highest catalyst efficiency. The tendency of naked ZnO to degrade into soluble zinc ions, under photo-degradation experiments, was studied under different pH values. Catalyst recovery and reuse experiments were conducted on both systems. The catalytic activity of the recovered systems was only slightly lower than the fresh system in each case. The fourth time recovered catalysts showed up to 50% efficiency loss in each case, presumably due to ZnO degradation and leaching out. However, fresh and recovered catalyst systems caused complete degradation of contaminants after enough time. Temperature showed a slight effect on rate of reaction, with immeasurably small activation energy value. Details of effects of other parameters on reaction rate and catalyst efficiency are described. Using CdS as sensitizing dye failed to enhance ZnO efficiency under direct solar light. The screening effect and tendency of CdS to leach out limit its use as ZnO sensitizer. Tendency of ZnO to leach out zinc ions into solution is discussed. The naked ZnO and AC/ZnO systems are promising photo-catalysts in future water purification technologies by direct solar light.

Chapter 7 - A water purification process is integrated into a compression and an absorption heat pump with energy recycled. In the compressor heat pump, the heat delivered in the condenser was recycled to the heat pump evaporator and the excess heat to balance the heat pump was used to preheat the impure water supply. The advantage of this system is that it does not use waste heat to operate—because the energy recycling happens at the boiling water temperature. In the absorption heat pump also a fraction of the heat obtained by the heat pump absorber and condenser is recycled to the evaporator by the integration of the auxiliary condenser of the water purification process. This system also has the advantage that it does not use waste heat to operate because of the energy recycling. A developed thermodynamic model shows that the proposed water purification process integrated into a heat transformer is capable of increasing the original COP values more than 120% due to the energy recycling from the water purification process. Therefore, the proposed integration of the water purification process to the heat pump allows increasing the actual COP values with any working fluid. In addition, a mathematical model based in the neural network is proposed to obtain the optimum COP value considering the energy recycling of the auxiliary condenser in the water purification process.

Chapter 8 – While explaining vapor pressure lowering in solutions, many general scholars mistakenly depend on rate of evaporation lowering, or on relative values of attraction forces between solvent and solute molecules. This is in spite of the correct explanations based on thermodynamics, as presented in physical chemistry references. Unfortunately, there seems to be no qualitative model available so far to explain vapor pressure lowering. The misconception is escalating, and an end needs to be made. A qualitative understandable model, to explain vapor pressure lowering, at least in solutions of nonvolatile solutes in volatile solvents, is needed. For this purpose, the authors propose here a *new qualitative model*, to explain vapor pressure lowering in such solutions. The new model is based on purely old thermodynamic concepts.

Chapter 9 - The problem of quality water improvement can be solved through phased water purification by electrolytic processing and then adding plant medicinal substances of an antibacterial action. Due to a low-cost electrochemical activation of running water the salts of heavy metals and chlorine-containing compounds disintegrate. The essences of well-known medical plants, extracted in an unusual way, improve taste and hygienic properties of water.

Chapter 10 - Phototrophic and chemotrophic microorganisms and their consortia with water plants (aquatic fern *Azolla*, water hyacinth *Eichonia cassipens*) are capable of accumulating metal ions of Ni, Pt, Ru, Cu, Cr, Pb, Zn, Si, Au and other pollutants. This ability enables using them for purification of agricultural and industrial waste water from toxic heavy metals, hydrocarbon and to obtain rare trace metals. For this purpose it is possible to use growing cultures and immobilized cells or enzyme hydrogenase. Purple bacteria (*Rhodobacter* spp., *Rhodopseudomonas* spp.) are able to accumulate Cu, Zn, Ni and Hg., showing various resistance to these metals. Green algae *Chlorella* spp. and water fern *Azolla*, duckweed Lemna spp. and water hyacinth showed higher ability for biosorption of metal ions.

In: Water Purification   ISBN 978-1-60741-599-2
Editors: N. Gertsen and L. Sønderby   © 2009 Nova Science Publishers, Inc.

Chapter 1

ADVANCED TECHNOLOGIES BASED ON OZONATION FOR WATER TREATMENT

Leonardo Morais Da Silva, Débora Vilela Franco, Ismael Carneiro Gonçalves and Lindomar Gomes de Sousa

Federal University of Jequitinhonha and Mucuri's Valley – UFVJM, Rua da Glória 187, 39100-000, Diamantina, Brazil

ABSTRACT

This chapter first describes the background information about the environmental concern caused by the inadequate release of several different toxic antropic substances in different water bodies. The environmental demand for the pollution-free waste discharge to different receiving water bodies is discussed and the development of integrated treatment process using the advanced chemical oxidation technology in combination with the well-established biochemical processes is introduced as a top challenge in order to provide an adequate pollution control in residuary waters. Ozone application aiming to provide removal of pathogenic agents and degradation of refractory pollutants is presented in order to emphasize its potentialities when compared to other oxidation processes based on application of chlorine and its derivates. An overview concerning the advanced chemical oxidation technologies based on ozonation and other advanced oxidation processes is presented, as well as some guidelines for application of these alternative processes in order to perform a rational analysis for the large-scale operation in water treatment facilities. Modern ozonation systems are presented and discussed in the light of the technological advances gained, mainly, in the last three decades. Emphasis is placed on ozonation devoted to drinking water treatment and several different aspects concerning the peroxone applications are discussed. The performance presented by different advanced oxidation technologies for the wastewater treatment is discussed taking into account target industrial wastewaters containing important refractory pollutants. Removal of pharmaceuticals present in water using ozonation is discussed placing emphasis on compounds classified as endocrine disruptors.

ABBREVIATIONS

BAT	Best available technology
ITP	Integrated treatment processes
DBPs	Disinfection byproducts
THMs	Trihalomethanes
THMFP	THM-formation potential
HAAs	Haloacetic acids
MIB	Methylisobornel
ED	Endocrine disruptors
PIE	Pharmaceuticals in the environment
APIs	Active pharmaceutical ingredients
STP	Sewage treatment plants
PCBs	Polychlorinated biphenyls
NOM	Natural organic matter
AOX	Absorbable organic halogens
USEPA	U.S. Environmental Protection Agency
MCL	Maximum concentration levels
TOC	Total organic carbon
COD	Chemical oxygen demand
BOD	Biochemical oxygen demand
BOD_5	5-day biochemical oxygen demand
BDOC	Biodegradable dissolved organic carbon
VOCs	Volatile organic compounds
AOT	Advanced oxidation technologies
AOP	Advanced oxidation processes
O_3/H_2O_2	Peroxone
EOP	Electrochemical ozone production
TCF	Totally chlorine free bleaching technology
SPE	Solid polymer electrolyte
MEA	Membrane electrode assembly
RZ	Reaction zone
CISTR	Continuous ideally stirred tank reactor
Ha	Hatta number
E_i	Instantaneous enhancement factor
E	Enhancement factor
$k_L a$	Volumetric mass transfer coefficient
UV	Ultraviolet radiation
UV_{254}	Ultraviolet radiation at 254 nm
RY 143	Reactive Yellow 143
RB 264	Reactive Blue 264
RB 5	Reactive Black 5
IDC	Initial dye concentration
TPCBP	Transient persistent colored byproducts

1. WATER POLLUTION

Clean water is so basic to human life that water droplets, bubbling brooks, and waterfalls are enduring symbols of the life force [1,2]. Obtaining an adequate supply of clean water has likely always been a challenge for much of humanity [1]. Despite the scientific and technological advancements of the modern society and, ironically, sometimes because of them, clean water is becoming an increasingly scarce and coveted resource [1-11]. From these considerations, one has that water security is now a critical environmental issue that touches the life of every human being [4].

The increasing world population with growing industrial demands has led to a situation where protection of the environment has become a major issue and a crucial factor for several industrial processes, which will have to meet the requirements of the sustainable development [5-7].

The rapid expansion of chemical industry since William Henry Perkin discovered and commercialized in 1856-1857 the first synthetic dye, mauveine, has resulted in the potential for release to the environment of approximately 80,000 xenobiotic compounds that are not natural components of the organisms exposed to them [1,2,5,6,12]. These "alien" chemical compounds have worked their way into our lives, usually as uninvited guests and too often with deleterious effects, from a variety of technology sources, including pesticides, personal care products, cleaning materials, building materials, houseware, laboratory work, pharmaceuticals, etc., which become "essential" to every sector of our chemically based economy [4,12].

Importantly, even in highly developed countries, we are only a few generations or less into widespread public exposures for many individual xenobiotics, including several different pharmaceuticals. New nanotech materials, biologics, genetic therapies, and genetically modified foods are more recent newcomers to this anthropogenic "chemical soup" [4]. The addition to water of a cocktail of trace quantities of pharmaceuticals, compounds that are designed to exhibit potent physiological activity, is arguably an important emerging water issue [4,13,14].

Growing intensification of land and water use for industry and agriculture has increased the need to reclaim wastewater for reuse, including supplementing the drinking water supply [1-3]. Concurrently this has increased the risk of water resource contamination. The contaminants can leach from the contaminated watercourses into the groundwater aquifers and appear at trace-level concentrations in drinking water. Because of their polar structure, several pharmaceuticals are not significantly adsorbed in the subsoil, thus reaching the groundwaters, which constitute a major source of drinking water [1,2].

In order to combine industrial activities with preservation of the environment (sustainable development), there is a tendency in most countries to adopt rigid environmental legislation where the well-known Green Technology plays a key role [5]. In this context, chemical processes, where the Best Available Technology (BAT) not entailing excessive cost and aspiring to performance without considerable impact, are preferred in substitution of the classical ones [5,12].

This environmentally friendly strategy minimizes (or even eliminates) the use or generation of hazardous substances in the design, manufacture, and application of chemical products [5,8,12]. Approaches and principles of Green Chemistry were presented by Anastas and Williamson [12], who discussed the terms environmentally benign chemical syntheses, alternative pathways to prevent pollution, and benign chemistry.

2. WATER TREATMENT: ELIMINATION OF PATHOGENICS AND HAZARDOUS CHEMICALS

The quality of drinking water has been an important concern for mankind. For instance, drinking water polluted by fecal wastes and hazardous chemicals is very dangerous since it constitutes a potential source for several waterborne diseases and mutagenic processes [1-5,9].

Therefore, water pollution control and the development of clean technologies for water purification comprise a very important issue for our society. In fact, in industrialized countries improvements have been achieved in water treatment technologies, which led to a near eradication of acute health hazards caused by waterborne diseases [2,5,9]. Thus, adoption of clean technological processes and water reuse for several industrial activities are imperative issues for our society in order to ensure preservation of water resources [5].

Nowadays, the occurrence of epidemics caused by infectious agents present in drinking water is occasional in most countries. Waterborne diseases are now well controlled by virtue of the use of disinfectant agents (e.g. chlorine and ozone) [9]. Unfortunately, the control of occurrence in water and wastewater of the potential hazardous chemicals constitutes a very difficult task from the applied point of view [3,5,8,11].

The antropic occurrence of the refractory hazardous chemicals in water bodies is a major concern for the population in different countries. Since the initial of the past century the industrialized society is mainly marked by a considerable growth in the manufacture and the use of several different synthetic chemicals. In addition, the activities pertaining to our metal based civilization has provoked the antropic release in the environment of different heavy metals.

These pollutants can be found in both surface and groundwater sources (rivers, lakes and streams) [1-3]. The main contamination sources concerning organics include wastes from industrial chemical production, textile industry, tannery and pulp bleaching processes and pesticide runoff from agricultural activities [1-3,5,9]. Major water contaminants include several organics (e.g. chlorinated hydrocarbons, endocrine disruptors, synthetic dyes, pesticides, etc.) and inorganics (e.g. heavy metals, cyanide, etc.) species.

The widespread use of inefficient technologies for the wastewater treatment has resulted in concentration of the pollutants in facilities as landfills, storage lagoons, treating ponds, etc., thus creating a potential source of contamination for groundwater and soil [1-3,5-7]. Thus, the development of more efficient technologies for the wastewater treatment comprises a great challenge for our society in order to ensure safety environment conditions for future generations. Based on these considerations, one can remark the understanding of the different pollutant sources, as well their interactions with the environment, is imperative in order to provide the pollution control under acceptable environmental conditions.

In the context of the classical treatment technologies several pollutants, particularly those which are oxygen-demanding substances (e.g. oil, grease, etc.), are removed by primary and secondary treatment processes [1-3]. However, in most of cases other pollutants such as salts, heavy metals and refractory (recalcitrant) organics remain unchanged in the waste after the treatment [3,5,8].

Conventional wastewater treatment facilities are basically constituted by next stages [2,3,15]:

(1) *Primary Waste Treatment* - removal from water of insoluble matter such as suspended solids, grease, etc. Solids are collected on screens and separated for subsequent disposal;

(2) *Secondary Waste Treatment* - removal of Biochemical Oxygen Demand (BOD) to an acceptable level usually by taking advantage of the same kind of biological processes; the action of microorganisms provided with added oxygen leading to degradation of the organic matter present in solution or in suspension. The main processes used in this case are the fixed film systems (trickling bed filters) and suspended growth systems (activated sludge processes);

(3) *Tertiary Waste Treatment* (or *Advanced Waste Treatment*) – processes carried out in order to provide disinfection and an additional treatment of the secondary waste treatment. The major target contaminants are pathogenic agents and the substances classified as dissolved organic compounds and dissolved inorganic materials.

The inadequate disposal of these "treated' effluents can result in several environmental problems. For instance, municipal treated sewage typically contains about 0.1% of solids after the treatment, which settle out in the downstream water [2]. So, when these solids are deposited in the ocean floor, aggregation takes place assisted by dissolved salts present in seawater, thus promoting the formation of sludge-containing sediment.

Millions of tons of organic compounds are manufactured in the world per year and a great part of these species are constantly released in the environment [1-3]. Most of the synthetic compounds, specially those classified as refractory, are non environmentally friendly substances to which the natural living organisms present in water and soil have not yet been exposed for a long time. Therefore, their effects upon these organisms are not known, particularly for long-term exposures at very low levels [4].

Recalcitrant organics are compounds of most concern in water and wastewater since these substances are poorly biodegradable by natural living organisms. Table 1 shows some biorefractory (recalcitrant) organics frequently found in water and wastewaters.

Considering that application of the biochemical (biological) process alone (e.g. activated sludge) does not constitute an efficient technology for removal of recalcitrant compounds, one has that water contaminated with these compounds must be treated using Integrated Treatment Processes (ITP), which are based on combination of physical and chemical processes such as air stripping, solvent extraction, carbon adsorption, ozonation and Advanced Oxidation Processes (AOP) (e.g. O_3/UV, $O_3/UV/H_2O_2$, etc.) [5,8]. According to this technological approach the complexicity of the organic matter is reduced in order to facilitate the further treatment using an specific biological process.

Table 1. Biorefractory organics frequently found in water and wastewaters

Benzene	Dichloroethyl ether	Styrene
Bromobenzene	Dinitrotoluene	Tetrachloroethylene
Bromochlorobenzene	Ethylbenzene	Trichloroethane
Butylbenzene	Ethylene dichloride	Toluene
Chloroform	2-ethylhexanol	1,2-dimethoxybenzene
Chloromethylethyl	Isocyanic acid	Endocrine Disruptors
ether	Isopropylbenzene	Textile dyes
Chloronitrobenzene	Methylbiphenyl	Pharmaceuticals
Chloropyridine	Methyl chloride	
Cibromobenzene	Nitrobenzene	
Dichlorobenzene		

3. DRINKING WATER TREATMENT: CHLORINE VS. OZONE

Chlorination may be performed using chlorine or other chlorinated compound which can be applied in liquid or solid form [3,9]. Since chlorine and its derivates are stable compounds, these oxidant species are commonly produced "off site". Chlorination is mainly applied in water treatment facilities in order to provide disinfection. The general trend of increasing chlorine disinfection difficulty is bacteria < viruses < protozoa [1-3,9,15].

Chlorine, Cl_2, is frequently added to water at levels below 1.0 g dm^{-3} [15]. When chlorine is added to water, it rapidly hydrolyzes producing hypochlorous acid, HOCl, which is a weak acid that dissociates forming the anion OCl^-. The concentration of Cl_2 is negligible under equilibrium conditions at pH > 3 when chlorine is added at levels below 1.0 g dm^{-3} [2]. Sometimes, hypochlorite salts are substituted for chlorine gas as a disinfectant, since hypochlorites are safer to handle than gaseous chlorine; calcium hypochlorite, $Ca(OCl)_2$, is commonly used in this case [1-3].

The two chemical species formed by chlorine in water, HOCl and OCl^-, are known as *free available chlorine*, which is very effective in killing bacteria. In the presence of ammonia the formation of monochloramine, dichloramine, and trichloramine takes place. The chloramines are called *combined available chlorine*, which are species more readily retained as a disinfectant throughout the water distribution system [2,3].

However, high concentrations of NH_3 in water are undesirable since it requires a higher demand for Cl_2. The Cl:N molar ratio at which a small amount of HOCl and OCl^- remain unreacted in solution resulting in formation of NCl_3 is denoted the breakpoint for chlorination process [2].

Despite of the benefits presented by chlorination, one has that several problems can arise due to formation of potential carcinogenic byproducts. These hazardous compounds known as Disinfection Byproducts (DBPs) are formed during chlorination of the Natural Organic Matter (NOM) and different organics present in water [1-3,5].

During chlorination in water treatment facilities free chlorine is normally applied directly as a primary disinfectant [15]. So, when added to the water, free chlorine reacts with NOM and bromide to form DBPs, primarily Trihalomethanes (THMs), and some Haloacetic Acids (HAAs) [3,9]. Factors affecting the formation of these halogenated DBPs include type and concentration of organic compounds, chlorine form and dose, time, bromide ion concentration, pH, organic nitrogen concentration, and temperature. Organic nitrogen significantly influenced the formation of nitrogen containing DBPs, including haloacetonitriles, halopicrines, and cyanogen halides [16,17].

Chlorine dioxide, ClO_2, is a disinfectant which in the absence of Cl_2 does not lead to formation of DBPs [1-3,5]. Despite of this fact, ClO_2 does not react with ammonia or other nitrogen-containing compounds in order to ensure a residual disinfectant power in the water distribution system. Besides, some concern has been raised over possible health effects during ClO_2 applications due to formation of ClO_3 [8,9,11].

In principle, DBPs could be controlled by four different ways: (1) removing the DBP precursors; (2) modifying the chlorination strategy; (3) changing disinfectants, or (4) removing the DBPs itself. However, since DBPs are difficult to remove, control strategies are mainly based on the first three procedures.

The elimination of DPBs from water through alternative methods of disinfection is strongly desirable. So, the use of ozonation and other alternative AOP ozone-based systems in order to provide disinfection indeed comprises an environmentally friendly strategy for treatment of drinking water [5].

Since the discovery of ozone by Schönbein in 1840, the ozonation process has proved to be efficient for the water purification [5,18-20]. The first experiment on water disinfection via ozonation was carried out by De Meritens in 1886. Since then, there has been a great interest in the use of ozone as a primary oxidant in replacement of chlorine for water disinfection in order to eliminate/minimize the formation of DBPs [5,9].

The use of ozone in water purification in a great scale was first proposed by Werner Siemens in 1889, but it took about ten years to actually employ ozone in order to provide a pure water supply for cities and villages surrounded by contaminated areas [20]. Ozone was first used for drinking water treatment in 1893 in the Netherlands. While being used frequently in Europe for drinking water disinfection and oxidation, it was slow to transfer to the United States [9].

Recent studies revealed that ozone is one of the most potent and effective germicide and viricide for water treatment [1-3,9,20]. Ozonation process presents the following advantages: (i) it is effective against bacteria, viruses, and protozoan cysts; (ii) inactivation efficiency for bacteria and viruses is not considerable affected in the 6 to 9 pH range, and (iii) ozone disinfection efficiency increases on increasing water temperature.

Ozone can be also used to remove taste, odor, and color caused by different substances present in water. Literature reports concerning water treatment [5,21] revealed that ozone loads of 2.5 to 2.7 mg dm^{-3} and 10 min of contact time (ozone residual of 0.2 mg dm^{-3}) significantly reduced taste and odors.

Ozone is sparingly soluble in water (12.07 mg dm^{-3} at 20 °C) and, therefore, the effective ozone concentration used in water treatment limits the mass transfer driving force of gaseous ozone into the water [3,22]. Consequently, typical concentrations of ozone found during water treatment range from < 0.1 to 1.0 mg dm^{-3}, although higher concentrations can be attained under optimum conditions [18,20].

Ozone is capable to provide disinfection with less contact time and concentration than all weaker disinfectants, such as chlorine, chlorine dioxide, and monochloramine [23].

According to the U.S. Environmental Protection Agency (USEPA), ozone is the most powerful disinfectant available [9].

Inactivation of bacteria and virus by ozone is attributed to an oxidation reaction [24,25]. The first site to be attacked appears to be the bacterial membrane. In addition, ozone disrupts enzymatic activity of bacteria by acting on the sulfhydryl groups of certain enzymes. Beyond the cell membrane and cell wall, ozone may act on the nuclear material within the cell [9]. In the case of virus the first site of action leading to virus inactivation is the virion capsid, which is used for virus fixation on cell surface.

It is important to mention that ozone can only be used as a primary disinfectant since it cannot maintain a residual effect in the distribution system [15]. Therefore, ozone disinfection should be coupled with a secondary disinfectant in order to ensure disinfection in the entire distribution system.

Two separate disinfectants can be used to provide primary and secondary disinfection. By separating the primary and secondary disinfection functions, the processes can be optimized for maximum inactivation and minimum DBPs formation. Interactive disinfection (using synergism between two disinfectants to enhance inactivation) can serve as a primary disinfectant. DBPs formation is in general reduced by using combined disinfectants. Specifically, continued use of chlorine in combination with other disinfectants can reduce DBPs formation [3,9].

The principal benefit of using ozone for controlling DBPs formation is that it allows free chlorine to be applied later in the treatment process after precursors have been removed and at lower doses, thereby reducing DBPs. Besides, water treatment technologies based on the ozonation process present the additional advantage of removing heavy metals and phosphate [1-3,5].

According to the literature [2,5,9], the main concern associated with ozone application in drinking water production is the presence of the bromide ion. Thus, to attain the levels required by environmental legislation both O_3 dose and residence time during water treatment should be previously evaluated in cases where raw water contains bromide ions in order to maintain the bromate, a potential carcinogen, formed during ozonation at concentrations lower than 10 mg dm^{-3}. Early work on oxidation of DBPs precursors seemed to indicate that the effects of ozonation, prior to chlorination, were quite site-specific and, therefore, this issue deserves a special attention for each particular case [26].

The major variables influencing the efficiency of the ozonation process are O_3-dose, alkalinity, pH, and, mainly, the nature of the organic matter present in water. At low pH values, oxidation via molecular ozone is quite effective. However, above some critical pH value the ozonation process is less effective. For instance, during oxidation of most humic substances the critical pH is 7.5, which is the approximate pH-value at which decomposition of ozone to hydroxyl free radicals, $HO^{\bullet}$, increases rapidly, thus increasing organic oxidation rates [18-20].

As bicarbonate alkalinity increases, it has a beneficial effect on THM-Formation Potential (THMFP) during ozonation [27,28]. This is because alkalinity scavenges any hydroxyl free radical formed during ozonation, leaving molecular ozone as the sole oxidant, which is only capable of oxidizing organic precursors to a lower oxidation sequence than does the hydroxyl free radical. Given neutral pH and moderate levels of bicarbonate alkalinity, THMFP level reductions of 3 to 20% have been shown at ozone doses ranging from 0.2 to 1.6 mg O_3 per mg carbon [29].

Characterization of raw waters for the ozone application was reported by Park et al. [30], who investigated the effects of particulates, ozone load, and sequential ozone injection. This study confirmed ozonation is a very efficient technology for drinking water treatment.

The ozonation kinetics plays a key role in the feasibility of treating organic compounds present in water. Since the work of Hoigné et al. [31-34] rate constants for several different compounds in aqueous media have been measured [35-40]. Differences related with the direct (via molecular ozone) or indirect (via $HO^{\bullet}$ radical) pathway of the ozonation process were presented by Yao and Haag [37], who reported the direct ozonation kinetics for several contaminants whose Maximum Concentration Levels (MCL) have been established by USEPA.

The differentiation between the two oxidative pathways taking place during ozonation is important since it permits evaluation of the nature of the product types in different natural waters, and application of the rate constants to other systems, including biological media [5,8,9]. In general, compounds presenting pseudo second order kinetic rate constant higher than 100 L mol^{-1} s^{-1} will be consumed significantly by either direct ozonation and via indirect processes, while compounds presenting a strong recalcitrant nature ($k \ll 100$ L mol^{-1} s^{-1}) will be destroyed mainly by the hydroxyl radical [18-20].

4. WATER RECYCLING AND THE ADVANCED OXIDATION TECHNOLOGIES FOR THE WASTEWATER TREATMENT

The water recycling process is a growing need for a variety of industries faced with increasing water costs and environmental constraints [1-3]. Several industrial finishing activities (e.g. printed circuit board, the dying process, wood-pulp bleaching, etc.) generate large volumes of rinse water contaminated with several different inorganic and organic compounds [3,5,41-55].

Physical separation of suspended solids, oils and greases followed by a biological treatment have been shown to be very economical and successful in the cases of municipal wastewater, food and farm processing water [1-3,15]. However, these classical wastewater technologies based upon phase transfer merely concentrate the pollutants, which then have to be eliminated in a further treatment [3,5,8].

In fact, there are several cases in which the effectiveness of these classical wastewater treatments drops away (e.g. effluents containing recalcitrant compounds and toxic substances for biological treatment) and, therefore, Advanced Oxidation Technologies (AOT) must be used as a coadjutant in order to perform a more efficient final treatment of these effluents [42-50].

AOT is considered as a complete technology for the degradation of several organics in order to meet the regulations for toxicity as well as Total Organic Carbon (TOC) reduction, thus comprising one of the few processes which can destroy hazardous and toxic recalcitrant organics on site [8,18-20,56]. AOT can be more beneficial than other available treatment technologies that act by shifting the problem (hauling and landfilling) or transferring the problem to another receptor (air stripping) [5].

An ideal environmentally acceptable oxidant used in AOT-systems should possess the following features [5,12]: (i) reactivity toward the target compound; (ii) neither produce nor leave undesirable byproducts during the course of the reaction; (iii) readily available, and (iv) reasonably inexpensive to purchase. In the light of these considerations, the most commonly used oxidants are ozone, hydrogen peroxide, peroxone (O_3/H_2O_2), chlorine dioxide, and potassium permanganate.

The degree to which recalcitrant organics are degraded during chemical oxidation process can be ascribed as [5,54,57,58]: (i) *Primary Degradation*, where a structural change in the parent compound occurs, thus allowing it to be eliminated more easily by other processes (e.g. activated sludge process); (ii) *Acceptable Degradation*, which involves the decomposition of the parent compound to the extent that its toxicity is reduced; (iii) *Ultimate Degradation*, the last step comprising the mineralization of the organic compound (its conversion to simple inorganic substances such as CO_2 and H_2O).

Taking into account the wastewater characteristics, final requirements, and economical aspects, some AOT based on combination of chemical oxidation and some biological process are more suitable than others for the wastewater treatment [54]. New environmental trends accounting for pollution control in water are mainly based on the so-called Integrated Treatment Processes (ITP), which comprises, for the most of cases, the application of chemical oxidation as a pre-treatment step followed by classical biological processes (e.g. activated sludge) in order to provide transformation of recalcitrant pollutants to more easily biodegradable intermediates [4,5,12,54,56].

Combination of chemical and biological processes for wastewater treatment can indeed result in a cheaper option for toxic wastewater treatment, where each process is affected by some intrinsic effluent characteristic [54]. During the chemical oxidation step the total amount of recalcitrant organics, pH, and temperature will affect the reaction rate, while the biological processes will be mainly affected by pH, occurrence of toxic substances, temperature, redox potential and dissolved oxygen.

Therefore, the use of ITP for pollution control in water can lead to different combinations, which can be arranged for the particular effluent and for each of them there is an optimum set-up condition. Besides, changes in composition of the inlet effluent are easily resolved changing parameters in each of the stages, thus permitting a broad set of possibilities and a flexible design for the whole treatment process [5].

The performance of an AOT devoted to wastewater reclamation of secondary municipal effluents and the study of the environmental quality of treated effluents were studied by Petala et al. [59]. In this particular case, the secondary effluent obtained from a conventional activated sludge process was fed to an AOT based on the sequence combination: *filtration* (moving-bed sand filter) → *adsorption* (granular activated carbon) → *disinfection/oxidation* (ozonation process). The efficiency of the AOT plant was measured by various physicochemical, microbiological and ecotoxicological parameters and the operation capacity of each process was evaluated for the removal of specific pollutants [59].

Analysis of these experimental findings revealed the continuous up-flow sand filtration unit resulted in the removal of suspended solids and the reduction of turbidity (45%) and organic matter, while the dissolved phosphorus and ammonium nitrogen concentrations were not substantially affected. Besides, it was found the activated carbon column was effective for the removal of organic compounds and turbidity; the effluent from the carbon adsorption column had TOC values lower than 10 mg dm^{-3}. In this case the effect of ozonation on organic matter removal was studied by the application of different ozone doses, between 7.1

and 26.7 mg dm^{-3}. Further removal of turbidity, up to 80%, and reduction of UV$_{254}$ absorption to 60% were observed at the highest ozone dosage.

According to Petala et al. [59] ozonation was very effective for the removal of microorganisms, since the complete removal of pathogenic microorganisms was verified. The cost of reclaimed water was estimated to 0.24 € m^{-3}; annual costs included construction cost, depreciation, operation and maintenance expenses and the monitoring cost of effluent quality. Power consumption, consumables and labor expenses, contributed mainly to the reclaimed water cost.

5. Ozone Technology for Water Treatment: Trends and Strategies

Nowadays steps are being taken, mainly due to increasing economic, social, legal, and environmental pressures, to avoid further degradation of the environment [5-7,54,59-61]. Therefore, there is an increasing demand for alternatives technologies for treatment of several different wastewaters.

In the special case of the wastewater treatment several environment advantages can be achieved based on the concept of ITP, where ozone or peroxone can be used in order to provide the major oxidation process accounted for removal of the recalcitrant organic matter. For instance, the substitution of chlorine dioxide by ozone as a bleaching agent permitted the development of the so-called Totally Chlorine Free (TCF) bleaching technology, thus reducing significantly the pollution burden and potential health hazard of chlorine derivatives of this industrial activity [3,5,60,61].

In the case of drinking water treatment, one has that treatment of water sources containing significant bromide ion concentrations, site-specific engineering evaluations should be completed prior to using ozone as a primary disinfectant due to the potential formation of bromate ion or brominated organic byproducts. In systems containing negligible concentrations of bromide ion, replacement of pre-chlorination with ozonation may allow for the point of chlorine addition to be moved beyond clarification or filtration [3,5].

An operational consideration of the ozone/chlorine system is the application point for chlorine; the most promising treatment strategy for preventing the enhancement of DBPs during ozonation is to locate ozonation after sedimentation and follow it by biologically active filtration.

Once ozone enters into water, it becomes highly unstable and rapidly decomposes through a complex series of reactions [18-20]. Ozone chemistry in water is very important since the O$_3$ molecule decomposes in water as follows [5,27,45]:

$$O_3 + HO^- \rightarrow HO_2^{\bullet} + O_2^{-\bullet} \tag{I}$$
$$HO_2^{\bullet} \leftrightarrow O_2^{-\bullet} + H^+ \tag{II}$$
$$O_3 + O_2^{-\bullet} \rightarrow O_3^{-\bullet} + O_2 \tag{III}$$
$$O_3^{-\bullet} + H^+ \rightarrow HO_3^{\bullet} \tag{IV}$$
$$HO_3^{\bullet} \rightarrow HO^{\bullet} + O_2 \tag{V}$$
$$HO^{\bullet} + O_3 \rightarrow HO_4^{\bullet} \tag{VI}$$

$$HO_4^{\bullet} \rightarrow HO_2^{\bullet} + O_2^{\bullet} \quad \text{(VII)}$$
$$HO_4^{\bullet} + HO_4^{\bullet} \rightarrow H_2O_2 + 2O_3 \quad \text{(VIII)}$$
$$HO_4^{\bullet} + HO_3^{\bullet} \rightarrow H_2O_2 + O_3 + O_2 \quad \text{(IX)}$$

According to this mechanism hydroxide ions (HO^-) initiate a chain of reactions when ozone enters into water. The chain reaction is sustained by $HO_2^{\bullet}$ formation, which can then initiate further reactions. The hydroxyl radical ($HO^{\bullet}$ - see step (V)) is the most important species formed during ozone decomposition.

Therefore, ozone can react in an aqueous medium directly with substrates, such as molecular ozone, or indirectly, via radical-intermediates formed during ozone decomposition in aqueous media [38,39,45]. Depending on the ozone behavior in the aqueous medium, the selectivity of ozonation can be high (direct reaction) or low (indirect reaction) [37,45,54]. Since the $HO^{\bullet}$ radical is not selective and possesses a very high oxidative potential, this species is therefore a much more effective oxidant than ozone itself.

Depending on the influence of the partial pressure of the dissolved oxygen, the following semi-reactions are representative of the standard redox potentials [5,54,62]:

$$O_{3(aq)} + 6H^+_{(aq)} + 6e^- \rightarrow 3H_2O_{(l)} \quad E^o = 1.51 \text{ V} \quad (1)$$

$$O_{3(aq)} + 2H^+_{(aq)} + 2e^- \rightarrow H_2O_{(l)} + O_{2(aq)} \quad E^o = 2.07 \text{ V} \quad (2)$$

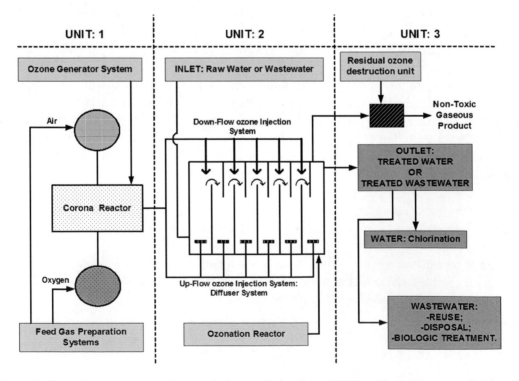

Figure 1. Flow diagram representing a typical ozonation system: UNIT: 1 (Feed Gas Preparation Systems and Ozone Generator System – *Corona Reactor*); UNIT: 2 (Inlet of the Raw Water/Wastewater and Ozonation Reactor - *Gas-Liquid Contactor System*), and UNIT: 3 (Residual Ozone Destruction Device and Distribution System).

According to these semi-reactions, even in cases where the reaction represented by eq. 2 is considered, the hydroxyl radical ($E^o = 2.80$ V) is by far a stronger oxidant than ozone.

An ozonation system consists at least of four unit processes [9,20,54]: (i) feed gas preparation system; (ii) ozone generator; (iii) gas-liquid contactor (reactor system), and (iv) off-gas treatment system. Each of these unit processes adds to ozonation costs and, therefore, optimization of these individual parameters is always necessary. The main parameters to be adjusted are the O_3-concentration in the gas phase and the ozone mass transfer rate (O_3 concentration in water), which is related with components (ii) and (iii), respectively.

Figure 1 shows a schematic representation of a typical ozonation system.

Feed Gas Preparation System

Most of the feed gas preparation systems used for ozone production is classified as using air, high purity oxygen or mixture of the two [19,20,27]. Figure 2 presents a flow diagram describing the different unit processes used for preparation of the gas feed for ozone production using the silent electric discharge technology (Corona reactor).

High purity oxygen can be purchased and stored as a liquid, or it can be generated "on site" through either a cryogenic process, with vacuum swing adsorption, or with pressure swing adsorption. Cryogenic generation of oxygen is a complicated process and is feasible only in large systems.

Pressure swing adsorption is a process whereby a special molecular sieve is used under pressure to selectively remove nitrogen, carbon dioxide, water vapor, and hydrocarbons from air, producing an oxygen rich (80–95% O_2) feed gas. The components used in pressure swing adsorption systems are similar to high pressure air feed systems in that both use pressure swing molecular absorption equipment. Low pressure air feed systems use a heat reactivated desiccant dryer.

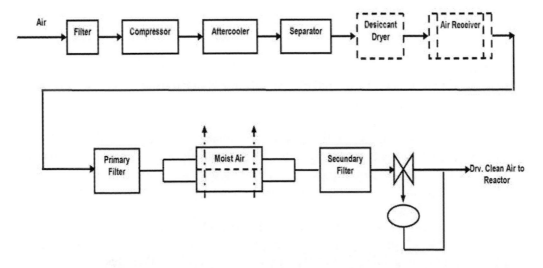

Figure 2. Flow diagram describing the different unit processes used for preparation of the gas feed for ozone production using the silent electric discharge technology (adapted from ref. [9]).

Liquid oxygen feed systems are relatively simple, consisting of a storage tank or tanks, evaporators to convert the liquid to a gas, filters to remove impurities, and pressure regulators to limit the gas pressure to the Corona reactor. On contrary, air feed systems for Corona reactors are fairly complicated as the air should be properly conditioned to prevent damage to the ozonizer. Air should be clean and dry, with a maximum dew point of -60 °C and free of contaminants [9]. So, small particles and oil droplets should be removed by filtration [27]. Air preparation systems typically consist of air compressors, filters, dryers, and pressure regulators.

If hydrocarbons are present in the feed gas, granular activated carbon filters should follow the particulate and oil filters. Moisture removal can be achieved by either compression or cooling (for large-scale system), which lowers the holding capacity of the air, and by desiccant drying, which strips the moisture from the air with a special medium. Desiccant dryers are required for all air preparation systems. Large or small particles and moisture cause arcing which damages generator dielectrics [20,27].

Typically, desiccant dryers are supplied with dual towers to allow regeneration of the saturated tower while the other is in service. Moisture is removed from the dryer by either an external heat source or by passing a fraction (10 to 30%) of the dried air through the saturated tower at reduced pressure. Formerly, small systems that require only intermittent use of ozone, a single desiccant tower is sufficient, provided that it is sized for regeneration during ozone decomposition time [9].

Air preparation systems can be classified by the operating pressure: ambient, low (less than 30 psig), medium, and high (greater than 60 psig) pressure. The distinguishing feature between low and high pressure systems is that high pressure systems can use a heatless dryer. A heatless dryer operates normally in the 100 psig range, rather than the 60 psig range [27].

Air receivers are commonly used to provide variable air flow from constant volume compressors. Oil-less compressors are used in modern systems to avoid hydrocarbons in the feed gas.

Ozone Generator Systems

Reports about ozone generation can be traced back to 1785 when van Marum, a Dutch physicist, found electric discharge in the air results in a characteristic ozone odor. In 1801 the same odor was observed during a water electrolysis experiment carried out by Schönbein [5-7].

Because ozone is a highly reactive gas under ordinary conditions, it has to be generated "on site" [5,20,54]. Nowadays, the "on site" ozone generation is carried out in great scale using the *silent electric discharge process* via Corona reactors, while in the case of medium and small scale applications ozone can be also generated via water electrolysis using specially designed Electrochemical reactors [5-7,54,57].

The main advantage presented by Electrochemical Ozone Production (EOP) is the high concentration achieved in the gaseous phase ($O_2 + O_3$), which can range from 10 up to 35 wt% [63-80]. EOP is gaining popularity due to a couple of features that are not achieved using the conventional Corona process (e.g. production of the ultra pure ozone at very high concentrations) [54]. Besides, specially designed Electrochemical reactors operating in electrolyte-free water permit the direct application of ozone into water streams, thus eliminating problems concerned with the mass transport from gas to the condensed phase [54,78,79].

Costs associated with ozone production using Corona reactors have dropped by almost 50% in the last two decades and, therefore, a great number of new industrial applications has appeared in recent years [5,54]. Depending on the particular ozone generation system, ozone generation consumes power at a rate of 8 to 17 Wh g^{-1} O_3 [5,9,54].

Ozone generation via Corona process - where a dry gas, either air or pure oxygen, is subjected to a silent electrical discharge, a reaction between the oxygenated species ($O^{\bullet} + O_2$), assisted by the free energetic electrons that are created from an electric spark, takes place in the gas phase generating the O_3-molecule [7,54,81,82].

Figure 3 presents a scheme describing a Corona reactor and the elementary processes leading to ozone formation, which take place during the silent electric discharge inside the discharge chamber.

The voltage required to produce ozone by corona discharge is proportional to the pressure of the source gas in the generator and the width of the discharge gap. Theoretically, the highest yield (ozone produced per unit area of dielectric) would result from a high voltage, a high frequency, a large dielectric constant, and a thin dielectric [9,20,54].

However, there are practical limitations to these parameters. As the voltage increases, the electrodes and dielectric materials are more subject to failure. Operating at higher frequencies produces higher concentrations of ozone and more heat requiring increased cooling to prevent ozone decomposition. Thin dielectrics are more susceptible to puncturing during maintenance. So, the design of any commercial Corona reactor requires a balance of ozone yield with operational reliability and reduced maintenance.

Two different geometric configurations for the electrodes are used in commercial ozone generators [9,27,81,82]: (i) concentric cylinders and (ii) parallel plates. The parallel plate configuration is commonly used in small generators and can be air cooled. The glass dielectric/high voltage electrode in commercial generators resembles a fluorescent light bulb and is commonly referred to as a "generator tube" [81,82].

Most of the electrical energy input to an ozone generator (about 85 percent) is lost as heat [54]. Because of the adverse impact of temperature on the production of ozone, adequate cooling should be provided to maintain generator efficiency. Excess heat is removed usually by water flowing around the stainless steel ground electrodes. The tubes are arranged in either a horizontal or vertical configuration in a stainless steel shell, with cooling water circulating through the shell.

Ozone generators are classified by the frequency of the power applied to the electrodes. Low frequency (50 or 60 Hz) and medium frequency (60 to 1,000 Hz) generators are the most common found in the water industry, however some high frequency generators are available [54]. Medium frequency generators are efficient and can produce ozone economically at high concentrations, but they generate more heat than low frequency generators and require a more complicated power supply to step up the frequency supplied by utility power. New installations tend to use medium or high frequency generators [82].

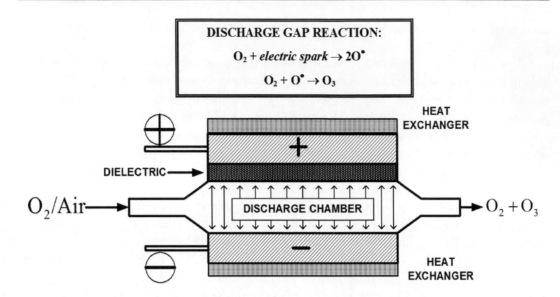

Figure 3. Corona reactor and the elementary processes taking place during the silent electric discharge (adapted from ref. [54]).

Although the corona technology requires a lower specific power consumption the concentration of the O_3 in the gaseous phase ($O_2 + O_3$) presented by conventional devices is low (~2.5 to 7.5 wt%) [81,82], thus restricting its use in several important applications involving degradation of the recalcitrant pollutants. In very special cases the O_3-concentration furnished by a Corona device can reach a maximum of ~15 wt% [9,54].

In principle, ozone generation can be also carried out via Photochemical technology, where pure oxygen or air, when irradiated by the UV light (λ = 254 nm) inside a photochemical reactor, produces ozone [7]. However, the photochemical ozone generation presents a very high specific energy demand (~ 1 kWh g^{-1}) due its low efficiency and, therefore, it is very expensive when compared with the Corona process [5,20,54].

Electrochemical Ozone Production (EOP) – EOP presents a couple of features that are not achieved with the Corona process, thus making it an interesting alternative for several medium and small ozone applications: (i) investment costs (per unit mass of produced O_3) are considerably lower than for the conventional Corona technology and (ii) concentrations of O_3 in the product gas that can be achieved are higher [5,7,54].

Several technological advances related to EOP were achieved in recent decades [54,57,67,79]. Electrolytic ozonizers based on Solid Polymer Electrolyte technology (SPE), which operate in electrolyte-free water and under ambient temperature conditions, permit ozone application directly into water streams ready for various oxidizing and/or disinfectant applications. In this case the total energy demand is minimized, since O_3-production is carried out at ambient temperature (refrigeration is not necessary) and the high ozone mass transfer rate obtained in this case avoids the accessories such as gas diffusers and pumping systems [54].

EOP can be also carried out employing electrolytic ozonizers using specially designed electrolytes [67]. This technology furnishes a very high efficiency ($\geq$ 35 wt%), which makes this electrochemical ozonizer rather competitive with the corona technology in several different applications where a higher O_3-concentration in the gaseous phase is necessary. In this case the cathodic process is the reduction of the oxygen present in the air and, as a consequence, the specific energy demand is very close to that presented by a conventional Corona device [54,57,67].

As previously discussed, the reaction mechanism leading to ozone generation during water electrolysis is a rather complex one [6,75]. This electrode process is less favorable from the thermodynamic point of view when compared to the oxygen evolution reaction. Therefore, one has that EOP can only be carried out using specially design electrocatalysts [54]. Fundamental and applied aspects concerned with the EOP process were reported by Da Silva et al. [5-7,54,75-77].

The fundamental process taking place in the membrane electrode assembly (MEA – see Fig. 4), which comprises a "sandwich" of gas diffusion electrodes (anode and cathode) pressed against a solid polymer electrolyte (e.g. Nafion 117 from Dupont$^{®}$) using porous current collectors is presented below:

ANODIC PROCESSES (+): Oxygen Evolution vs. Ozone Production

$H_2O_{(aq.)} + S \rightarrow S\text{---}HO^{\bullet}_{(ads.)} + H^+_{(aq.)}\text{---}SPE + e^-$(current collector)

$S\text{---}HO^{\bullet}_{(ads.)} \rightarrow S\text{---}O^{\bullet}_{(ads.)} + H^+_{(aq.)}\text{---}SPE + e^-$(current collector)

$2S\text{---}O^{\bullet}_{(ads.)} \rightarrow 2S\text{---}O_{2(ads.)} \rightarrow O_{2(g)}$ (*oxygen evolution* $- E^o = 1.23$ V)

$2S\text{---}O_{2(ads.)} + S\text{---}O^{\bullet}_{(ads.)} \rightarrow O_{3(g)}$ (*ozone production* $- E^o = 2.07$ V))

CATHODIC PROCESS (–):

$2H^+_{(aq.)}\text{---}SPE + 2e^-$(current collector) $\rightarrow H_{2(g)}$ (*hydrogen production*),

where: **S** is an active surface site and *SPE* is the solid polymer electrolyte.

Figure 4 presents the design of an electrochemical ozonizer based on the SPE technology recently developed by Da Silva and Jardim (Patent – EGT/FAPEMIG, Brazil).

The main difference between the ozone generation process using the electric silent discharge and the electrolysis processes is the fact that ozone are produced in the bulk gas phase and at the solid/aqueous interface, respectively. While in the case of the bulk gas phase reaction the electric spark can dissociate the O_3-molecule inside the Corona (ozone decomposition step), thus limiting the efficiency of this process for ozone production, in the case of the surface reaction taking place at the electrode/solution interface inside the Electrochemical reactor the electric potential gradient does not lead to O_3-degradation; the ozone molecule is constantly removed from the electrode surface by forced convection and then spontaneously dissolved in water.

Therefore, EOP permits production of ozonated water containing a very high ozone concentration [54]. Besides, the electrolysis of water (condensed phase) resulting in formation of the O_2-O_3 mixture (gaseous phase) eliminates the use of compression systems in order to deliver the oxidant gas in water [54].

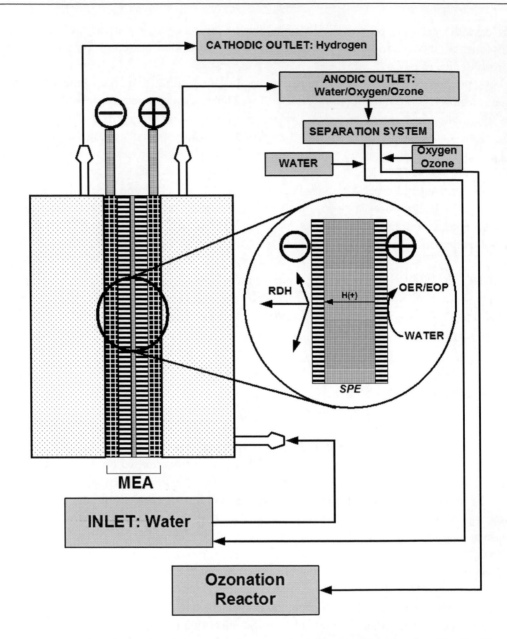

Figure 4. Design of an electrochemical ozonizer based on the SPE technology developed by Da Silva and Jardim.

Since pure hydrogen is generated during EOP, one has that this clean fuel can be in principle used in a hydrogen fuel cell in order to reduce the total energy demand dissipated in the whole process [54]. Figure 5 shows a scheme representing a hybrid environmentally friendly electrochemical system proposed for ozone production using the electrochemically generated hydrogen as an auxiliary energy source.

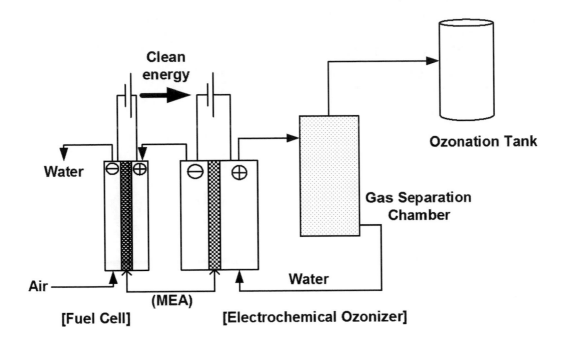

Figure 5. Scheme representing the hybrid environmentally friendly electrochemical system proposed for ozone production using the electrochemically generated hydrogen as an auxiliary energy source (adapted from ref. [54]).

Ozonation Reactor: Bubble Production and the Ozone Mass Transfer in Water

An ozonation reactor is at least a two phase system, consisting of the gas phase containing ozone and the fluid or product phase (generally contained in a liquid), where the ozone must be adequately transferred for chemical reaction [5,9,27].

In the case of mixed gas-liquid reactors, the absolute partial pressure of ozone is the main design parameter. So, the mixing energy and whether this energy has a macroscopic, or turbulent, or a microscopic, or laminar, character as well as the hydrodynamic pattern (complete-mix, plug-flow, concurrent, countercurrent) of the ozonation reactor, are the main design and control parameters for the task assigned to ozone [5,9,18,27,54].

Effectiveness of ozone as either a disinfectant or an oxidant is affected by ozone reaction kinetics and the ozone mass transfer rate from the gas phase to the liquid phase [20,27]. So, the ozonation process depends considerably on partial ozone concentration in the gaseous phase (O_2/O_3) and the ozone mass transfer rate [54]. While the former is limited by the ozone generator system, the later depends on ozone distribution in the liquid phase provided by the gas-liquid contactor system.

The most effective way to increase the ozone mass transfer rate is to increase the interfacial area available for mass transfer by decreasing the size of the gas bubbles dispersed in solution and increasing their residence time [5]. Thus, the mass transfer rate will be a maximum when a given mass of the gaseous mixture is introduced in the liquid forming a great number of very small bubbles, thus increasing to a maximum the reaction zone area located in the gas/liquid interface [5,27,54]. In this case, the two important parameters for the gas-liquid contactor mixed-gas reactor are the *gas-liquid dispersion volume*, V and the *gas-liquid specific contact area, a*. The value of a is related to the gas volume fraction (void fraction), ε, and the *Sauter mean bubble diameter*, d_S, of the gas.

If the gas phase consists of N spherical bubbles of different diameters, d, then [5]:

$$a = \frac{N\pi\langle d^2 \rangle}{V}, \qquad (3)$$

where $\langle d^2 \rangle$ is the average of the squares of the diameters. Similarly, the volume of the gas is given by the relation:

$$\varepsilon V = \frac{N\pi\langle d^3 \rangle}{6} \qquad (4)$$

Consequently, the specific contact area a is given by:

$$a = \frac{6\varepsilon\langle d^2 \rangle}{\langle d^3 \rangle} = \frac{6\varepsilon}{d} \qquad (5)$$

The importance of the bubble size distribution in the liquid phase is apparent from analysis of equation 5, which reveals that a is inversely proportional to the average bubble size of the dispersed gas. The average bubble size depends not only on the properties of gas and liquid but also, to a large extent, on mechanisms responsible for the breaking up and mutual coalescence of bubbles, that is, on reactor hydrodynamics [5,54].

In a gas-liquid interface, where gas absorption is followed by an irreversible chemical reaction (e.g. degradation reaction), two steps control the overall reaction rate [5,54]: (i) the mass transfer from the gas phase to the liquid phase and (ii) the chemical reaction in the liquid process. The simplest and most commonly used model to treat this gas-liquid system is the film theory [20,38,39,54], where it can be considered, assuming that the oxygen/ozone mixture is not very soluble in water, that no mass transfer limitation is observed within the gas phase and only the liquid phase mass transfer resistance needs to be considered.

In this context, it is feasible to assume that there is a liquid film of average thickness, dL, between the liquid bulk and the gas-liquid interface. Also, the mass transport is at a steady-state within the film and, therefore, there is no mass accumulation. Figure 6 shows a simplified sketch representative of a gas-liquid interface where the film theory can be applied [5].

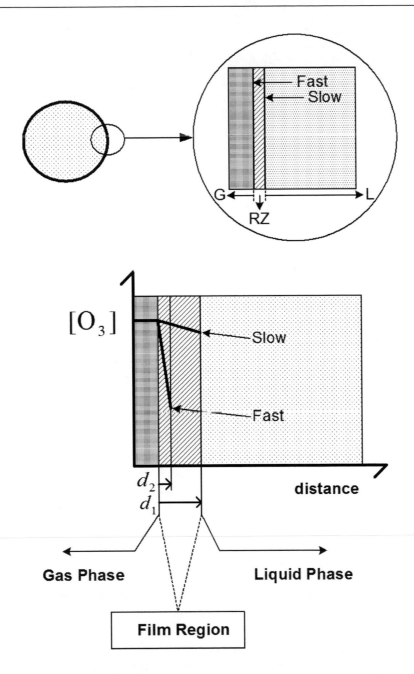

Figure 6. Simplified sketch representative of a gas-liquid interface where the film theory can be applied (adapted from ref. [5]).

According to the sketch presented in figure 6, an elementary bubble surface region constitutes a gas/liquid interface, which comprises the Reaction Zone (RZ). According to the literature [5], both the RZ-thickness and the ozone concentration profile ([O_3] vs. distance) depend on the nature of the chemical reaction occurring at the outer surface of the gas phase. In the light of the film model, the nature of the chemical reaction can be analyzed by two limiting cases corresponding to slow and fast reactions.

The film theory is mainly based on two dimensionless parameters, namely, the Hatta number (Ha) and the instantaneous enhancement factor (E_i), which describe the different phenomena occurring in the reaction zone [20,38,39]. From a theoretical point of view if the ozonation is characterized as *Slow* ($Ha < 0.3$ and $E_i = 1$) the reaction rate is controlled by chemical reactions (O_3-diffusion in the film region is not important). In this case, the ozone concentration in the film region suffers a slight decrease and the chemical reaction remains in the more external region of the film, which can be denoted by d_1 (see Fig. 6).

On contrary, if the reaction is characterized as *Fast* ($Ha > 3$ and $E_i > 1$) the reaction rate is limited by the ozone mass transfer rate. In this case the ozone concentration suffers a strong depletion in the film region, which possesses a thickness represented by d_2. Intermediate cases where $0.3 < Ha < 3.0$ are representative of ozonation reactions where both chemical kinetics and the mass transfer rate are important. A model for this intermediate case was presented by Benbelkacem and Debellefontaine [38], who validated their model experimentally by investigating the ozonation of fumaric acid in a semi-batch bubble column reactor. According to these authors the ozonation of fumaric acid occurs mainly within the film region.

Because ozone and, or hydroxyl radical react with several recalcitrant organics very rapidly, from a practical point of view the ozone mass transfer rate is very important for several different degradation processes [54]. Studies dealing with the ozonation of effluents of the textile industry reveal that ozone mass transfer can control the degradation reaction rate, presenting rates for the O_3-mass transfer in the 0.245×10^{-3} to 0.774×10^{-3} mg mol dm^{-3} s^{-1} interval [5,54]. Depending on the efficiency presented by the gas diffuser system, the average size of microbubbles changes from 0.2 to 4.0 mm [5,54,83-85].

In a Continuous Ideally Stirred Tank Reactor (CISTR), where the reaction mixture is considered to be uniform throughout the tank, any increase in the O_3-mass transfer rate will lead to an increase in CISTR performance. An efficient mass transfer rate also significantly decreases O_3-dosage as a consequence of the increase in the ozone utilization efficiency and the increase in the decomposition rate of the pollutants or pathogenic micro-organism dispersed in the contaminated aqueous medium.

In the case of ozonation of potable water, for a given dosage the initial ozone demand (i.e. the amount of ozone required before its concentration reaches saturation) uses efficient diffusers leading to a high O_3-mass transfer, ensuring very good decontamination efficiency. From the above discussion, may be observed that the use of ozone in water and wastewater treatment strongly depends on optimization of the amount of the ozone added into the liquid phase (nominal load) and the contact time, which, by means of a reactor tank, will permit the effective load to be as close as possible of the nominal ones (O_3-residual ≈ 0).

Gas-liquid contactors currently used in water and wastewater systems include submersible orifices, such as, nozzles, sintered metal or sintered-glass plates, sprinklers, rubber membranes and electrostatic spraying microbubble generators [5]. In combination with these bubble generation system devices, providing mechanical agitation can be used to increase the contact between phases. Such devices provide kinetic energy to the bulk fluid in order to provide efficient bubble dispersion [52].

The use of gas diffusers in tanks under pressure promotes operational conditions with a very high efficiency, thus reducing the O_3-dosage and contact time. On the contrary, open tanks (under atmospheric pressure) present a lower disinfectant efficiency when compared with pressurized tanks and present considerable O_3 losses to the atmosphere (utilization efficiency of ozone << 100%) [54].

The bubble diffuser contactor is by far the most commonly used device in order to promote the ozone mass transfer from gas into the liquid phase [27]. This method offers the advantages of no additional energy requirements, high ozone transfer rates, process flexibility, operational simplicity, and no moving parts.

Ozone can be introduced in the liquid phase using a countercurrent flow configuration (ozone and water flowing in opposite directions), an alternating cocurrent/countercurrent arrangement, or a cocurrent flow configuration (ozone and water flowing in the same direction) [5,9,85].

The bubble diffuser contactor must be designed in order to provide plug flow hydraulic conditions [85]. The contactor volume is determined in conjunction with the applied ozone dosage and estimated residual ozone concentration to satisfy the treatment process (disinfection and, or oxidation of organics).

As recently reported [5,38,39,54], the global kinetic process concerning ozonation of long-chain organic compounds is very complex, since the oxidative process taking place in aqueous solutions containing organic pollutants involves the ozone mass transfer process from gas to liquid phase, which is combined with the ozone consumption (chemical reaction and/or O_3 self-degradation) at the liquid side of the gas/solution interface.

This combined process is normally described using the parameter denoted as Enhancement Factor, E, which describes the acceleration of the ozone transference across the liquid film caused by the chemical reaction located on the solution side of the interface [38,54].

From a practical point of view, E can be calculated based on the experimental determination of the volumetric mass transfer coefficient in the absence, $k_L a$, and in the presence, $k_L a^*$, of a chemical reaction ($E \equiv k_L a^*/k_L a$) using the unsteady state method [54]. This experimental approach is based on analysis of the transient behavior presented by the ozone concentration in pure water (or aqueous solutions) until the O_3-saturation is reached for a given temperature [55].

Considering that a stationary condition prevails under ozone saturation and that the O_3-self decomposition process is negligible ($k_d \cong 0$) the ozone mass transference from gas to the water phase is describe by next equation [55]:

$$\ln\left(\frac{[O_3]_L^{\text{sat.}} - [O_3]_L}{[O_3]_L^{\text{sat.}}}\right) = -k_L a \cdot t, \tag{6}$$

where $[O_3]_L$ and $[O_3]_L^{\text{Sat.}}$ are the instantaneous and the stationary ozone concentration in water for a given temperature.

Figure 7 shows the transient behavior describing the ozone mass transference from gas to the water phase towards the saturation conditions in water and 25 $^{\circ}$C, and the liner dependence predicted by equation 6.

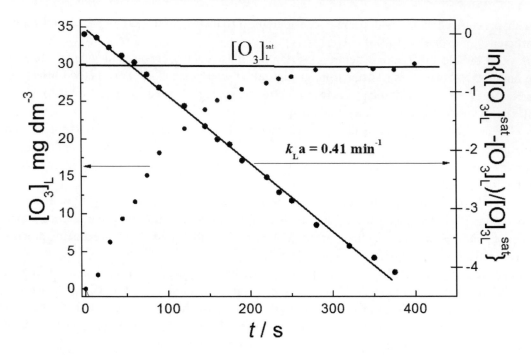

Figure 7. Transient behavior describing the ozone mass transference from gas to the water phase towards the saturation conditions in water and 25 °C, and the liner dependence predicted by equation 6. Ozone dose: 0.35 ± 0.02 g h^{-1} (adapted from ref. [55]).

Analysis of data presented in figure 7 revealed that $k_L a$ is 0.41 ± 0.03 min^{-1} (25 °C). This $k_L a$-value indicates a rather good mass transference from gaseous to the water phase, which is in good agreement with the literature reports concerning the ozone dissolution in water using porous plate diffuser and moderate volumetric flow rate values [38,39,85].

Bubble diffuser contactors are typically constructed with 4.50 to 7.00 m water depths to achieve 85 to 95% ozone transfer efficiency [9,20,27]. Since all the ozone is not transferred into the water, the contactor chambers are covered to contain the off-gas. The concentration of ozone in the off-gas from a contactor is usually well above the fatal concentration. For example, at 90% transfer efficiency, a 3% ozone feed stream will still contain 3,000 ppm of ozone in the off-gas [9]. Off-gas is collected and the ozone converted back to oxygen prior to release to the atmosphere. Ozone is readily destroyed at high temperature (> 350 °C or by a catalyst operating above 100 °C) to prevent moisture buildup. The off-gas destruct unit is designed to reduce the concentration to 0.1 ppm of ozone by volume, which is the current limit for worker exposure in an eight hour shift [27]. A blower is used on the discharge side of the destruct unit to pull the air from the contactor, placing the contactor under a slight vacuum to ensure that no ozone escapes.

The optimal operating conditions can be achieved in practice by means of an automated system using devices that permit maintaining rigid control of the residual ozone [5]. Other variables are controlled in order to obtain optimal operational conditions: (i) water (effluent) flow rate; (ii) O_3-concentration in the gaseous phase and (iii) the ozone load. Specific software especially developed for optimization of these variables is available where, for a considered pollutant charge distributed in a particular effluent, it is possible to optimize the diffuser system, ozone load and the choice of the ozone generator and the reactor tank [5,54].

6. ADVANCED OXIDATION PROCESSES: OZONATION OF POLLUTED WATERS CONTAINING REFRACTORY ORGANIC COMPOUNDS

The purpose of all AOP is to produce hydroxyl radicals ($HO^{\bullet}$ - E^o = 2.80 V), which is highly reactive oxidizing agent that reacts with and destroys most organic pollutants in water [5,54]. There are a variety of AOP commercially available [86]. Each has a scope of wastewater treatment applications that it is best suited for. The "in situ" hydroxyl radical production offers full-scale treatment systems employing different types of AOP. With these various technologies, hydroxyl can deliver systems that employ UV light as an energy source to create reactions or dark AOP that do not depend upon UV light to provide treatment [3,9,86].

As a rule, AOP efficiencies depend on the chemical nature of the effluent subjected to the treatment, such as, pH, turbidity, Chemical Oxygen Demand (COD) and the presence of radical scavengers. A project aiming at evaluation of the economic viability of AOP for a given acceptable level of contaminants was described by Luck et al. [87].

In the case o ozonation, these free radicals are can be produced during the spontaneous decomposition of ozone (dark AOP) [86]. By accelerating the ozone decomposition rate in alkaline conditions, the hydroxyl radical concentration is elevated, which increases the oxidation rate. This procedure increases the contribution of indirect oxidation over direct ozone oxidation [5].

Ozone applications in ITP present at least two important features [54]: (i) biodegradability increases of the dissolved organics and (ii) introduction of a considerable amount of oxygen into the water, thus creating excellent conditions during the biologically active filtration process. Ozonation alone can be efficiently applied in order to purify water loops used in the pharmaceutical and electronic industries and to provide an adequate wood pulp bleaching without further release of carcinogenic compounds normally originated when the complex natural organic matter is oxidized using chorine and its derivates [9,86].

A review dealing with alternative technologies for wastewater treatment involving application of several different AOP systems was presented by Gogate and Pandit [47]. In the special case of AOP involving ozonation, literature reports have pointed out some considerations involved in wastewater treatment [5,86]. For example, it must be ensured that O_3 is not depleted during the course of the reaction, since the radicals released in the various peroxyl radical reactions generate further $HO^{\bullet}$ radicals by the rapid reaction with O_3. Promising AOP are mainly expected from hybrid processes as O_3/UV_{254}, peroxone (O_3/H_2O_2), O_3 at high pH, and H_2O_2/UV_{254} [5,47,54,86].

The peroxone process requires an ozone generation system and a hydrogen peroxide feed system. The process involves two essential steps: (i) ozone dissolution and (ii) hydrogen peroxide addition. Hydrogen peroxide can be added after ozone (thus allowing ozone oxidation and disinfection to occur first) or before ozone (i.e., using peroxide as a pre-oxidant, followed by hydroxyl radical reactions) or simultaneously [87,88].

There are two major effects from the coupling of ozone with hydrogen peroxide [47]: (i) oxidation efficiency is increased by conversion of ozone molecules to hydroxyl radicals and (ii) ozone transfer from the gas phase to the liquid is improved due to an increase in ozone reaction rates. In this case ozonation can be described as occurring in two stages. In the first stage, ozone rapidly destroys the initial oxidant demand present, thereby enhancing the ozone transfer rate into solution from the gas phase. Addition of hydroxyl free radicals to the first stage should be minimized since the hydrogen peroxide competes with ozone-reactive molecules (i.e., initial demand) for the ozone present.

In the second stage, organic matter is oxidized, taking place much slower than in the first stage. Adding hydrogen peroxide during the second stage makes it possible to raise the overall oxidation efficiency, since the reaction of hydrogen peroxide with ozone produces hydroxyl radicals enhancing chemical reaction rates. In practice, the addition of hydrogen peroxide to the second stage of ozonation can be achieved by injecting the hydrogen peroxide into the second chamber of an ozone contactor [9].

Energy consumption of the peroxone process includes that for ozone generation and application, plus for metering pumps to feed peroxide. The peroxide addition step does not require any more training from an operator than any other liquid chemical feed system. Systems should be checked daily for proper operation and for leaks. Storage volumes should also be checked daily to ensure sufficient peroxide is on hand, and to monitor usage [86,87].

A key difference between ozone and peroxone processes is that the ozone process relies heavily on the direct oxidation of aqueous ozone at pH < 7.0, while peroxone relies primarily on oxidation with hydroxyl radical. In the peroxone process, the ozone residual is short lived because the added peroxide greatly accelerates the ozone decomposition. However, the increased oxidation achieved by the hydroxyl radical greatly outweighs the reduction in direct ozone oxidation because the hydroxyl radical is much more reactive. The net result is that oxidation can more reactive and much faster in the peroxone process compared to the ozone molecular process.

Peroxone has found a niche in oxidizing difficult-to-treat organics, such as taste and odor compounds including geosmin and Methylisobornel (MIB) [89]. In addition, peroxone and other AOP have been shown to be effective in oxidizing halogenated compounds such as 1,1-dichloropropene, trichloroethylene, 1-chloropentane, and 1,2-dichloroethane [90,91]. Peroxone is also used for the destruction of herbicides (e.g. atrazine), pesticides, and Volatile Organic Compounds (VOCs).

6.1. Degradation Kinetics of Recalcitrant Organic Compounds

The removal of organics from wastewaters and effluents has become an increasingly important issue because of steadily increasing industrial activities [3]. Important sources of contamination are the pulp and paper mill industry, the petrochemical industry, the food-processing industry, the runoff from urban areas, etc. [3,9,20].

Ozonation of organics depends on pH, chemical reaction (intrinsic kinetics), ozone mass transfer and ozone load in the gaseous phase [18-20,27,54]. The driving force for ozone mass transfer, comprising the difference between the dissolved and the equilibrium ozone concentration at the gas-liquid interface, is system-dependent. The dependence of this driving force for ozone mass transfer on the wastewater characteristics results from the fact that the concentration of the dissolved ozone varies considerably with the rate of the O_3-self degradation and the nature of the chemical reaction.

Da Silva and Jardim [5] recently presented a simplified scheme for representing the ozonation of organic pollutants taking place at the *gas-liquid* interface. According to this scheme, ozone absorption in the liquid phase is followed by the irreversible degradation reaction, as follows:

$$O_{3(g)} \rightarrow O_{3(l)} \ (mass\ transfer\ step) \tag{7}$$

$$[O_{3(l)} \Leftrightarrow \Sigma(Rad)_{(l)}] + \nu X_{(l)} \rightarrow Products_{(l)} \ (degradation\ step) \tag{8}$$

where: $O_{3(g)}$ and $O_{3(l)}$ represent ozone present in the gaseous and liquid phases, respectively; $[O_{3(l)} \Leftrightarrow \Sigma(Rad)_{(l)}]$ describes the O_3-decomposition that yields oxygenated free radicals (e.g. $HO^\bullet$); ν is the stoichiometric coefficient, and X is a given target organic compound.

In water, the direct reaction of dissolved ozone can hardly be separated from other reactions that arise from decomposition of aqueous ozone to secondary oxidants that are much more reactive than ozone itself [9,18-20,27]. As described by equation 8, one has the overall oxidation process taking place during ozonation of the organic pollutants will depend on the parallel reactions taking place simultaneously in the water phase, which is described below [5,63]:

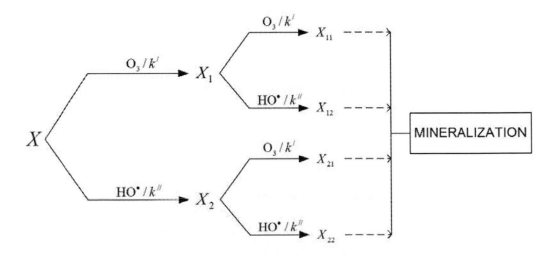

Considering a particular case where ozonation comprises an irreversible process which is not controlled by mass transfer ($E \equiv k_L a^*/k_L a \cong 1$), one has the degradation rate of the parent organic compound, X, can be expressed as follows [7,38,39]:

$$\frac{d[X]}{dt} = -\{k'[X][O_{3(aq.)}] + k''[X][HO^\bullet_{(aq.)}]\} \tag{9}$$

where: $[X]$ is the concentration of the parent compound; $[O_{3(aq.)}]$ and $[HO^\bullet_{(aq.)}]$ are the concentrations in water phase of ozone and the hydroxyl radical, respectively; $k'\ (= n_1 k_1)$ and

k'' ($= n_2 k_2$) are the apparent kinetic rate constants comprising the product of the stoichiometric coefficients (n), which reflect oxidations of daughter products (X_{11}, X_{12},...) and the actual rate constant (k_1 and k_2).

According to equation 9, slow degradation kinetics depends considerably on the total concentration of the effective oxidants present in solution (O_3 and/or $HO^\bullet$). Also, the competition of direct and indirect kinetic pathways will depend on pH and on the presence of free radical scavengers (e.g. carbonate).

Considering that pseudo-first order conditions will prevail in the semi-batch mode (constant ozone application in the ozonation reactor), the kinetic rate law described by equation 9 can be simplified resulting in the next equation:

$$\frac{d[X]}{dt} = -k_{obs}[X], \tag{10}$$

where $k_{obs} = k'[O_{3(ads.)}] + k''[HO^\bullet]$.

Integration of equation 10 furnishes the next expression for degradation of organics for the particular conditions specified above:

$$\ln\left(\frac{[X]}{[X]_0}\right) = -k_{obs} \cdot t \tag{11}$$

Analysis of different ozonation processes using readily measurable parameters in order to represent the concentration of the parent compound ($[X]$) (e.g. absorbance, COD, TOC, etc.) revealed that the reaction model described by equation 11 can be applied in order to design ozonation reactor for the wastewater treatment.

6.2. The Textile Effluent

Ozonation is very efficient in order to provide discoloration. Therefore, ozonation comprises an important process for treatment of several different colored wastewaters.

Based on the EPA's toxic release inventory [10], approximately 2,200 ton of four hazardous dyes are discharged annually into publicity owned treatment works. The classical methods employed for treating textile wastewaters include various combinations of the biological (activated sludge), physical and chemical processes [61].

The textile effluent is an important case for AOP processes since the use of classical methods for discoloration does not furnish satisfactory results [92-102]. Dye molecules are highly structured polymers, which are very difficult to break down biologically and, therefore, they cannot be treated efficiently by simple classical methods. Besides, the main drawback of these processes is the generation of a large amount of sludge or solid waste, resulting in high operational costs for sludge treatment and disposal.

Different alternative technologies based on ozonation can be used in combination with biological process aiming to provide an efficient discoloration and a TOC/COD reduction of the textile wastewater [93-98].

Textile waste effluents are one of the wastewaters that are very difficult to treat satisfactorily because they are highly variable in composition and contain several different recalcitrant compounds [101,102]. Wastewaters that are generated at various stages of the dyeing process differ in compositions and temperature. The high pollution load is mainly caused by spent dyeing baths. Their constituents are unreacted dyeing compounds, dispersing agents (surfactants), salts and organics washed out of the material which undergoes dyeing [61].

For low strength dye waste effluents, ozonation alone is sufficient to totally eliminate the color and reduce the turbidity. However, for medium and high strength waste effluents, ozonation is found to be sufficient to reduce the color, but not enough to reduce the turbidity. Hence, coagulation of the textile effluent using aluminum sulphate (~ 60 mg dm^{-3} for a TOC value of ~ 600-1000 ppm) or especially designed polymers can be necessary [92-102].

As reported by Tzitzi et al. [48], ozonation of the wastewater after coagulation-precipitation process, under the same conditions as the raw wastewater ozonation, exhibited more efficient discoloration (> 90%) and COD reduction (> 30%), while the biodegradability was found to increase. After this initial step of degradation the residual organic carbon generated during partial mineralization can be further effectively degraded using the activated sludge process. Thus, the combination of ozonation with proper chemical coagulation and an activated sludge process is a promising alternative technology for dealing with textile industry effluent, which considerable reduces the sludge disposal.

According to Shu and Huang [95], who investigated the chemical oxidation of non-biodegradable azo dyes by ozonation and photo-oxidation process in a pilot scale using a photochemical UV/O$_3$ reactor, application of the UV-light did not significantly enhance the degradation ability of the ozonation reaction.

As previously discussed by Franco et al. [54], there is a distinction between total discoloration of the dye solution and the total degradation (mineralization) of the aqueous solution containing refractory dyes. The discoloration process via ozonation takes place when the chromophore bond(s) is(are) removed, while many colored byproducts of the parent dye molecule may remain stable in solution [52,54,55]. Therefore, discoloration may be the initial step in the degradation route of a dye molecule, which is not necessarily accompanied by quantitative carbon removal (considerable degree of mineralization), thus requiring a lower oxidant dose than mineralization.

Discoloration Kinetics

During the curse of the ozonation reaction is possible that the daughter products (colored byproducts) may compete with the parental dye molecules for the oxidant (ozone and/or hydroxyl radical). Besides, taking into account the complexity nature of the ozonation process of long-chain organic molecules, the overall pseudo-first order rate constant, k_{obs}, contains the effect of the intrinsic kinetics and may reflect more than one mass transfer-chemical regime [54,55].

In the light of these considerations, the kinetic of the discoloration process carried out in the semi-batch mode (pseudo-first order conditions) can be adequately described by the following rate law [54,55,103]:

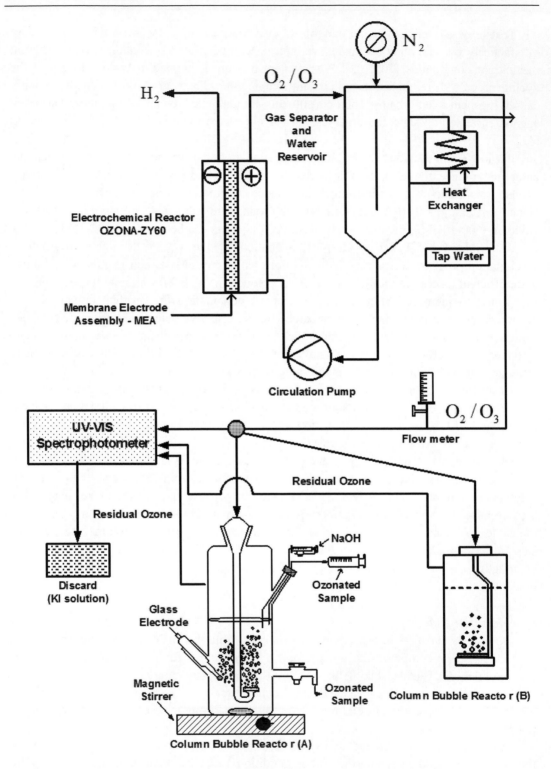

Figure 8. Flow diagram representing the experimental set-up used for ozone generation and its application for discoloration/degradation of textile dyes (adapted from ref. [54]).

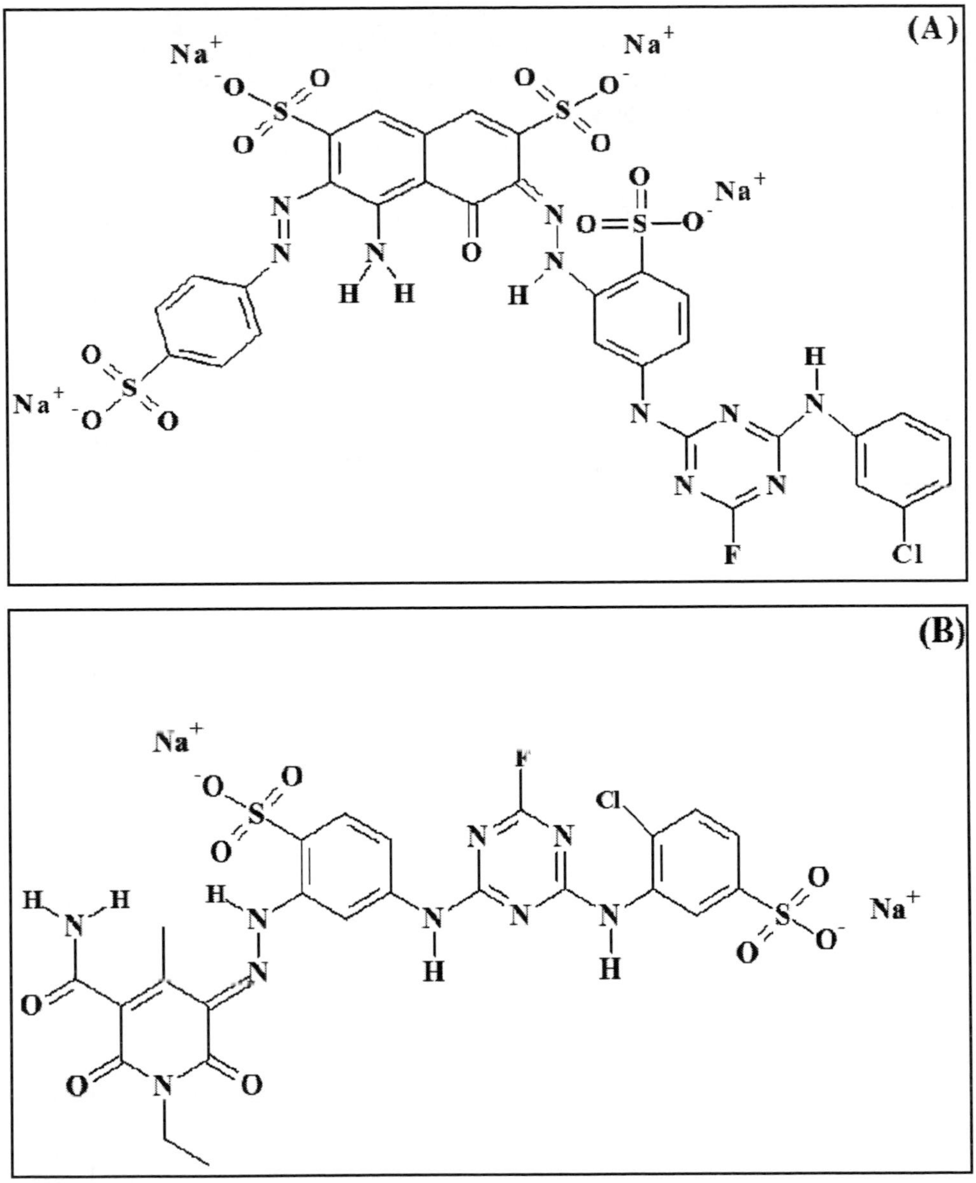

Figure 9. Chemical structures of the commercial textile dyes: (A) Reactive Blue 264 (RB 264 - CAS Number: 70528-89-1 - $C_{31}H_{18}ClFN_{10}Na_4O_{13}S_4$) and (B) Reactive Yellow 143 (RY 143 - CAS Number: 75268-65-4 - $C_{24}H_{19}ClFN_9Na_2O_9S_2$).

$$\ln\left(\frac{A}{A_0}\right) = -\{\alpha(\nu_{EOP})^x ([HO^-]^y ([cro]_0 [HO^\bullet])^z\} \cdot t = -k_{obs} \cdot t \tag{12}$$

where: A/A_0 represents the normalized absorbance measured at a fixed wavelength; $[cro]_0$ is the initial chromophore concentration; ν_{EOP} is the ozone application rate; $[OH^\bullet]$ is the hydroxyl radical concentration and [OH⁻] is the hydroxyl anion concentration. α, x, y and z are the empirical constants for the particular ozonation process.

Figure 8 presents a flow diagram representing the experimental set-up used for EOP and its application for discoloration/degradation of textile dyes under controlled pH conditions [54,55]. In this case ozone was generated using the electrochemical reactor OZONA-ZY60 developed by Da Silva and Jardim (Patent – EGT/FAPEMIG, Brazil).

Figures 9A and 9B show the molecular structures for the commercial textile dyes Reactive Yellow 143 (RY 143) and Reactive Blue 264 (RB 264).

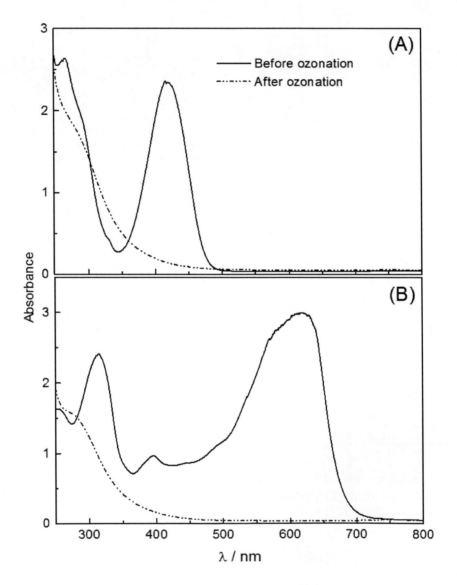

Figure 10. UV-VIS spectra for textile dyes RY 143 (A) and RB 264 (B) before and after ozonation at pH 12. IDC = 100 mg dm⁻³. Ozone dose: 0.35 ± 0.02 g h⁻¹ (adapted from ref. [54]).

Table 3. Dependence of t_d values on pH for RY 143 and RB 264. IDC = 100 mg dm^{-3}, V = 1 dm^3 (adapted from ref. [54])

pH	RY143	RB 264
	t_d/min	t_d/min
2 ± 0.05	25	14
7 ± 0.05	23	12
12 ± 0.05	20	11

Figure 10 shows the UV-VIS spectra obtained for RY 143 and RB 264 before and after ozonation ($t = 1$ h and pH 12) using a low ozone dose of 0.35 ± 0.02 g h^{-1}. This study was carried out as a function of the solution pH ($2 - 12$) using the Column Bubble Reactor A (see Fig. 8) [54]. Both dyes show a maximum absorbance in the visible region, corresponding to 421 and 619 nm for RY 143 and RB 264, respectively. Initial Dye Concentration (IDC) was 100 mg dm^{-3} and final discoloration was reported as absorbance reduction higher than 99 % obtained for a given ozonation time defined as t_d.

Analysis of figure 10 reveals that ozonation leads to a quantitative removal of the absorbance in the visible region. Besides, analysis of the absorbance bands in the UV-region revealed the total amount of aromatics decreased about 15 and 40% for both RY 143 and RB 264, respectively [54]. Chu et al. [104] also reported the dependence of the UV absorption bands on the ozonation time in order to account for degradation of the Reactive Black 5 (RB5) and found decay of the absorbance at 311 nm as evidence of the degradation of aromatic groups in the dye molecules and their intermediates.

Table 3 shows the total discoloration time, t_d, values as functions of the pH for RY 143 and RB 264, where is found that total discoloration ($\geq$ 99%) is slightly attained faster in alkaline solution as a result of the influence of the indirect oxidative pathway via hydroxyl radical on discoloration process. Discoloration also depends on the nature of the dye ($t_{d(RB264)} < t_{d(RY143)}$), thus revealing that the destruction of the chromophore centers in RB 264 is more pronounced.

Figure 11 shows representative kinetic profiles for discoloration, while Table 4 gathers k_{obs} as a function of the pH for RY 143 and RB 264. A rather good linear behavior ($r > 0.998$) was verified in all cases using the pseudo-first order kinetic model described by equation 12.

Two linear segments in the pseudo-first order profile were obtained in all cases. The presence of two k_{obs} in the kinetic profile supports the existence of two corresponding half-life time constants, $\tau_{1/2}$, for discoloration and formation of intermediate colors before total discoloration is achieved (production of Transient Persistent Colored Byproducts - TPCBP) [54,55,104].

The existence of two linear segments in the kinetic profile was also verified by Hsu et al. [52], who proposed that slope changes in the kinetic profile, after a given threshold t-value, were due to changes in the chemical nature of primary substances in solution, forming TPCBP.

Ozonation of the RB5 solution carried out by Chu et al. [104] revealed that that dye solution gradually bleached from black, to brown, to yellow, and then to colorless. Besides, these findings revealed that the lower the IDC, the higher the discoloration rate. This dependency of discoloration rate on the IDC was attributed to the production of more than one kind of intermediate during the ozonation processes [54,55].

Analysis of Table 4 clearly reveals the pH does not lead to a significant change in discoloration kinetics, thus indicating that the "direct oxidative process" comprises the main oxidative process leading to discoloration.

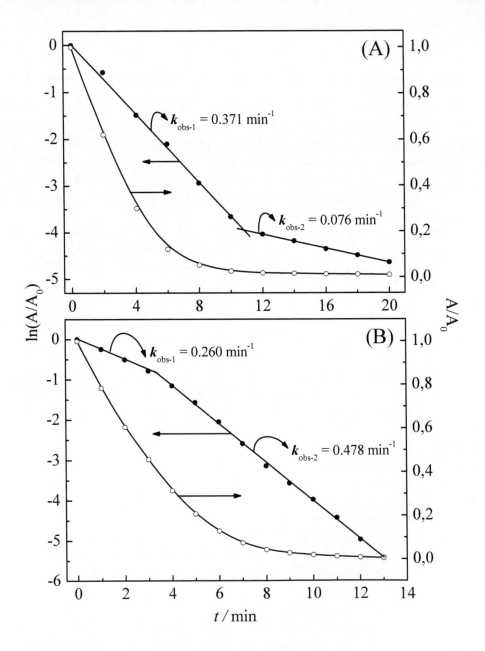

Figure 11. Kinetic profiles representative of discoloration via ozonation. (A) RY 143, pH 2; (B) RB 264, pH 2. IDC = 100 mg dm^{-3}. T = 24 °C. Ozone dose: 0.35 ± 0.02 g h^{-1}. V = 1 dm^3 (adapted from ref. [54]).

Table 4. Dependence of k_{obs} on pH for RY 143 and RB 264. IDC = 100 mg dm^{-3} (adapted from ref. [54])

DYES	pH 2 ± 0.05		pH 7 ± 0.05		pH 12 ± 0.05	
	$k_{obs\text{-}1}$/ min^{-1}	$k_{obs\text{-}2}$/ min^{-1}	$k_{obs\text{-}1}$/ min^{-1}	$k_{obs\text{-}2}$/ min^{-1}	$k_{obs\text{-}1}$/ min^{-1}	$k_{obs\text{-}2}$/ min^{-1}
RB 264	0.260	0.478	0.210	0.380	0.284	0.581
RY 143	0.371	0.076	0.360	0.070	0.202	0.099

The linear behavior in the pseudo-first order profiles supports that ozonation of parental dye molecule and TPCBP are both slow chemical processes. Considering that v_{EOP}, $[cro]_o$, [OH$^-$] and [OH$^\bullet$] (see eq. 12) are both constants in the present case, one can propose that changes in k_{obs} as functions of the ozonation time and the dye nature, can be attributed to modifications suffered in the intrinsic kinetics between the oxidant (O$_3$ and/or HO$^\bullet$) and the different chromophore centers present in the parental and non-parental (TPCBP) dye molecules.

The discoloration kinetics can be divided in three stages [54,55]: (i) primary attack: the discoloration process is governed by chemical reaction involving the oxidant (O$_3$ and/or HO$^\bullet$) and the more reactive chromophore centers of the parental molecule; (ii) secondary (transient) attack: color removal rate is influenced by changes suffered in the intrinsic nature of the oxidation process as a consequence of the competition between the new chromophore centers present at the TPCBP structure, which are formed after the primary attack, and (iii) tertiary attack (last stage of discoloration): color removal takes place via oxidation of the remaining chromophore centers present in TPCBP.

Therefore, changes in k_{obs} as functions of the ozonation time can be attributed to modifications in the [dye]/[TPCBP] ratio, which results in a considerable modification in the intrinsic kinetics between the oxidant (O$_3$ and/or HO$^\bullet$) and the chromophore centers [54].

In order to account for influence of mass transfer on ozonation process, it was proposed an empirical linear relation for the discoloration process, which correlates, for a given temperature, the enhancement factor (E) with the initial dye concentration ([IDC]) and the ozone application rate ($z(v_{EOP})$). This relation was denoted as [105]:

$$E = x + y[\text{IDC}] + z(v_{EOP}), \tag{13}$$

where x, y and z are experimental parameters determined for each dye system.

Analysis of equation 13 reveals that E increases with [IDC] due to the chemical kinetics, since ozonation of dye is pseudo-first order with respect to O$_3$ and the dye. Therefore, the E-value should increase with the concentration of both the dye and the dissolved ozone. Besides, any increase in v_{EOP} leads to a concomitant increase in turbulence at the gas/liquid interface, thus enhancing the mass transfer at the reactive zone (see model for bubble/liquid interface presented in Fig. 6). From these considerations, one has that keeping constant the v_{EOP} under semi-batch conditions the ozonation becomes pseudo-first order with respect to the dye, and the E-parameter increases linearly only with [IDC].

It was found that k_{obs} declines logarithmically with [IDC]. This behavior was empirically described by the next relation [105]:

$$k_{obs} = w[IDC]^{-m}, \tag{14}$$

where w and m are empirical constants. It was verified by Wu and Wang [105] that the linear $\log(k_{obs})$-$\log([IDC])$ relationship can be always applied for discoloration of azo dye solution regardless the co-existence of other compounds.

Degradation of eight commercial reactive azo dyes with different structures containing different substituted groups was studied by ozonation individually and in mixture [106], where was proved that ozonation carried out in alkaline conditions (pH 10) using an appropriate ozone dose is indeed effective for discoloration and COD removal. The influence of the molecular structure was evidenced in this study; even for dye solutions decolorized up to 95-99%, it was verified that COD removal was different for each dye. Contrary to discoloration process, this behavior indicates the complexicity of the molecular structure plays a key role during COD removal [106].

Oxidation and cleavage of substituent groups were evidenced by the release of chloride, nitrate and sulfate during ozonation. Increase in biodegradability was observed after ozonation, as measured by the BOD_5/COD ratio [106].

6.3. The Pulp and the Paper Mill Effluents

The pulp and paper industry has historically been one of the world's largest consumers of freshwater resources and producers of wastewater discharges, which consumes around 200 m^3 per ton of cellulose produced [7,47,61].

The pulp and paper wastewater can cause serious environmental problems if it is discharged into rivers and lakes without a previous treatment. Application of the conventional biological processes on treatment of pulp and paper industry wastewater eliminates important organic compounds present in the waste [7,61]. However, these biochemical decomposition processes do not comprise an efficient technology for removal of the recalcitrant organic matter in order to satisfy the discharge limits [5,7,54].

Most of the persistent compounds present in pulp and paper wastewater exhibits a high molar mass (> 1 kDa) and are resistant (recalcitrant) to conventional biological degradation requiring, therefore, an alternative coadjutant technology for treatment of these wastewaters. A very large amount of colored pollutant compounds (200 to 300 kg ton^{-1} of pulp), mainly chlorolignin, is produced during the conventional bleaching of softwood kraft pulp [61].

Therefore, development of new technologies based on the concept of ITP for treatment of pulp and paper mills wastewaters is a very important issue due to the urgent necessity to meet current environmental legislation in different countries. Special attention has to be paid to the so-called TCF-technologies, which comprise the use of alternative oxidative processes in substitution of ClO_2 in order to provide a better, safer, effluent treatment [7].

In this context, coadjutant oxidation processes based on application of ozone and, or hydrogen peroxide is very promising for treatment of the pulp and paper mills wastewaters. Particularly, ozonation has proved to be an efficient technology that has matured for this particular wastewater treatment [5].

Table 5. Operation costs for ozonation studies (adapted from ref. [111])

Parameter	Ozone	Ozone + Peroxide	
Ozone dose, mg dm^{-3}	175	110	100
Ozone generator capacity, kg h^{-1}	18.2	11.5	10.4
Peroxide dose, mg dm^{-3}	—	50	100
Waterwater Flow, m^3 day^{-1}	2,500	2,500	2,500
Operationg cost, $ m^{-3}			
Energy	0.3	0.19	0.17
Maintenance	0.03	0.02	0.02
Chemical (H$_2$O$_2$)	—	0.04	0.08
Total Operation Costs, $ m^{-3}	0.33	0.25	0.27

In fact, the use of advanced effluent treatment in the pulp and paper industry comprising the combination of ozone with fixed bed biofilm reactors is one of the most efficient tertiary effluent treatment processes to give maximum elimination of COD, color and AOX (Absorbable Organic Halogens) using a minimum ozone dose. This technology provides an effective elimination of polluting and disturbing substances (instead of merely separating them as is done with precipitation) and transfers the oxidative residue into a small amount of biological excess sludge [5].

The first studies on the ozonation of pulp and paper effluents were carried out in the 1970s. With the recent advances in ozone-generating technologies and the development of ITP processes, the application of ozone for pulp mill effluent treatment becomes not only technically feasible, but also economically viable [107-111].

Laboratory studies and pilot tests concerning different mill effluents presented by Helbe et al. [108] revealed biological effluent treatment with relatively low BOD after the ozonation requires biofilm technologies, *i.e.*, biofilters, for the further biological treatment, since classical biological processes such as activated sludge systems cannot be operated well with this expected low loaded effluent. Also, if COD removal efficiencies higher than 50% are required, advanced treatment with ozone and biofiltration must be undertaken in two stages. Pilot tests with COD removal efficiencies higher than 80% were achieved using an ozone consumption of 0.6 to 0.8 kg per kg of COD eliminated after biofiltration [5].

Cost-effective analysis revealed the added costs for a paper mill using ozonation technology represent an increase on the order of 1% of the manufacturing cost of the overall paper production process [5,111]. Operating costs for ozonation studies were reported by Sevimli is presented in Table 5 [111].

The use of AOP, including ozone [108,112,113], peroxone [114] and Fenton's reagent [115,116], has been successfully applied for the removal of refractory organics and color from pulp and paper effluents. Literature reports [109,117] have shown that ozonation is mainly effective in removing color, total phenols, and acute toxicity of bleaching effluents.

Studies were carried out by Assalin et al. [107] in order to optimize the ozonation conditions, such as pH, temperature, and applied ozone dose, revealed that ozonation of pulp mill effluent is mainly controlled by the mass transfer of ozone from the gas phase to the wastewater; the overall oxidation process is system dependent since the driving force for ozone transference varies considerably with the rate of self-decomposition of ozone in wastewater, which in turns depends on the effluent's characteristics.

Assalin et al. [107] reported the degradation kinetics for the all parameters studied at pH 12 was lower than those carried out at pH 10. Ozonation carried out at pH 10 and 12 resulted in a mineralization degree of 32.4% and 13.2%, respectively. Under the same ozonation conditions the treatment furnished a discoloration degree of 67.4% and 59.3%, respectively.

Removal of COD, color, and UV_{254} of the paper mill effluent was carried out by Sevimli [111] using three different chemical oxidation methods: (i) ozonation; (ii) peroxone, and (iii) Fenton's reagent. It was found in this study that ozonation lead to a decrease of COD to the value of 270 mg dm^{-3} after 45 min of treatment, which corresponded to 195 mg dm^{-3} of the consumed ozone; the COD removal efficiency was 43%. Besides, it was also found that peroxone application, ozone with 50 and 100 mg dm^{-3} H_2O_2, lead to final COD levels of 182 and 170 mg dm^{-3}, by consuming 210 and 330 mg dm^{-3} ozone, respectively, in the same ozonation period. These final COD levels corresponded to 61 and 64% COD removal efficiencies, respectively.

Discoloration study revealed that the wastewater previously presenting a dark brown color before treatment, presented a coloration turned into light yellow after ~ 15 min of ozonation [111]. The color removal efficiencies obtaining after ozonation and peroxone application (O_3+100 mg dm^{-3} H_2O_2), during 15 min at pH of 7.1, were 75 and 93%, respectively. It was found that an additional ozonation step during 30 min lead to a final discoloration of 91 and 97%, respectively.

The absorbance at 254 nm can be used in order to represent the degree of aromaticity and unsaturated compounds in water effluents. In the light of this approach, Sevmili [111] reported a rapid decrease in UV_{254} for both ozone and peroxone applications, which indicates that these processes can lead to a considerable degradation of the persistent organic matter. In the case of peroxone the UV_{254} removal rate was maximum during the first 15 min. Application of O_3 and peroxone (O_3+100 mg dm^{-3} H_2O_2) carried out during 45 min lead to a UV_{254} removal of 75 and 84%, respectively.

6.4. Pharmaceuticals in Water

Several pharmaceutical agents are among a growing body of anthropogenic chemicals that lead to undesirable interferences in the environment. Of the three often-overlapping toxicological endpoints for chemicals, namely, the killing of cells, the mutation of DNA in ways that may lead to cancer, and the disruption of chemical signaling mechanisms controlling cellular development, it is well known the least about the third [4]. This last area of toxicity is known as *endocrine disruption*; these chemical species can interfere with the hormonal command of cellular development and the result can be severe impairment of growing creatures [4,13,43].

Pharmaceuticals are generally absorbed by humans or animals after intake and are then attacked by metabolic degradation processes. However, significant fractions of the original substances often are excreted in unmetabolized form or as active metabolites via urine or feces to be emitted into raw sewage, which may or may not be treated [118,119]. Some pharmaceutical pollutants escape degradation in waste treatment plants and enter the environment [118,120]. In addition to metabolic excretion, disposal by flushing of unused or expired medication and drug-containing waste from manufacturing facilities can also contribute to environmental contamination [121]. Flushing unused medicines down the toilet appears to be of minor importance, while patient excretion following therapy is widely considered to be the primary pathway to the environment [122]. Even posthumously, the drugs administered in the closing phases of our lives likely leach into cemeteries and groundwater [123].

Pollution from pharmaceuticals in surface and groundwaters is becoming recognized as an environmental concern leading to the area of study known as Pharmaceuticals in the Environment (PIE) [4]. The pharmaceutical industry is attaining more effective Active Pharmaceutical Ingredients (APIs) by designing for increased potency, bioavailability and degradation resistance. APIs show a wide range of persistence in aquatic environments, and some are highly persistent. In such cases, the pharmacologically valuable properties of degradation resistance and bioavailability return as hazards because they translate into unwelcomed exposures of humans and the environment to bioactive anthropogenic compounds.

It has been shown recently that even in these infinitesimal concentrations, some pharmaceuticals have the potential to interfere detrimentally with the normal development of aquatic life. The low concentrations of pharmaceuticals are unlikely to elicit acute health toxic effects. What is not well enough known is whether more subtle effects such as growth, fertility, sex ratios of higher organisms, or reproductive behavior are potentially being impacted in aquatic life by the long-term, low-level exposure to pharmaceutically active compounds [124].

Moreover, in ecotoxicity testing, a few surrogates are used to represent the diversity of wild species, while several different species, for example, rat, mouse, rabbit, and guinea pig, are being used as surrogates for human risk assessment [125]. This raises considerable uncertainties when extrapolating data obtained in the laboratory on a few surrogates to species exposed in the field, which are often not closely related to them and may differ considerably in their sensitivity to the chemical being tested. Not surprisingly, large safety margins are often invoked when carrying out environmental risk assessment using such data [125].

The persistence of pharmaceuticals in the environment plays a major role in determining their potential for producing adverse environmental effects. The bad news side of this is that the industry is actively working to make pharmaceuticals more resistant to degradation chemistries suggesting that the problems will only get worse with time. The good news is that this overarching fact presents society with opportunities to obviate the problems. Either by restricting or by better controlling pharmaceutical releases and by developing more effective technologies for their rapid destruction in water, we should be able to substantially ameliorate the problems.

In the discussion of what is already available technology for water treatment, emphasis is placed upon important aspects of degradation technologies including the degree of pharmaceutical degradation, the identity and characteristics of the degradation intermediates and byproducts, and the possible degradation pathways [4]. In the light of these considerations, ozonation and AOP ozone-based systems, such as O_3/H_2O_2, O_3/UV_{254}, and $O_3/H_2O_2/UV_{254}$, which are already underpin significant technologies for the treatment of wastewaters, become an interesting emerging technology in order to promote removal of these contaminants from water [5,54].

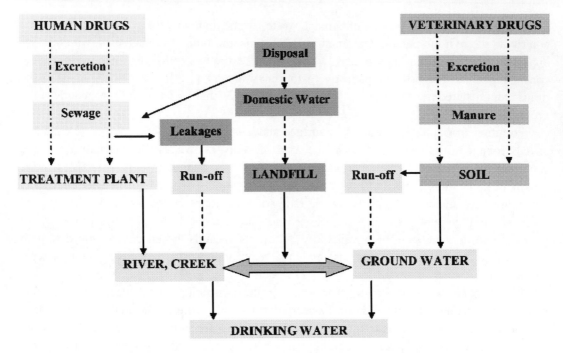

Figure 12. Fate of pharmaceuticals in the environment (adapted from ref. [118]).

The fates of human and veterinary drugs after urinary and fecal excretions are quite different. The excreted human pharmaceuticals pass through Sewage Treatment Plants (STP) prior to entering rivers or streams. Veterinary drugs are more likely to directly contaminate soil and groundwater without any sewage treatment. Figure 12 presents the fate of pharmaceuticals in the environment [118].

Many pharmaceutical compounds pass, at least in part, through sewage treatment plants to end up in environmental waters since STPs were often not designed to handle pharmaceuticals.

Seasonal variation in temperature and light intensity are considered to be the factors determining the fate of pharmaceuticals in surface waters [126,127]. Boreal winter climate conditions with low temperatures and low daylight hours may lead to decreased bio-and photodegradation of pharmaceuticals compared with summer. These processes are likely to be even less effective in rivers covered by ice and snow. Several reports have described the temporal and the spatial distributions of pharmaceuticals in surface water systems [4,127].

In surface water, the main elimination processes are biodegradation, sorption, and photodegradation. In general pharmaceuticals have a designed resistance to biodegradation, thereby inhibiting one of the major removal mechanisms. The sediment type has been shown to significantly affect the sorption of pharmaceuticals such as ibuprofen and diclofenac, thus implying that the efficiency of elimination by sorption is site specific and cannot be relied upon on a global basis.

Despite the ability of natural processes to break down many pharmaceuticals as discussed above, there is an increasing need for strategies to mitigate organic micropollutants in water as the demand for clean water is steadily rising. Compounds that are particularly susceptible to oxidation often contain heteroatoms with lone pairs of electrons (oxygen, nitrogen, sulfur), thus suggesting that APIs should be susceptible to oxidative treatment [4].

Several chemical methods have been employed for the degradation of pharmaceuticals, including ozonation and several AOP. While the effectiveness of direct photolysis is governed by the absorption spectra of the contaminant and the quantum yield, the dominant mechanism taking place in AOP involves highly reactive hydroxyl radicals, often significantly lowering the UV influence (dose) required for oxidation compared with direct photolysis [128-131]. Oxidative treatment of clofibric acid, ibuprofen, and diclofenac has been reported with O_3, H_2O_2/UV, and O_3/H_2O_2.

Ozonation was effective in the degradation of diclofenac with complete conversion of the chlorine into chloride ions and 32% mineralization [4]. A 90 min treatment of diclofenac with H_2O_2/UV resulted in 39% mineralization [129]. The combined application of O_3/H_2O_2 degraded all three compounds to more than 98% at 5.0 mg dm^{-3} O_3 and 1.8 mg dm^{-3} H_2O_2 [130].

Oxidative treatment with both H_2O_2/UV and O_3 completely removes the toxicity of a mixture consisting of carbamazepine, clofibric acid, diclofenac, sulfamethoxazole, ofloxacin, and propranolol in a moderately hard synthetic medium within 1 min of treatment [131].

Ozone treatment of biologically purified water from wastewater treatment plants is reported to reduce concentrations of many pharmaceuticals below detection limits [4]. The results are generally based on the disappearance of the parent compounds. This treatment would be useful in cases where the wastewater use poses ecotoxicological risks, such as irrigation in agriculture or dilution into surface waters [132].

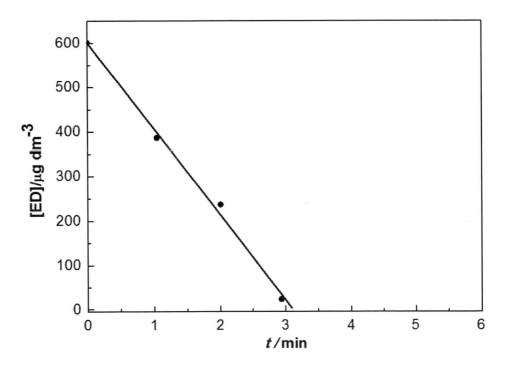

Figure 13. Removal of ED from water using ozonation. Composition: diclofenac, ibuprofen, bisphenol-A, estrone and 17α-ethinylestradiol, resulting in a total ED concentration of 600 μg dm^{-3} (adapted from ref. [54]).

Ozonation of the macrolid antibiotic lincomycin results in reduction of the toxicity of the treated solution toward the alga *S. leopoliensis* compared with untreated solutions containing the antibiotic [131]. Ozone treatment in synthetic wastewater at a dose rate of 2.96 mg dm^{-3} has also been reported to degrade in 1 h the fluoroquinole veterinary antibiotic enrofloxacin [133].

Ozonation and membrane filtration have achieved removal rates of >95% for many pharmaceutical compounds compared with an average of 60% for secondary wastewater treatment plants [4].

The degradation of different Endocrine Disruptors (ED) via ozonation was carried out by Franco et al. [54] and Deborde et al. [41].The aqueous ozone-induced oxidation of several ED was studied by Deborde et al. [41], who found that at pH > 5 ozone reacted to the greatest extent with dissociated ED forms. Besides, for all ED considered, O$_3$ exposures of only ~2 µg min dm^{-3} were calculated to achieve ≥ 95% removal efficiency. In the light of these findings the authors argued the ozonation process could thus highly oxidize the studied ED under water treatment conditions [41].

The fate of ED, pharmaceutical, and personal care product chemicals during simulated drinking water treatment processes was investigated by Westerhoff et al. [42], who found that ozone is capable to oxidize steroids containing phenolic moieties (e.g. estradiol, estrone, etc.) more efficiently than those without aromatic or phenolic moieties (e.g. progesterone, testosterone, etc.).

Ozonation of different ED under semi-batch was carried out by Franco et al. [54]. Figure 13 shows the dependence of total ED concentration on ozonation time. The quantitative HPLC analysis of the ozonated samples, comprising the summation of the several different peak areas, revealed that ozone application leads to total degradation of the ED mixture after ~ 3 min of ozonation. These results are in agreement with literature reports [41-43], and confirm that ozonation is an efficient process for treatment of waters containing ED. The treatment of the experimental findings obtained at short ozonation times ($t \leq 3$ min) furnished an average degradation rate of 30 µg dm^{-3} min^{-1} [54].

6.5. Pesticides in Water

The widespread application of DDT in the last century marked the beginning of a period of very rapid growth in pesticide use for many different purposes [1-3]. For instance, U.S. agriculture used in 90s of the last century about 365 million kg of pesticides per year, whereas about 900 million kg of insecticides were used in nonagricultural applications including forestry, landscaping, gardening, food distribution, and home pest control [1-2].

The potential exists for large quantities of pesticides to enter water either directly, in applications such as mosquito control or indirectly, primarily from drainage of agricultural lands. The harmful nature of organochlorines is due to their persistence, toxicity and a capacity to accumulate in living issues [134-143]. Taken together, these properties make organochlorines arguably the most damaging group of chemicals to which natural systems can be exposed. Most organochlorines are extremely stable.

The carbon-chlorine bond is commonly very strong and resists being broken down by physical processes. As a result many organochlorines remain in the environment for long periods of time. Moreover, when organochlorines do breakdown, they usually produce other organochlorines, the carbon-chlorine bond remaining intact as part of another compound, and these are sometimes more toxic and far more hazardous than the original substance [137].

When organochlorines enter the aquatic environment, their behavior depends very much upon their physical and chemical properties. Solvents such as chloroform and carbon tetrachloride are generally volatile and tend to evaporate from the water column leaking to the atmosphere. Less volatile compounds (semi-volatiles) such as chloroethane, chlorobenzene, chlorophenols and Polychlorinated Biphenyls (PCBs) tend to become bound up in sediments and enter the food chain [143].

Previous studies have shown that ozonation is effective at removing contaminants such as triazine herbicides, phenylureas and chlorophenoxy acids [135,144]. There is very little reported evidence of ozonation being an effective process for the removal of chlorinated solvents. However, some treatment plants have incorporated combinations of ozone with other oxidants and UV irradiation (AOP) in order to produce a higher concentration of hydroxyl radicals in the water, compared to ozone alone.

As the hydroxyl radicals are highly reactive they are capable of oxidising organic contaminants to carbon dioxide and mineral acids. Therefore, an increase in the radical concentration will heighten the possibility of the oxidation of unreactive organochlorine compounds such as chlorobenzene and chloroethane [135].

As previously discussed AOP based on ozonation may be used with several different wastewaters containing pesticides [7]. The treatment and removal of these compounds are among the main goals of the water and wastewater treatment industries and, as a result, constant advances are being made to improve the technologies of removal.

The literature dealing with the ozonation of more than 30 pesticides was reviewed by Reynolds [145], who reported that organophosphorus insecticides are generally more readily oxidized by ozonation than are organochlorine pesticides, while most organonitrogen and phenolic pesticides are readily oxidized via ozonation processes. More recently, an extensive revision concerning the degradation of several different pesticides by ozonation and ozone-based AOP was presented by Ikehata and El-Din [135,136].

7. OZONATION: BIODEGRADABILITY AND TOXICITY

In the case of heavily contaminated effluents containing a considerable amount of recalcitrant organics, where a biodegradability enhancement of the effluent is necessary before its introduction into the biological reactor, the use of AOT is imperative in order to reduce toxicity.

Ozonation processes are very effective in order to promote partially degradation of complex organic molecules which can be further eliminated via biologically active filtration process [5]. Indeed, when ozonation process is placed upstream of the biological filtration, and environmental conditions such as dissolved oxygen, pH and temperature are favorable, microbiological activity is increased in the filter and biodegradability of dissolved organic carbon is enhanced [5,9,54].

Ozone addition not only increases the biodegradability of the dissolved organics, but also introduces large amounts of oxygen to the water, thus creating an excellent environment for biological growth on the filtering media.

Different studies involving the investigation of biodegradability enhancement for several different industrial effluents after ozone application have been reported in the literature [5,7,86]. Rationalization of these findings permitted to conclude that biodegradability enhancement after ozone application depends on several parameters: (i) intrinsic properties of the effluent to be treated; (ii) reaction time and (iii) ozone dose. Therefore, it can conclude these variables play a key role in biodegradability enhancement and, therefore, optimization of the operating conditions adopted in the effluent treatment is very important in order to ensure biodegradability conditions.

Ozonation typically increases the biodegradability of NOM in water because many large organic molecules are converted into smaller organic molecules that are readily biodegradable. This increase in Biodegradable Dissolved Organic Carbon (BDOC) can lead to accelerated bacterial growth and regrowth in the distribution system if not removed in the treatment plant.

The biodegradability of the dye wastewater was evaluated by monitoring changes in 5-day Biochemical Oxygen Demand (BOD_5) with respect to COD [146]. Analysis of these findings indicate that the wastewater biodegradability increased with an increase in ozonation time, suggesting that partial oxidation of the dye occurred, not total mineralization, resulting in lower molecular weight compounds that were more amenable to microbial oxidation.

Figure 14 shows the feasibility of oxidation of the soluble organic matter during ozonation of the textile dye Reactive Orange 122 at different pH-values [55].

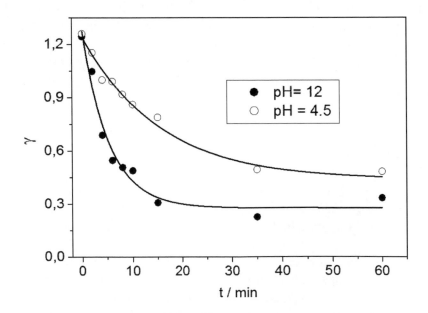

Figure 14. Influence of the ozonation process carried out in acid and alkaline solutions on feasibility of oxidation (γ = COD/TOC) of the soluble organic matter. Dye: Reactive Orange 122. IDC = 300 mg dm$^-$^{3} (adapted from ref. [55]).

Analysis of figure 14 clearly reveals that $\gamma(\equiv COD/TOC)$ decreases with increasing ozonation time, especially in alkaline conditions, thus indicating that ozonation leads to an increase in the oxidation feasibility of the soluble organic matter. As previously discussed by Franco et al. [54], this behavior is an experimental evidence that ozonation can constitutes, mainly in alkaline medium, an efficient pre-treatment step for ITP devoted for the wastewater treatment, where chemical oxidation and biologic processes are combined in order to provide an efficient and economical technology for remediation of recalcitrant pollutants.

8. Conclusion

The environmental aspects presented by ozonation and the advanced oxidation processes using ozone as a source for the "in situ" hydroxyl radical production, allied with the considerable efficiency presented by these processes for removal of pathogenic agents and several different classes of hazardous compounds, were presented in this chapter in order to justify the application of these alternative process on water and wastewater treatment.

Advances gained in the field of the ozone generation using the electric silent discharge and water electrolysis processes were presented. Besides, the well-know ozone generation using Corona ozonizers, it was reported the use of specially designed Electrochemical ozonizers can permit in some particular cases the adequate water treatment, thus resulting in several environmental and economical benefits.

It was emphasized that water and wastewater treatments using advanced oxidation processes must be carried out taking into account the specific characteristic of the system in order to avoid the release of strange substances in the environment. Thus, each water or effluent containing refractory hazardous substances deserves a special attention.

When correct applied, the advanced oxidation processes based on ozonation can indeed provide total discoloration, taste and odor removal, and toxicity reduction. Finally, upon the concept of the integrated treatment processes, one has the application of the advanced oxidation processes based on ozonation can furnish an increase in the oxidation feasibility of the dissolved organic matter, which can be further treated using well-established biochemical process.

References

[1]   Baird, C. *Environmental Chemistry*, 3th ed., W. H. Freeman and Company: New York, NY, 2004.

[2]   Manahan, S. E. *Environmental Chemistry*, 8th ed., CRC Press LLC: New York, NY 2005.

[3]   Tchobanoglous, G.; Burton, F. L. In *Wastewater engineering: treatment, disposal, and reuse*; Clark, B. J.; Morriss, J. M., eds.; 3rd ed.; Metcalf and Eddy, Inc.: Singapore, SG, 1991.

[4]   Khetan, S.K.; Collins T. J. Human pharmaceuticals in the aquatic environment: a challenge to green chemistry. *Chem. Rev.* 2007, *107(6)*, 2319-2364.

[5]   Da Silva, L. M.; Jardim, W. F. Trends and strategies of ozone application in environmental problems. *Quim. Nova* 2006, *29(2)*, 310-317.

[6]   Da Silva, L. M.; de Faria, L. A.; Boodts, J. F. C. Green processes for environmental application: electrochemical ozone production. *Pure Appl. Chem.* 2001, *73(12)*, 1871-1884.

[7]   Da Silva, L. M.; Santana, M. H. P.; Boodts, J. F. C. Electrochemistry and green chemical processes: electrochemical ozone production. *Quim. Nova* 2003, *26(6)*, 880-888.

[8]   Jr. Eckenfelder, W. W. *Industrial Water Pollution Control*, 2nd ed., McGraw-Hill: New York, NY, 1989.

[9]   EPA (1999). Alternative Disinfectants and Oxidants Guidance Manual, www.epa.gov/ogwdw000/mdbp/alternative_disinfectants_guidance.pdf

[10]   EPA (1995). *Toxics Release Inventory, Public Data Release*. www.denix. osd.mil/portal/page/portal/content/environment/ARC/FY1996/Toxics%20Release%20I nventory%20Public%20Data%20Report%201995.pdf

[11]   Benefield, L. D.; Judkins Jr., J. F.; Weand, B. L. *Process Chemistry for Water and Wastewater Treatment*. Prentice-Hall: New Jersey, NJ, 1982.

[12]   Anastas, P.; Williamson, T. *Green Chemistry, Theory and Practice*, Oxford University Press: Oxford, UK, 1998.

[13]   Deborde, M.; Rabonau, S.; Dughet, J.-P.; Legube, B. Kinetics of aqueous ozone-induced oxidation of some endocrine disruptors. *Environ. Sci. Technol.* 2005, *39(16)*, 6086-6092.

[14]   Mckenzie, K. S.; Sarr, A. B.; Mayura, K.; Bailey, R. H.; Miller, D. R.; Rogers, T. D.; Norred, W. P.; Voss, K. A.; Plattner, R. D.; Kubena, L. F.; Phillips, T. D. Oxidative degradation and detoxification of mycotoxins using a novel source of ozone. *Food Chem. Toxicol.* 1997, *35(5)*, 807-820.

[15]   Drinan, J. E. *Water & Wastewater Treatment: A Guide for the Nonengineering Professional*. CRC Press LLC: New York, NY, 2001.

[16]   Reckhow, D.A., Singer, P. C.; Trussell, R. R. *Ozone as a coagulant aid*. Seminar proceedings, Ozonation, Recent Advances and Research Needs, AWWA Annual Conference, Denver, CO, 1986.

[17]   Hoigné J.; Bader, H. The formation of trichloronitromethane (chloropicrin) and chloroform in a combined ozonation/chlorination treatment of drinking water. *Water Res.* 1988, *22(3)*, 313-317.

[18]   Hoigné, J. *The Chemistry of Ozone in Process Technologies for Water Treatment*, Plenum-Press: New York, NY, 1988.

[19]   Diaper, E. W. J. In *Ozone in Water and Wastewater Treatment*; Evans, F. L., Ed.; Ann Arbor Science: Michigan, MI, 1972.

[20]   Hoigné, J. In *Handbook of Ozone Technology and Applications*, Rice, R. G.; Netzer, A., Eds.; Ann Arbor Science: Michigan, MI, 1982.

[21]   Suffet, I. H.; Anselme, C.; Mallevialle, J. *Removal of Tastes and Odors by Ozonation*. Conference proceedings, AWWA Seminar on Ozonation: Recent Advances and Research Needs, Denver, CO, 1986.

[22]   Kinman, R. N. Water and wastewater disinfection with ozone: a critical review. *Crit. Rev. Environ. Contr.* 1975, *5(3)*, 141-152.

[23] DeMers, L. D.; Renner, R. C. *Alternative Disinfection Technologies for Small Drinking Water Systems.* AWWARF and AWWA: Denver, CO, 1992.

[24] Bringmann, G. Determination of the lethal activity of chlorine and ozone on *E. coli. Z. f. Hygiene* 1954, *139(2)*, 130-139.

[25] Chang, S. L. Modern concept of disinfection. *J. Sanit. Eng. Division.* 1971, *97(1),* 689-707.

[26] von Gunten, U.; Hoigné, J. Ozonation of Bromide-Containing Waters: Bromate Formation through Ozone and Hydroxyl Radicals. *Disinfection By-Products in Water Treatment*; Minear, R.A.; G.L. Amy, eds; CRC Press, Inc.: Boca Raton, FL, 1996.

[27] Langlais, B.; Reckhow, D. A.; Brink, D. R. *Ozone in Water Treatment: Application and Engineering*, CRC: Boca Raton, FL, 1991.

[28] Pontius, F. W. *Water Quality and Treatment*, American Water Works Association, McGraw-Hill: New York, NY, 1990.

[29] Georgeson, D. L.; Karimi, A. A. Water quality improvements with the use of ozone at the Los Angeles water treatment plant. *Ozone Sci. Eng.* 1988, *10(3)*, 255-276.

[30] Park, H-S.; Hwang, T-M.; Kang, J-W.; Choi, H.; Oh, H-J. Characterization of raw water for the ozone application measuring ozone consumption rate. *Water Res.* 2001, *35(11)*, 2607-2614.

[31] Hoigné, J.; Bader, H. The role of hydroxyl radical reactions in ozonation processes in aqueous solutions. *Water Res.* 1976, *10(5)*, 377-386.

[32] Hoigné, J.; Bader, H. Rate constants of reactions of ozone with organic and inorganic compounds in water—I: Non-dissociating organic compounds. *Water Res.* 1983, *17(2)*, 173-183.

[33] Hoigné, J.; Bader, H. Rate constants of reactions of ozone with organic and inorganic compounds in water—II: Dissociating organic compounds. *Water Res.* 1983, *17(2)*, 185-194.

[34] Hoigné, J.; Bader, H.; Haag, W. R.; Staehelin, J. Rate constants of reactions of ozone with organic and inorganic compounds in water—III. Inorganic compounds and radicals. *Water Res.* 1985, *19(8)*, 993-1004.

[35] von Gunten, U. Ozonation of drinking water: Part I. Oxidation kinetics and product formation. *Wat. Res.* 2003, *37(3)*, 1443-1467.

[36] von Gunten, U. Ozonation of drinking water: Part II. Disinfection and by-product formation in presence of bromide, iodide or chlorine. *Wat. Res.*2003, *3(3)*, 1469-1487.

[37] Yao, C. C. D.; Haag, W. R. Rate constants for direct reactions of ozone with several drinking water contaminants. *Wat. Res.* 1991, *25(7)*, 761-773.

[38] Benbelkacem, H.; Debellefontaine, H. Modeling of a gas-liquid reactor in batch conditions. Study of the intermediate regime when part of the reaction occurs within the film and part within the bulk. *Chem. Eng. Proc.* 2003, *42(3)*, 723-732.

[39] Benbelkacem, H.; Cano, H.; Mathe, S.; Debellefontaine, H.; Maleic acid ozonation: reactor modeling and rate constants determination. *Ozone Sci. Eng.* 2003, *25(1)*, 13-24.

[40] Whitlow, J. E.; Roth, J. A. Heterogeneous ozonation kinetics of pollutants in wastewater. *Environ. Progress* 1988, *7(1)*, 52-57.

[41] Watkinson, A. J.; Murby, E. J.; Constanzo, S. D. Removal of antibiotics in conventional and advanced wastewater treatment: Implications for environmental discharge and wastewater recycling. *Wat. Res.* 2007, *41*, 4164-4176.

[42]   Westerhoff, P.; Yoon, Y.; Snyder, S.; Wert, E. Fate of endocrine-disruptor, pharmaceutical, and personal care product chemicals during simulated drinking water treatment processes. *Environ. Sci.Technol.* 2005, *39(17)*, 6649-6663.

[43]   Huber, M. M.; Canonica, S.; von Gunten, U. Oxidation of pharmaceuticals during ozonation and advanced oxidation processes. *Environ. Sci.Technol.* 2003, *37(5)*, 1016-1024.

[44]   Ciardelly, G.; Ranieri, N. The treatment and reuse of wastewater in the textile industry by means of ozonation and electroflocculation. *Wat. Res.* 2001, *35(2)*, 567-572.

[45]   Masten, S. J. S.; Davies, H. R.; The use of ozonation to degrade organic contaminants in wastewaters. *Environ. Sci. Technol.* 1994, *28(4)*, 181-185.

[46]   Szpyrkowicz, L.; Juzzolino, C.; Kaul, S. N. A comparative study on oxidation of disperse dyes by electrochemical process, ozone, hypochlorite and Fenton reagent. *Wat. Res*. 2001, *35(9)*, 2129-2136.

[47]   Gogate, P.; Pandit, A. B. A review of imperative technologies for wastewater treatment II: hybrid methods. *Adv. Environ. Res*. 2004, *8(3-4)*, 553-597.

[48]   Tzitzi, M.; Vayenas, D. V.; Lyberatos, G. Pretreatment of textile industrial wastewaters with ozone. *Wat. Sci. Technol.* 1994, *29(9)*, 151-160.

[49]   Churchley, J. H. Ozone for dye waste color removal: four years experience at Leek STW. *Ozone Sci. Eng.* 1998, *20(2)*, 111-120.

[50]   Eremektar, G.; Selcuk, H.; Meric, S. Investigation of the relation between COD and the toxicity in the textile finishing industry wastewater: Effect of preozonation. *Desalination* 2007, *211*, 314-320.

[51]   Konsowa, A. H. Decolorization of wastewater containing direct dye by ozonation in a batch bubble column reactor. *Desalination* 2003, *158(1)*, 233-240.

[52]   Hsu, Y. C.; Chen, J. T.; Yang, H. C.; Chen, J. H. Decolorization of dyes using ozone in a gas-induced reactor. *AIChE J.* 2001, *4(1)*, 169-176.

[53]   Libergott, N.; van Lierop, B.; Nolin, A.; Faubert, M. G.; Laflamme, J. K. Modifying the bleaching process to decrease AOX formation. *Pulp Paper Can.* 1991, *92(3)*, 84-89.

[54]   Franco, D. V.; Boodts, J. F. C.; Jardim, W. F.; Da Silva, L. M. Electrochemical ozone production as an environmentally friendly technology for water treatment. *Clean* 2008, *36(1)*, 34-44.

[55]   Santana, M. H. P., Da Silva, L. M., Freitas, A. C., Boodts, J. F. C., Fernades, K. C., De Faria, L. A. Application of electrochemically generated ozone to the discoloration and degradation of solutions containing the dye Reactive Orange 122. *J. Hazard. Mat.* 2008, *in press*.

[56]   Marco, A.; Esplugas, S.; Saum, G. How and why combine chemical and biological processes for wastewater treatment. *Wat. Sci. Technol.* 1997, *35(3)*, 321-325.

[57]   Scott K. *Electrochemical Processes for Clean Technology*. Hartnolls: Cornwall, UK, 1995.

[58]   Tatapudi, P.; Fenton, J. M. *In Environmental oriented electrochemistry*. Sequeira, C. A. C, ed. New York, NY, 1994.

[59]   Petala, M; Tsiridis,V; Samaras, P.; Zouboulis, A.; Sakellaropoulos, G. P. Wastewater reclamation by advanced treatment of secondary effluents. *Desalination* 2006, *195(3)*, 109-119.

[60] Karnik, B. S.; Davies, S. H.; Baumann, M. J.; Masten, S. J. The effects of combined ozonation and filtration on disinfection by-product formation. *Wat. Res.* 2005, *39(3)*, 2839-2850.

[61] Hao, O. J.; Kim, H.; Chiang P. C. Decolorization of wastewater. *Crit. Rev. Environ. Sci. Technol.* 2000, *30(4)*, 449-505.

[62] Zoubov, N.; Pourbaix, M. In *Atlas of Electrochemical Equilibria in Aqueous Solutions*; M. Pourbaix, ed.; NACE International: Texas, TX, 1974.

[63] Da Silva, L. M.; Franco, D. V.; Forti, J. C.; Jardim, W. F.; Boodts, J. F. C. Characterization of a laboratory electrochemical ozonation system and its application in advanced oxidation processes. *J. Appl. Electrochem.* 2006, *36(3)*, 523-530.

[64] Suarasana, I.; Mudurab, M.; Chirab, R.; Andreia, G.; Munceleanua, I.; Morara, R. A novel type ozonizer for wastewater treatment. *J. Electrostat.* 2005, *63(2)*, 831–836.

[65] Wendt, H.; Kreysa, G. *Electrochemical Engineering: science and technology in Chemistry and Other Industries*, Springer-Verlag: Berlin, GE, 1999.

[66] Foller, P. C.; Tobias, C. W. The Anodic Evolution of Ozone. *J. Electrochem. Soc.* 1982, *129(3)*, 506-515.

[67] Foller, P. C.; Kelsall, G. H. Ozone generation via the electrolysis of fluoboric acid using glass carbon anodes and air depolarized cathodes. *J. Appl. Electrochem.* 1993, *23(4)*, 996-1010.

[68] Foller, P.; Goodwin, M. L. The electrochemical generation of high concentration ozone for small-scale applications. *Ozone Sci. Eng.* 1984, *6(2)*, 29-36.

[69] Han, S.-D.; Kim, J. D.; Singh, K. C.; Chaudhary, R. S. Electrochemical generation of ozone using solid polymer electrolyte – State of the art. *Indian J. Chem.* 2004, *43(8)*, 1599-1614.

[70] Park, S.-G. Stable ozone generation by using boron-doped diamond electrodes. *Russ. J. Electrochem.* 2003, *39(3)*, 321-322.

[71] Katoh, M.; Nishiki, Y.; Nakamatsu, S. Polymer electrolyte-type electrochemical ozone generator with an oxygen cathode. *J. Appl. Electrochem.* 1994, *24(3)*, 489-494.

[72] Arihara, K.; Terashima, C.; Fujishima, A. Electrochemical production of high-concentration ozone-water using freestanding perforated diamond electrodes. *J. Electrochem. Soc.* 2007, *154(4)*, E71-E75.

[73] Onda, K.; Ohba, T.; Kusunobi, H.; Takezawa, S.; Sunakawa, D.; Araki, T. Improving characteristics of ozone water production with multilayer electrodes and operating conditions in a polymer electrolyte water electrolysis cell. *J. Electrochem. Soc.* 2005, *152(10)*, D177-D183.

[74] Wang, Y.-H.; Cheng, S.; Chan, K.-Y. Synthesis of ozone from air *via* a polymer-electrolyte-membrane cell with a doped tin oxide anode. *Green Chem.* 2006, *8(3)*, 568-572.

[75] Da Silva, L. M.; de Faria, L. A.; Boodts, J. F. C. Electrochemical ozone production: influence of the supporting electrolyte on kinetics and current efficiency. *Electrochim. Acta* 2003, *48(6)*, 699-709.

[76] Franco, D. V.; Da Silva, L. M.; Jardim, W. F.; Boodts, J. F. C. Influence of the Electrolyte Composition on the Kinetics of the Oxygen Evolution Reaction and Ozone Production Processes. *J. Braz. Chem. Soc.* 2006, *17(4)*, 746-757.

[77]   Da Silva, L. M.; Franco, D. V.; de Faria L. A.; Boodts, J. F. C. Surface, kinetics and electrocatalytic properties of Ti/(IrO$_2$ + Ta$_2$O$_5$) electrodes, prepared using controlled cooling rate, for ozone production. *Electrochim. Acta* 2004, *49(3)*, 3977-3988.

[78]   Stucki, S.; Baumann, H.; Christen, H.; Kötz, R. J. Performance of a pressurized electrochemical ozone generator. *J. Appl. Electrochem.* 1987, *1(3)*, 773-778.

[79]   Stucki, S.; Theis, G.; Kötz, R.; Devantay, H.; Christen, H. J. In situ production of ozone in water using a membrel electrolyzer. *J. Electrochem. Soc.* 1985, *132(2)*, 367-371.

[80]   Tatapudi, P. and Fenton, J. M.; Synthesis of Ozone in a Proton Exchange Membrane Electrochemical Reactor. *J. Electrochem. Soc.*1993, *140(12)*, 3527-3530.

[81]   Fang, Z.; Qiu, Y.; Sun, Y.; Wang, H.; Edmund, K. Experimental study on discharge characteristics and ozone generation of dielectric barrier in a cylinder-cylinder rector and a wire-cylinder reactor. *J. Electrostatics* 2008, *66(3)*, 421-426.

[82]   Park, S. -L.; Moon, J. -D.; Lee, S. -H.; Shin, S. -Y. Effective ozone generation utilizing a meshed-plate electrode in a dielectric-barrier discharge type ozone generator. *J. Electrostatics* 2006, *64(2)*, 275-282.

[83]   Gao, M. T.; Hirata, M.; Takanashi, H.; Hano, T. Ozone mass transfer in a new gas-liquid contactor-Karman contactor. *Sep. Purif. Technol.* 2005, *42(3)*, 145-149.

[84]   Koide, K.; Kato, S.; Tanaka, Y.; Kubota, H. Bubbles Generated from Porous Plate. *J. Chem. Eng. Japan* 1968, *1(1)*, 51-56.

[85]   Westerterp, K. R.; Wijngaarden, R. J. In *Ullmann's Encyclopedia of Industrial Chemistry*; Vol. B4; Elvers, B.; Hawkins, S.; Schulz, G., eds.; 5th Edition, VCH Publishers: Cambridge, UK, 1992.

[86]   Tang, W. Z. *Physicochemical Treatment of Hazardous Wastes*, CRC Press LLC: New York, NY, 2004.

[87]   Luck, F.; Djafer, M.; Vel Leitner, N. K.; Gombert, B.; Legube, B. Destruction of pollutants in industrial rinse waters by advanced oxidation processes *Water Sci. Technol.* 1997, *35(4)*, 287-292.

[88]   Duguet, J. P.; Tsutsumi, Y.; Bruchet, A.; Mallevialle, A. Chloropicrin in Potable Water: Conditions of Formation and Production during Treatment Processes. In *Water Chlorination: Chemistry, Environmental Impact and Health Effects*. Lewis Publishers: Chelsea, MI, 1985; Vol.5.

[89]   Ferguson, D. W.; McGuire, M. J.; Koch, B.; Wolfe, R. L.; Aieta, E. M. Comparing peroxone and ozone for controlling taste and odor compounds, disinfection byproducts, and microorganisms. *J. AWWA* 1990, *82(4)*, 181-191.

[90]   Masten, S. J.; Hoigné, J. Comparison of ozone and hydroxyl radical-induced oxidation of chlorinated hydrocarbons in water. *Ozone Sci. Eng.* 1992, *14(3)*, 197-214.

[91]   Aieta, E. M.; Reagan, K. M.; Lang, J. S.; McReynolds, L.; Kang, J-K.; Glaze, W. H. Advanced oxidation processes for treating groundwater contaminated with TCE and PCE: pilot-scale evaluations. *J. AWWA.*1988, *88(5)*, 64-72.

[92]   Meriç, S.; Kaptan, D.; Ölmez, T. Colour and COD removal from wastewater containing Reactive Black 5 using Fenton's oxidation process. *Chemosphere* 2004, *54(3)*, 435-441.

[93]   van der Zee, F. P.; Villaverde, S. Combined anaerobic-aerobic treatment of azo dyes – A short review of bioreactor studies. *Wat. Res.* 2005, *39(4)*, 1425-1440.

[94]   Hassemer, M. E. N.; Sens, M. L. *Revista Saneamento Ambiental* 2002, *7(1)*, 30-36.

[95]   Shu, H. Y.; Huang, C. R. Degradation of commercial azo dyes in water using ozonation and UV enhanced ozonation processes. *Chemosphere* 1995, *31(8)*, 3813-3825.

[96]   Arslan, I.; Balcioglu, I. A.; Tuhkanen, T.; Bahnemann, D. W. H_2O_2/UV-C and Fe^{2+}/H_2O_2/UV-C vs. TiO_2/UV-A treatment for reactive dye wastewater. *J. Environ. Eng.* 2000, *126(3)*, 903-910.

[97]   Lopez, A.; Ricco, G.; Mascolo, G.; Tiravanti, G.; Di Pinto, A. C.; Passino, R. Biodegradability enhancement of refractory pollutants by ozonation: a laboratory investigation on an azo-dyes intermediate. *Wat. Sci. Technol.* 1998, *38(4-5)*, 239-245.

[98]   Forgacs, E.; Cserháti, T.; Oros, G. Removal of synthetic dyes from wastewaters: a review. *Environ. Inter.* 2004, *30(3)*, 953-971.

[99]   Neamtu, M.; Yediler, A.; Siminiceanu, I.; Macoveanu, M.; Kettrup, A. Decolorization of disperse red 354 azo dye in water by several oxidation processes – a comparative study. *Dyes Pig.* 2004, *60(1)*, 61-68.

[100] Torrades, F.; Montaño, J. G.; Hortal, J. A. G.; Domènech, X.; Peral, J. Decolorization and mineralization of commercial reactive dyes under solar light assisted photo-Fenton conditions. *Solar Energy* 2004, *77(5)*, 573-581.

[101] Meriç, S.; Selçuk, H.; Belgiorno, V. Acute toxicity removal in textile finishing wastewater by Fenton's oxidation, ozone and coagulation-flocculation processes. *Wat. Res.* 2005, *39(6)*, 1147-1153.

[102] Pierce, J. Colour in textile effluents – the origin of the problem. *J. Soc. Dyers Colourists* 1994, *110(4)*, 182-186.

[103] Sevimli, M. F.; Sarikaya, H. Z.; Yazgan, M. S. A new approach to determine the practical ozone dose for color removal from textile wastewater. *Ozone Sci. Eng.* 2003, *25(1)*, 137-143.

[104] Chu, L.-B.; Xing, X.-H.; Yu, A.-F.; Zhou, Y.-N.; Sun, X.-L.; Jurcik, B. Enhanced ozonation of simulated dyestuff wastewater by microbubbles. *Chemosphere* 2007, *68(7)*, 1854–1860.

[105] Wu, J.; Wang, T.; Ozonation of aqueous azo dye in a semi-batch reactor *Wat Res* 2001, *35(4)*, 1093-1099.

[106] Sarayu, K.; Swaminathan, K.; Sandhya, S. Assessment of degradation of eight commercial reactive azo dyes individually and in mixture in aqueous solution by ozonation. *Dyes Pig.* 2007, *75(2)*, 362-368.

[107] Assalin, M. R.; Rosa, M. A.; Durán, N. Remediation of kraft effluent by ozonation: effect of applied ozone concentration and initial pH. *Ozone: Sci. Eng.* 2004, *26(1)*, 317-322.

[108] Helbe, A.; Schlayer, W.; Liechti, P. A.; Jenny, R.; Möbius, C. H. Advanced effluent treatment in the pulp and paper industry with a combined process of ozonation and fixed bed biofilm reactors. *Water Sci. Technol.* 1999, *40(11-12)*, 343-350.

[109] Freire, R. S.; Kubota, L. T.; Durán, N. Remediation and toxicity removal form Kraft paper mill effluent by ozonization. *Environ. Technol.* 2001, *22(3)*, 897-904.

[110] Nakamura, Y.; Sawada, T.; Kobayashi, F.; Godliving, M. Microbial treatment of Kraft Pulp wastewater pretreated with ozone. *Water Sci. Technol.* 1997, *35(2)*, 227-282.

[111] Sevimli, M. F. Post-treatment of pulp and paper industry wastewater by advanced oxidation processes. *Ozone Sci. Eng.* 2005, *27(3)*, 37-43.

[112] Kallas, J.; Munter, R. Post treatment of pulp and paper industry wastewaters using oxidation and adsorption processes. *Water Sci. Technol.* 1994, *29(5–6)*, 259-272.

[113] Oeller, H. J.; Demel, I.; Weinberger, G. Reduction in residual COD in biologically treated paper mill effluents by means of combined ozone and ozone/UV reactor stages. *Water Sci. Technol.* 1997, *35(2–3)*, 269-276.

[114] Gulyas, H.; von Bismark, R.; Hemmerling, L. Treatment of industrial wastewaters with ozone/hydrogen peroxide. *Water Sci. Technol.* 1995, *32(7)*, 127-134.

[115] Pérez, M.; Torrades, F.; Garcia-Hortal, J. A.; Doménech, J. Removal of organic contaminants in paper pulp treatment effluents under fenton and photo-fenton conditions. *Appl. Catal. B. Environ.* 2002, *36(1)*, 63-74.

[116] Rodriguez, J.; Contreras, D.; Parra, C.; Freer, J.; Baeza, J.; Durán, N. Pulp mill effluent treatment by fenton-type reactions catalyzed by iron complexes. *Water Sci. Technol.* 1999, 40(11–12), 351-355.

[117] Kunz, A.; Mansilla, H.; Durán, N. A degradation and toxicity study of three textile reactive dyes by ozone. *Environ. Technol.* 2002, *23(6)*, 911-918.

[118] Halling-Sorensen, B.; Nielsen, S. N.; Lanzky, P. F.; Ingerslev, F.; Lutzhoft, H. C. H.; Jorgensen, S. E. Occurrence, fate and effects of pharmaceutical substances in the environment-A review. *Chemosphere* 1998, *36(2)*, 357-361.

[119] Daughton, C. G.; Ternes, T. A. Pharmaceuticals and personal care products in the environment: agents of subtle change? *Environ. Health Perspect.* 1999, *107*(S6), 907-938.

[120] Kummerer, K. Resistance in the environment *J. Antimicrob. Chemother.* 2004, *54(2)*, 311-320.

[121] Wilson, C. J.; Brain, R. A.; Sanderson, H.; Johnson, D. J.; Bestari, K. T.; Sibley, P. K.; Solomon, K. R. Structural and functional responses of plankton to a mixture of four tetracyclines in aquatic microcosms. *Environ. Sci. Technol.* 2004, *38(23)*, 6430-6438.

[122] Heberer, T. Occurrence, fate, and removal of pharmaceutical residues in the aquatic environment: a review of recent research data. *Toxicol. Lett.* 2002, *131(1-2)*, 5-17.

[123] Daughton, C. G. Cradle-to-cradle stewardship of drugs for minimizing their environmental disposition while promoting human health. I. rationale for and avenues toward a green pharmacy. *Environ. Health Perspect.* 2003, *111(5)*, 757-774.

[124] Jones, O. A. H.; Voulvoulis, N.; Lester, J. N. Potential ecological and human health risks associated with the presence of pharmaceutically active compounds in the aquatic environment *Crit. Rev. Toxicol.* 2004, *34(5)*, 335-3450.

[125] Walker, C. H. Ecotoxicity testing of chemicals with particular reference to pesticides. *Pest Manag. Sci.* 2006, *62(7)*, 571-583.

[126] Loraine, G. A.; Pettigrove, M. E. Seasonal variations in concentrations of pharmaceuticals and personal care products in drinking water and reclaimed wastewater in southern California. *Environ. Sci. Technol.* 2006, *40(3)*, 687-695.

[127] Vieno, N. M.; Tuhkanen, T.; Kronberg, L. Seasonal variation in the occurrence of pharmaceuticals in effluents from a sewage treatment plant and in the recipient water. *Environ. Sci. Technol.* 2005, *39(21)*, 8220-8226.

[128] Rosenfeldt, E. J.; Linden, K. G. Degradation of endocrine disrupting chemicals bisphenol A, ethinyl estradiol and estradiol during UV photolysis and advanced oxidation processes. *Environ. Sci. Technol.* 2004, *38(20)*, 5476-5483.

[129] Vogna, D.; Marotta, R.; Napolitano, A.; Andreozzi, R.; d'Ischia, M. Advanced oxidation of the pharmaceutical drug diclofenac with UV/H$_2$O$_2$ and ozone. *Water Res.* 2004, *38(2)*, 414-422.

[130] Zwiener, C.; Frimmel, F. H. Oxidative treatment of pharmaceuticals in water.*Water Res.* 2000, *34(6)*, 1881-1885.

[131] Andreozzi, R.; Campanella, L.; Fraysse, B.; Garric, J.; Gonnella, A.; Lo Giudice, R.; Marotta, R.; Pinto, G.; Pollio, A. Effects of advanced oxidation processes (AOPs) on the toxicity of a mixture of pharmaceuticals. *Water Sci. Technol.* 2004, *50(5)*, 23-28.

[132] Ternes, T. A.; Stuber, J.; Herrmann, N.; McDowell, D.; Ried, A.; Kampmann, M.; Teiser, B. Ozonation: a tool for removal of pharmaceuticals, contrast media and musk fragrances from wastewater? *Water Res.* 2003, *37(8)*, 1976-1982.

[133] Balcioglu, I. A.; Otker, M. Treatment of pharmaceutical wastewater containing antibiotics by O$_3$ and O$_3$/H$_2$O$_2$ processes. *Chemosphere* 2003, *50(1)*, 85-95.

[134] Cooper, C.; Burch, R. An investigation of catalytic ozonation for the oxidation of halocarbons in drinking water preparation. *Wat. Res.* 1999, *33(18)*, 3695-3700.

[135] Ikehata, K.; El-Din, M. G. Aqueous pesticide degradation by ozonation and ozone-based advanced oxidation processes: A review (part II). *Ozone Sci. Eng.* 2005, 27(23), 173-202.

[136] Ikehata, K.; El-Din, M. G. Aqueous pesticide degradation by ozonation and ozone-based advanced oxidation processes: a review (part I). *Ozone Sci. Eng.* 2005, *27(2)*, 83-11.

[137] Acher, A. J.; Hapeman, C. J.; Shelton, D. R.; Muldoon, M. T.; Lusby, W. R.; Avni, A.; Waters, R. Comparison of formation and biodegradation of bromacil oxidation-products in aqueous-solutions. *J. Agric. Food Chem.* 1994, *42(9)*, 2040-2047.

[138] Adams, C. D.; Randtke, S. J. Ozonation By-products of Atrazine in Synthetic and Natural-Waters. *Environ. Sci. Technol.* 1992, *26(11)*, 2218-2227.

[139] Al Momani, F. A.; Gamal El-Din, M.; Smith, D. W.; Bhandari, A.; Hutchinson, S. L. Pesticides and herbicides. *Water Environ. Res.* 2004, *76(6)*, 1775-1856.

[140] Baldauf, G. Removal of pesticides in drinking-water treatment. *Acta Hydrochim Hydrobiol.* 1993, *21(4)*, 203-208.

[141] Beltrán, F.; Acedo, B.; Rivas, J. Use of ozone to remove Alachlor from surface water. *Bull. Environ. Contam. Toxicol.* 1999, *62(3)*, 324-329.

[142] Bolzacchini, E.; A. Brambilla, M. O.; Polesello, S.; Rindone, B. Oxidative pathways in the degradation of triazine herbicides - A mechanistic approach. *Water Sci. Technol.* 1994, 30(7), 129-136.

[143] EPA (1998). Status of Pesticides in Registration, Registration, and Special Review. http://www.epa.gov/oppsrrd1/Rainbow/98rainbo.pdf

[144] Acero, J. L.; Benitez, F. J.; Real, F. J.; Maya, C. Oxidation of acetamide herbicides in natural waters by ozone and by the combination of ozone/hydrogen peroxide: kinetic study and process modeling. *Ind. Eng. Chem. Res.* 2003, *42(23)*, 5762-5769.

[145] Reynolds, G. Pesticides: a review. *Ozone Sci. Eng.* 1989, *11(3)*, 339-340.

[146] Lackey, L. W.; Mines Jr., R. O.; McCreanor, P. T. Ozonation of acid yellow 17 dye in a semi-batch bubble column. *J. Hazard. Mat.* 2006, *B138*, 357–362.

In: Water Purification
Editors: N. Gertsen and L. Sønderby

ISBN 978-1-60741-599-2
© 2009 Nova Science Publishers, Inc.

Chapter 2

PARADIGM SHIFT FROM A CLEAN OCEAN TO A BOUNTIFUL OCEAN: AN ESSENTIAL VISION REVEALED BY ECOLOGICAL MODELING OF "TIDAL FLATS - CENTRAL BAY AREA COUPLING" AND "BENTHIC-PELAGIC ECOSYSTEM COUPLING"

Akio Sohma

Environmental, Natural Resources and Energy, Div.
Mizuho Info and Research Institute, Tokyo, Japan

ABSTRACT

A newly developed ecosystem model - the first model describing the ecological connectivity consisting of both benthic-pelagic and central bay-tidal flat ecosystem coupling while simultaneously describing the vertical micro-scale in the benthic ecosystem - was developed and applied to Tokyo Bay (Sohma *et al.* 2005a, 2008). The model permits the prediction/evaluation of the effects of environmental measures, such as tidal flat creation/restoration, sand capping, dredging, and nutrient load reduction from rivers, on the hypoxic estuary from the perspectives of (1) the whole estuary composed of temporal-spatial mutual linkage of benthic-pelagic or central bay-tidal flat ecosystems (holistic approach), and (2) each biochemical and physical process contributing to oxygen production/consumption (elemental approach). The model outputs demonstrated the significant ecosystem responses as follows. First, the oxygen consumption in the benthic system during summer was quite low due to the low level of dissolved oxygen (hypoxia), although reduced substances, Mn^{2+}, Fe^{2+}, and S^{2-}, were highly produced and accumulated in the pore water. This result denotes importance to use the oxygen consumption rate under the high level of dissolved oxygen as the index of hypoxia potential. Second, both the tidal flat creation and nutrient load reduction decreased the anoxic water volume and mass of detritus in Tokyo Bay. However, the creation of tidal flats led to the higher biomass of benthic fauna, while the nutrient load reduction led to the lower biomass of benthic fauna compared to the existing situation. This result clarifies the differences from a measure aimed at a "bountiful ocean; a non-hypoxic and rich production of higher level trophic biology" to a measure just aimed at a "clear ocean; a non-hypoxic and low level of particulate organic matter" and also the differences from a bountiful ecosystem to a higher water quality. Lastly, in the simulation, Tokyo Bay reproducing reclaimed tidal flats (earlier Tokyo Bay system) prevented the increase of oxygen consumption potential (hypoxia potential) and the decrease of higher trophic production to red tide, compared to the existing Tokyo Bay system with reclamation of tidal flats. This result demonstrates the higher ecosystem tolerance of the earlier Tokyo Bay to red tide, and the tidal flat's function of keeping an optimized ecological balance to environmental perturbation.

Keywords: coastal ecosystem, environmental restoration, hypoxia, tidal flat reproduction/ restoration, nutrient load reduction

1. INTRODUCTION

Eutrophication, a serious problem in Japanese estuaries such as Tokyo Bay, has been thought to direct its ecosystem to red tide, hypoxia, and eventually to the decrease in the number of species and biomass of living organism. In Japan, the nutrient load reduction from rivers has been conducted since the 1960s and the reduced values in the case of Tokyo Bay is now achieved at the value from one third to half in terms of nitrogen and phosphorus (Ministry of the Environment, 2006). As a result, the water quality in Chemical Oxygen Demand (COD), Total Nitrogen (TN), and Total Phosphorus (TP) has now recovered as the values of these dropped to lower compared to the worst time. However, the number of species and biomass of living organisms has not recovered. One of the reasons is thought to be the disappearance of shallow waters (tidal flats) due to reclamation. Nowadays, a specific vision/direction for how to recover the eutrophic estuaries is required.

In order to contribute this, an ecosystem model, Ecological Connectivity Hypoxia Model: ECOHYM ("ZAPPAI"[1] in Japanese) (Sohma et al, 2005a, 2008) was newly developed. The features of ECOHYM can be briefly described as (1) modeling both the benthic and pelagic ecosystems and their linkage, (2) modeling both tidal-flat and central bay areas and their linkage, and (3) describing a micro-scale vertical spatial resolution of the benthic ecosystem. The vanguard approach/attempt of achieving all three items, (1), (2), and (3) simultaneously, was not only the "technological" challenge of model development but also the "philosophical" challenge of how we should apprehend the estuarine environment.

In this document, firstly, the "philosophy" and "technology" on the development of ECOHYM are introduced. Secondly, the detailed description of ECOHYM is demonstrated. Lastly, how much the collaboration of "philosophy" and "technology" was achieved and what was revealed from it, are overviewed through interweaving the several analysis from the model implementation to Tokyo Bay.

---

[1] The Japanese model name, "ZAPPAI" is named after a type of Japanese poem born from the general public (commoner) during the samurai period (Edo period) (Miyata, 2003). "ZA" means "miscellaneous" and "PPAI" means "haiku" in Japanese.

2. "PHILOSOPHY" - HOW WE UNDERSTAND
THE ENVIRONMENTAL PROBLEM OF ESTUARIES

In this section, the significant visions on considering the series of eutrophication problems in the estuary are overviewed. The visions shown in this section are not independent but are linked to each other. Therefore, they may include duplication. However, all of the visions introduced here are thought to be a meaningful philosophy. They were the reasons why the new model, ECOHYM, was developed and were also the motivation towards the challenge of newly developing this model. It is now possible to estimate/predict some of the visions quantitatively as the result of developing ECOHYM.

2.1. Significance of the Ecosystem Chain Response - "Interaction between Tidal Flat and Central Bay Area" and "Interaction between Benthic and Pelagic Systems"

The importance of the roles of the tidal-flat ecosystem and benthic ecosystem are explained here on the topics of hypoxia in Tokyo Bay. Hypoxia is a serious environmental problem in the semi-closed coastal zone, Tokyo Bay in Japan, being chronic from May to September (Chiba Prefectural Fisheries Research Center, 2001-2006, Kanagawa Prefectural Fisheries Research Institute, 2005-2006). Hypoxia is related explicitly or implicitly with the phenomena of eutrophication, reclamation of tidal flats, red tide (rapid growth of phytoplankton), blue tide (upwelling of oxygen depleted water body from sea-bottom to sea-surface), and decrease of fishery biomass (Suzuki et al., 1998; Imao et al., 2004). Figure 1 shows the conceptual diagram on the linkage of hypoxia and the accompanied phenomena/processes with hypoxia (Sohma et al., 2003). Generally, a tidal flat is a high biological production area where many species of living organisms dominate (Odum, 1971). Tidal flats are usually located along the coastal line of the bay and have been targets of reclamation due to their allocations when considering the convenience and cost benefits. One of the causes believed for red tide/blue tide repetition is precisely the disappearances of tidal flats due to reclamation (Kikuchi, 1993; Ishida and Hara, 1996; Aoyama and Suzuki, 1997). Disappearance of tidal flats reduces the aquaculture of benthic fauna (bivalve etc.) dominated there and reduces the predation pressure to phytoplankton. The low predation pressure induces abrupt increase in the phytoplankton population (red tide) and the large amount of dead shape of phytoplankton settles out toward the seafloor. The accumulated non living organism (dead shape of phytoplankton) at the seafloor changes the metabolism around the sediment-water interface and increases the oxygen consumption at sea bottom (Furota, 1988; Suzumura et al. 2003). The oxygen depleted water body formulated at the sea bottom, due to accession of oxygen consumption, leads the mass mortality of living organisms which dominates not only at sea bottom of the central bay area (hypoxic area), but also at the sea surface of the central bay area or at the tidal flat area caused by upwelling/transverse flow. The flow transports hypoxic water body from the bottom to the surface or from the central bay to tidal flats. The mass mortality caused by oxygen depletion promotes further harmful effects in the estuary (Figure 1). In this way, generation mechanism of hypoxia or ripple effect of hypoxia is formulated with the association (hereinafter referred to as "ecosystem network") comprised of mutual interaction between the benthic and pelagic systems and between the central bay (hypoxic) and tidal flat areas.

Hereinafter, the chain response of the ecosystem in Figure 1, left, which started from the disappearance of tidal flat areas towards the environmental deterioration, is what I have defined to as "environmental deterioration spiral (negative spiral)" (Sohma et al, 2003). The environmental deterioration spiral has the possibility to lead survival of few living organisms which have high tolerance against the low level of oxygen. As the result, it is provisioned that the negative spiral leads to the ocean characterized by (1) low/poverty biodiversity, (2) difficulty in nutrient transition from lower to higher trophic level, and (3) low utilizable potential of stored nutrients in the estuary.

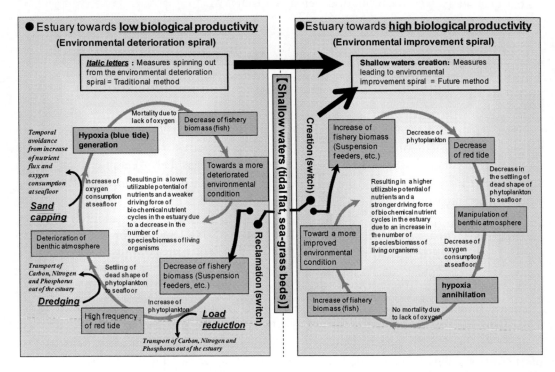

Figure 1. An example of the ecological chain responses in the environmental deterioration spiral and environmental improvement spiral accompanied by tidal flat reclamation and creation. The positions of other environmental measures (load reduction, dredging and sand capping) and the results of spirals are also illustrated.

2.2. Essence of Objectives of Environment Improvement Measures – "To Inhibit Eutrophication or to Recover the Bountiful Ecosystem" and "Positive Spiral and Negative Spiral"

Considering the objectives of environmental measures such as (1) nutrient load reduction, (2) sand capping, (3) dredging and (4) tidal flat reproduction/restoration from the perspective of the environmental deterioration spiral, nutrient load reduction, sand capping and dredging essentially resemble each other and differ from tidal flat reproduction/restoration. It means that the objectives of nutrient load reduction, sand capping and dredging are basically to escape/spinout from the processes/paths consisting of the environmental deterioration spiral, i.e. (a) increase of phytoplankton, (b) settling of dead shape of phytoplankton to seafloor, and (c) increase of oxygen consumption around the seafloor (Figure 1, left). The expectation of such measures is to stop and prevent the environmental deterioration spiral. In contrast, the tidal flat restoration is based on the thought of generating the driving force of environmental improvement and creating the reversed spiral to a negative spiral, rather than stopping/preventing the environmental deterioration spiral (Figure 1, right); viz. reproducing/creating the tidal flat as the comfortable dominant zone for living organisms artificially, and making living organisms resuscitated/established autonomously and making the lost bio-function returned. As a result, predation pressure to phyplankton increases and red tide is prevented. Preventing red tide leads to the reduction of sedimentation flux of dead shape of phytoplankton toward the seafloor and the benthic circumstances for living organisms is recovered. Hypoxia is prevented and the mortality of fish/shellfish from lack of oxygen does not occur. The ecosystem chain response explained above may recover the hypoxic estuary autonomously towards the estuary featuring (1) high/bountiful biodiversity, (2) facility in nutrient transition from lower to higher trophic level, and (3) higher utilizable potential of stored nutrients in the ocean. The reversed spiral to "environmental deterioration spiral" mentioned above is what I have defined as "environmental improvement spiral (positive spiral)" (Sohma et al, 2003). The measure, tidal flat reproduction/restoration, inducing the environmental improvement spiral may be said "the drastic measure" against the environmental deterioration spiral. In contrast, the measures stopping/preventing the environmental deterioration spiral, such as nutrient load reduction, sand capping and dredging, may be said as "the stopgap measure".

When considering the measures of nutrient load reduction, sand capping, dredging and tidal flat reproduction/restoration from the perspective of nutrients cycling (nitrogen and phosphorus cycling), nutrient load reduction, sand capping and dredging are the methods aiming at the transportation of the excess nutrients, which cannot be used effectively to a higher trophic level production, out of the system (estuary). In contrast, tidal flat reproduction/restoration is the method aiming to raise the potential of effective nutrient utilization through the recovery of a bountiful ecosystem and assimilation of the nutrients to living organisms of a higher trophic level. In simple terms, nutrient load reduction, sand capping and dredging are the measures of which the primary objective is "inhabitation of eutrophication (reduction of excess nutrients)", but the tidal flat reproduction/restoration is the measure of which the primary objective is, "recovery of a bountiful ecosystem (lifting up the potential of nutrient utilization)".

3. "TECHNOLOGY" – THE REQUIREMENTS FOR AN UNPRECEDENTED NEW MODEL; "MULTIPLE ECOSYSTEM" AND "VERTICAL MICRO SCALE MECHANISMS OF THE BENTHIC SYSTEM"

In order to conduct a coastal environmental management (recovery and development of the coastal zone) strategically, the prediction/estimation on the effect of environmental improvement measures (i.e. nutrient load reduction, sand capping, dredging and tidal flat reproduction/restoration) and/or the development (i.e. tidal flat reclamation) on the coastal ecosystem is required. It is more favorable if the prediction/estimation can demonstrate the ecosystem response (1) of both short and long time-scales, (2) quantitatively (which direction the ecosystem responds towards) and qualitatively (how much the effect is) and (3) mechanically to explain the causes of the effect. On the basis of the discussion described in section 2, the ideal method to perform such a prediction/estimation, is to embrace the whole estuary as a "multiple ecosystem" composed of temporal-spatial mutual linkage of each ecosystems between the tidal flat area and the central bay area, and ecosystems between the benthic system and the pelagic system.

In order to evaluate the multiple ecosystem on the theme of eutrophication problem, it is crucial to consider (1) the physical-biochemical processes and (2) their mechanical linkage; those that are thought to be significant for eutrophication effects on the ecosystem and those that occur inside and at the boundary of each ecosystem comprising a multiple ecosystem (i.e. tidal flat pelagic ecosystem, tidal flat benthic ecosystem, central bay pelagic ecosystem, and central bay benthic ecosystem). The reason is that the multiple ecosystem is composed of tangles of physical-biochemical processes and they propagate the effect of environmental measures, development, impact and disturbance. The ecosystem response is the result of spillover effects from the tangles. The positive spiral and negative spiral shown in Figure 1 may be just one example of the result from the tangles.

The ecosystem model describes each physical-biochemical processes and the ecological dynamics derived from the mutual interaction (tangles) of physical- biochemical processes. Therefore, it is a powerful tool to reveal both the ecosystem mechanisms and the ecosystem response to seasonal/daily or incidental changes of external/internal environment of the estuary, such as meteorology, nutrient load from rivers, red tide, sand capping, dredging and reproduction/reclamation of tidal flats etc. On the bases of these advantages of the ecosystem model, many series of the models have been developed. However, when the development of ECOHYM started, no ecosystem model satisfied all the following requirements simultaneously, which should be considered to estimate/predict the hypoxic estuary from the perspectives/background mentioned above.

(1) Requirements on Modeling the Benthic Ecosystem in the Central Bay Area

The main cause of hypoxia is thought to be the consumption of oxygen around the seafloor in the central bay area. This consumption originates from biochemical processes in the benthic sediment which change precipitously on a micro-scale in the vertical direction (Canfield et al., 1993). Thus, to demonstrate hypoxia dynamics accurately, describing the vertical profiles of biochemical processes in the micro-scale (mm scale pitch) is required (Rysgaard and Berg, 1996). One-dimensional benthic biochemical models with a vertical micro-scale have now reached such a level of sophistication and comprehensiveness that they can accurately reproduce benthic metabolism as well as carbon, nutrients, or oxygen cycling in the sediment or sediment-water interface (Soetaert et al., 1996a, 1996b, 2000; Boudreau, 1996; Sohma and Sayama, 2002; Dedieu et al, 2007).

(2) Requirements on Modeling the Tidal Flat Ecosystem

A tidal flat, where high potential for hypoxia improvement exists, is an oxygen producing area caused by (a) the photosynthesis of benthic algae, sea-grass and sea-weed, and (b) the accelerated aeration driven by emersion/submersion and nearshore waves. Oxygen dynamics in tidal flats are more complex and vary considerably within a daily scale (Kuwae et al., 2003). Furthermore, similar to the central bay area, metabolism in the tidal flat sediment changes precipitously within a micro-scale in the vertical direction (Revsbech et al., 1986; Kuwae et al., 2003). Therefore, in order to evaluate the dynamics of metabolism or oxygen production/consumption mechanisms in a tidal flat, an ecosystem model that simultaneously describes the time course of oxygen changes on a 1-day scale as well as vertical benthic metabolic profiles on a micro-scale is required. Heretofore, several ecosystem models applied to the tidal flat area have been developed (Baretta and Ruardij, 1988; Hata and Nakata, 1998; Sohma et al., 2000; Nakamura et al., 2004). Some models are able to calculate the dynamics of the boundary between the oxic (aerobic) layer and anoxic (anaerobic) layer in the benthic system, although they do not calculate the details of diagenetic processes along the vertical scale (Baretta and Ruardij, 1988; Hata and Nakata, 1998; Sohma et al., 2000). Moreover, some models have focused not only on the seasonal dynamics but also dynamics assessed on a daily scale (Sohma et al., 2000; Nakamura et al., 2004). However, to the best of our knowledge, no ecosystem model had existed yet which simultaneously described both the benthic vertical micro-scale metabolic mechanisms and daily dynamics in a tidal flat.

(3) Requirements on Modeling a Multiple Ecosystem

As mentioned above, hypoxia generation/annihilation and its ripple effects shown in Figure 1 are formulated by mutual interaction between benthic-pelagic ecosystems and between central bay-tidal flat ecosystems, i.e. a multiple ecosystem. Therefore, when evaluating the ecological response of the hypoxic estuary, the model is required to contain (a) the benthic and pelagic systems in both the central bay and tidal flat areas, and (b) mutual interactions between the benthic and pelagic systems, and between the central bay and tidal flat areas. In recent years, a number of benthic-pelagic coupling models have been developed (Baretta and Ruardij, 1988; Baretta et al., 1995; Baretta-Bekker and Baretta, 1997; Sohma et al., 2001; Sohma et al., 2004; Luff and Moll, 2004), with several models describing diagenetic (metabolic) processes in detail (Luff and Moll, 2004).

Furthermore, the requirements in items (1), (2) and (3) mentioned above are not independent from each other, because vertical micro scale mechanisms in the benthic system (items (1) and (2)), or the daytime scale dynamics in the tidal flat area (item (2)) is controlled by the ecosystem network in the multiple ecosystem (item (3)). Thus, the ecosystem model which treats all the requirements in items (1), (2) and (3) at the same time (i.e. describing both (a) the micro scale spatial resolution in benthic and (b) the ecosystem dynamics of daytime scale, plus (c) including the benthic-pelagic ecosystem both in the central bay and tidal flat simultaneously) is important/useful to evaluate the ecosystem dynamics of a hypoxic estuary. However, such models or model studies had not existed.

4. TOWARD THE COLLABORATION OF "PHILOSOPHY" AND "TECHNOLOGY"- OBJECTIVES AND SIGNIFICANCES OF THE RESEARCH; HOLISM AND REDUCTIONISM

On the basis of the philosophical and technological background mentioned above, a new ecosystem model, the Ecological Connectivity Hypoxia Model (ECOHYM), the first model to meet all the requirements in items (1), (2) and (3) simultaneously (i.e. describing both (a) the micro-scale spatial resolution in benthic and (b) the ecosystem dynamics of daytime scale, plus (c) including the benthic-pelagic ecosystem both in the central bay and tidal flat) was developed. With this, the challenges for the clarification of the mechanisms of hypoxia and for the prediction of the ecosystem response and tolerance to environmental measures, development, impact and disturbance were performed from two perspectives: (1) the whole estuary, composed of temporal-spatial mutual linkage of benthic-pelagic or central bay-tidal flat ecosystems (holism), and (2) each physical-biochemical process, contributing to oxygen production/consumption (reductionism).

For ECOHYM to complete these challenges, firstly, the selection of treating the physical and biochemical processes in the model is significant. It should be performed based on the confirmation of the known knowledge of the hypoxia and its related phenomena mechanically. Each selected physical-biochemical process should be formulated based on the latest scientific knowledge as much as possible. This approach is derived from reductionism. The additional requirement for ECOHYM is not the apprehension of each physical and biochemical processes fragmentary nor the description of the ecosystem by superposition (the stack) of them, but is the description of the mechanical linkage and interaction of the each process. Therefore, the numerical construction (Sohma, 2005b), which can describe the autonomous response/the feedback effect due to the entanglement of each process and can estimate the dynamics of the ecosystem as a whole, has to be applied to ECOHYM. Modeling the internal mechanisms of the benthic and pelagic ecosystems or the tidal flats and central bay ecosystems, and also linking each ecosystem by such a numerical construction enables us to regard the whole estuarine ecosystem as the ecosystem of temporal-spatial mutual linkage of the benthic and pelagic systems or the central bay and tidal flat areas. This approach is derived from holism.

The success of the development of such a model reveals where and how much each modeled physical and biochemical processes contribute relatively on the oxygen consumption and production, and leads to the clarification of hypoxic mechanisms. In addition, it enables us to predict the response of the whole ecosystem of the estuary to the environmental measures and perturbations quantitatively and qualitatively, while considering the ripple effects of the measures/perturbations through the entanglement of physical and biochemical processes. These results are also linked to establishing the foundation of the cost-performance evaluation on the environmental improvement technology, such as nutrient reduction, dredging, sand capping, and tidal flat restoration. Here, "quantitatively" means the direction/trend of the temporal and spatial dynamics.

On modeling, several assumptions have to be forced to the ambiguous/unknown/ uncertain processes. However, the fact that model outputs the result of mechanical interactions of many processes, suggests the possible breakthrough on some of the yet-to-be-defined processes reductively/reversely from the perspective of the whole ecosystem balance (the holistic approach), although they have not been clarified yet through the research of each process piece by piece. Such a method is sometimes effective on advancing the frontiers of science.

Because all existing models are the condensed/simplified description of the real system, any model includes some approximations and assumptions. However, if the political measure is argued by reference to the model output/results, and if the persons engaging in the decisions of political measures may discuss about the output/results with sufficient understanding of the approximations and assumptions imposed in the model, the model can be expected to be used in the consensus building on making the action plan of the coastal management.

5. MODEL DESCRIPTION

5.1. Construction of the Model

ECOHYM is composed of two models: a hydrodynamics model and an ecological model for the benthic and pelagic systems. The ecological model is generalized to enable its application to both the central bay and tidal flat area. The whole construction of ECOHYM is illustrated in Figure 2. The hydrodynamics model is calculated independently from the ecological model, whereas the ecological model receives input of the flow-temperature field from the hydrodynamics model. Therefore, the physical field calculated in the hydrodynamics model such as advection, eddy diffusion, and temperature, is not affected by the model variables (plankton, detritus, nutrients and dissolved oxygen etc.) treated in the ecological model, but model variables of the ecological model is moved passively by the physical field. More specifically, physical field transports the model variables of the ecological model and changes the biochemical balance among the model variables at any computational grid. As such, hydrodynamics changes the biochemical processes which are formulated by model variables. The concept, "eco-hydrodynamics", emphasizes the importance of this interaction among the physical field and biochemical processes. The interaction between the central bay and tidal flat areas is also performed through the physical field (flow field and eddy viscosity) calculated by the hydrodynamics model. The physical processes playing the transportation in the sediment or sediment-water interface such as molecular diffusion, irrigation, bioturbation and burial etc. are treated in the ecological model, while not considered in the hydrodynamics model.

Hereinafter, to clarify the definition of central bay and tidal flat areas in the model, the central bay area is defined as that area where the transparency/depth is so low/deep that it hinders benthic algae, sea-grass and sea-weed photosynthesis. The tidal flat area is defined as that area where the transparency/depth is so high/shallow that benthic algae, sea-grass and sea-weed photosynthesis occurs. Areas of submersion/emersion with tidal level shift are included in the tidal flat area. In the eutrophic estuary in Japan, the central bay area usually has a higher potential to become hypoxic than the tidal flat areas.

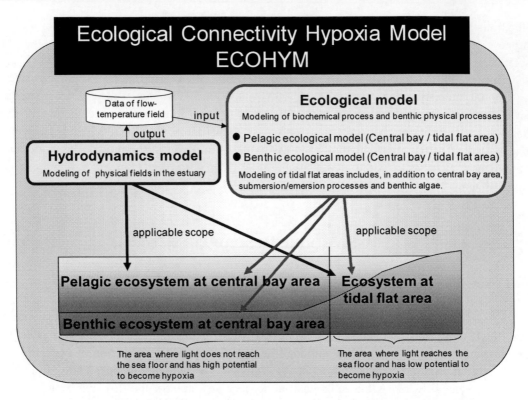

Figure 2. The construction of ECOHYM. The model is composed of two models, the hydrodynamics model and the ecological model for the benthic and pelagic systems. The ecological model is generalized to enable its application to both the central bay and tidal flat areas.

5.2. Hydrodynamics Model

The hydrodynamics model simulates the three dimensional physical field in the pelagic system of estuary and demonstrates the long term variability of flow field, salt and heat transport. The target area of the model is an estuary, i.e. meso-scale (1~100km) semi-closed coastal zone where the seawater is exchanged with external ocean and is diluted by the inflow of freshwater from rivers. The model equations and algorithms of the hydrodynamics model are well described by Nakata et al (1983a, 1983b). Thus, only the outline is described here. The model includes the tidal forcing, surface wind and local density gradient with realistic costal topography and bathymetry described by computational grids/mesh. Under hydrostatic and Boussinesq approximations on a rotating Cartesian coordinate system, the model employs the equations of fluid motion, flow continuity and conservation of heat and salt to determine the local distribution of model variables; i.e., mean velocity components, surface displacement, temperature and salinity. The model equations of motion referred to a Cartesian coordinate system (Figure 3) are as follows.

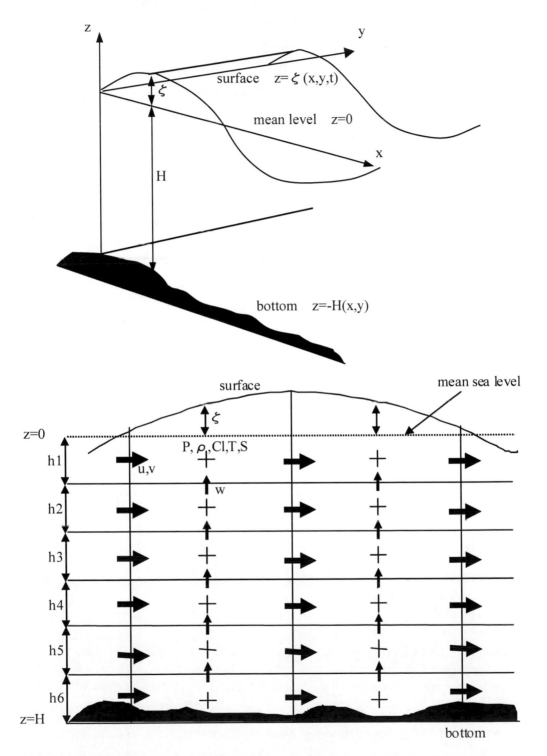

Figure 3. Hydrodynamics model coordinate (upper) and multi level layer in the vertical direction (lower).

(1) The equation of motion in the x direction

$$\frac{\partial u}{\partial t} = -u\frac{\partial u}{\partial x} - v\frac{\partial u}{\partial y} - w\frac{\partial u}{\partial z} + f_0 v - \frac{1}{\rho}\frac{\partial P}{\partial x}$$
$$+ \frac{\partial}{\partial x}(N_x \frac{\partial u}{\partial x}) + \frac{\partial}{\partial y}(N_y \frac{\partial u}{\partial y}) + \frac{\partial}{\partial z}(N_z \frac{\partial u}{\partial z})$$

(5.2.1)

(2) The equation of motion in the y direction

$$\frac{\partial v}{\partial t} = -u\frac{\partial v}{\partial x} - v\frac{\partial v}{\partial y} - w\frac{\partial v}{\partial z} + f_0 u - \frac{1}{\rho}\frac{\partial P}{\partial y}$$
$$+ \frac{\partial}{\partial x}(N_x \frac{\partial v}{\partial x}) + \frac{\partial}{\partial y}(N_y \frac{\partial v}{\partial y}) + \frac{\partial}{\partial z}(N_z \frac{\partial v}{\partial z})$$

(5.2.2)

(3) The equation of hydrostatic pressure approximation

$$-\frac{1}{p}\frac{\partial P}{\partial z} - g = 0$$

(5.2.3)

(4) The equation of continuity

$$\frac{\partial u}{\partial x} + \frac{\partial v}{\partial y} + \frac{\partial w}{\partial z} = 0$$

(5.2.4)

(5) The thermal equation

$$\frac{\partial T}{\partial t} = -u\frac{\partial T}{\partial x} - v\frac{\partial T}{\partial y} - w\frac{\partial T}{\partial z}$$
$$+ \frac{\partial}{\partial x}(K_x \frac{\partial T}{\partial x}) + \frac{\partial}{\partial y}(K_y \frac{\partial T}{\partial y}) + \frac{\partial}{\partial z}(K_z \frac{\partial T}{\partial z})$$

(5.2.5)

(6) The saline equation

$$\frac{\partial S}{\partial t} = -u\frac{\partial S}{\partial x} - v\frac{\partial S}{\partial y} - w\frac{\partial S}{\partial z}$$
$$+ \frac{\partial}{\partial x}(K_x \frac{\partial S}{\partial x}) + \frac{\partial}{\partial y}(K_y \frac{\partial S}{\partial y}) + \frac{\partial}{\partial z}(K_z \frac{\partial S}{\partial z})$$

(5.2.6)

(7) The equation of state

$$\rho = \rho(S, T)$$

(5.2.7)

Where, t = time (T), (x, y, z) = Cartesian coordinate system with x and y measured in the horizontal plane of undisturbed sea and z the distance above that surface (henceforth that surface is referred to as the mean sea level) (L), (u, v, w) = the velocities of x, y, and z directions (L/T), f_0 = Coriolis parameter (1/T), ρ = density of sea water (Mass/L^3), P = pressure (Mass/LT^2), g = acceleration of the earth's gravity (L / T^2), H = the depth from the mean sea level to the bottom (L), T = temperature of sea water (degree), S = salinity(`), (N_x, N_y, N_z), and (K_x, K_y, K_z) are eddy coefficients of viscosity and diffusion (L^2/T).

The assumptions implied in equation (5.2.1) to equation (5.2.7) are as follows:

(i) Eqation (5.2.4) assume incompressible flow.

(ii) Eqation (5.2.3) is the usual hydrostatic approximation involving the neglect of vertical accelerations.

(iii) For the equation of state (5.2.7), Kundsen's expression is adopted. That is

$$\rho = \frac{\sigma_t}{1000} + 1$$

$$\sigma_t = \Sigma_t + (\sigma_t + 0.1324)\{1 - A_t + B_t(\sigma_0 - 0.1324)\}$$

$$\sigma_0 = -0.069 + 1.4708S - 0.00157S^2 + 0.0000398S^3$$

$$\Sigma_t = -\frac{(T - 3.98)^2}{503.57} \cdot \frac{T + 283.0}{T + 67.26}$$

$$A_t = T(4.7869 - 0.098185T + 0.0010843T^2) \times 10^{-3}$$

$$B_t = T(18.03 - 0.8164T + 0.01667T^2) \times 10^{-6} \tag{5.2.8}$$

(iv) Coriolis parameter is constant (f -plane approximation).

From vertical integration of equation (5.2.3), the following equation is obtained.

$$P = P_0 + g \int_z^\zeta p\, dz' \tag{5.2.9}$$

Therefore, the terms of pressure gradient in equations (5.2.1) and (5.2.2) can be transformed as follows:

$$-\frac{1}{p}\frac{\partial P}{\partial x} = -g\frac{\partial \zeta}{\partial x} - \frac{g}{p}\int_z^\zeta \frac{\partial p}{\partial x}dz' - \frac{1}{p}\frac{\partial P_0}{\partial x} \tag{5.2.10}$$

$$-\frac{1}{p}\frac{\partial P}{\partial y} = -g\frac{\partial \zeta}{\partial y} - \frac{g}{p}\int_z^\zeta \frac{\partial p}{\partial y}dz' - \frac{1}{p}\frac{\partial P_0}{\partial y} \tag{5.2.11}$$

Where, P_0 = atmospheric pressure (Mass/LT^2), ζ = tidal level from the mean sea level (L).

From Equations (5.2.1), (5.2.4) and (5.2.10), equation (5.2.1) can be transformed into flux form as follows:

$$\frac{\partial u}{\partial t} = -\frac{\partial}{\partial x}(u^2) - \frac{\partial}{\partial y}(uv) - \frac{\partial}{\partial z}(uw) + f_0 v - g\frac{\partial \zeta}{\partial x} - \frac{g}{\rho}\int_z^0 \frac{\partial \rho}{\partial x}dz - \frac{1}{\rho}\frac{\partial P_0}{\partial x}$$

$$+ \frac{\partial}{\partial x}(N_x \frac{\partial u}{\partial x}) + \frac{\partial}{\partial y}(N_y \frac{\partial u}{\partial y}) + \frac{\partial}{\partial z}(N_z \frac{\partial u}{\partial z})$$

$$\tag{5.2.12}$$

In the same way as x axis, for y axis, equation (5.2.2) is transformed as follows:

$$\frac{\partial v}{\partial t} = -\frac{\partial}{\partial x}(uv) - \frac{\partial}{\partial y}(v^2) - \frac{\partial}{\partial z}(vw) + f_0 u - g\frac{\partial \zeta}{\partial y} - \frac{g}{\rho}\int_z^0 \frac{\partial \rho}{\partial y}dz - \frac{1}{\rho}\frac{\partial P_0}{\partial y}$$
$$+\frac{\partial}{\partial x}(N_x \frac{\partial v}{\partial x}) + \frac{\partial}{\partial y}(N_y \frac{\partial v}{\partial y}) + \frac{\partial}{\partial z}(N_z \frac{\partial v}{\partial z})$$

(5.2.13)

In addition, from equations (5.2.4), (5.2.5), and (5.2.6), the thermal and saline equations can be transformed as follows:

$$\frac{\partial T}{\partial t} = -\frac{\partial}{\partial x}(uT) - \frac{\partial}{\partial y}(vT) - \frac{\partial}{\partial z}(wT)$$
$$+\frac{\partial}{\partial x}(k_x \frac{\partial T}{\partial x}) + \frac{\partial}{\partial y}(k_y \frac{\partial T}{\partial y}) + \frac{\partial}{\partial z}(k_z \frac{\partial T}{\partial z})$$

(5.2.14)

$$\frac{\partial S}{\partial t} = -\frac{\partial}{\partial x}(uS) - \frac{\partial}{\partial y}(vS) - \frac{\partial}{\partial z}(wS)$$
$$+\frac{\partial}{\partial x}(K_x \frac{\partial S}{\partial x}) + \frac{\partial}{\partial y}(K_y \frac{\partial S}{\partial y}) + \frac{\partial}{\partial z}(K_z \frac{\partial S}{\partial z})$$

(5.2.15)

Also, the dynamic boundary condition: $\frac{D\zeta}{Dt} = 0$ at surface ($z = \zeta$), which means that water does not pass through the sea surface, and the dynamic boundary condition: $\frac{DH_b}{Dt} = 0$ at sea-bottom ($z = -H_b$), which means that no water passes through the sea bottom are satisfied. By using these boundary conditions, the following equation describing the dynamics of surface level can be obtained.

$$\frac{\partial \zeta}{\partial t} = -\frac{\partial}{\partial x}\left(\int_{-H}^{\zeta} u\, dz\right) - \frac{\partial}{\partial y}\left(\int_{-H}^{\zeta} v\, dz\right)$$

(5.2.16)

The seven equations (5.2.4), (5.2.8), (5.2.12), (5.2.13), (5.2.14), (5.2.15) and (5.2.16) are the model equations of the hydrodynamics model. The model is calculated by finite volume method (FVM) and by the explicit leap frog computational scheme, whose algorithm conserves mass of model compartments (temperature and salinity) and volume of sea water.

The vertical mixing process is parameterized with a turbulence model of second moment closure, which determines local distributions of the turbulent kinetic energy, k, and the mixing length, l, by means of well-established k-kl equations (e.g., Blumberg and Mellor, 1978; Mellor and Yamada, 1982).

As for the boundary conditions, the conditions at (1) land boundaries (coastal line), open sea (bay mouse), (3) air-water interface (sea-surface) and (4) sediment-water interface (sea-bottom) are set on the hydrodynamics model. At (1) land boundaries (coastal line), the free-slip condition is applied to the momentum (current), i.e.

$$\mathbf{u} \bullet \mathbf{n} = 0 , \quad \frac{\partial \mathbf{u}_t}{\partial n} = 0 \qquad\qquad (5.2.17)$$

For temperature and salinity at (1) land boundaries, the following condition (i.e., temperature and salinity are not transported at land boundaries) is applied.

$$\frac{\partial T}{\partial n} = \frac{\partial S}{\partial n} = 0 \qquad\qquad (5.2.18)$$

Where, $\mathbf{u}$ is the current vector [L/T] and $\mathbf{n}$ [-] is the normal vector to the coastline (outward directed from ocean area). n denotes the element of normal direction of coastal line. t denotes the element of tangential direction of coastal line.

At (2) open sea boundaries (bay mouse), the sea surface elevation, ζ, described by trigonometric function is given and free-stream condition is applied to the momentum (current). For the concentration of salinity and temperature, the value of their inflow from the outside to the inside of the calculation area are specified by prescribed functions and the values for the outflow from the inside to the outside set the free-stream condition. i.e.,

$$\zeta = \zeta_0 + \sum_i A_i \cos(\omega_i t - \delta) \qquad\qquad (5.2.19)$$

$$\frac{\partial \mathbf{u}_t}{\partial n} = 0 \qquad\qquad (5.2.20)$$

$$\frac{\partial T}{\partial n} = \frac{\partial S}{\partial n} = 0 \quad (\text{ for outflow side }) \qquad\qquad (5.2.21)$$

$$T = T_o, S = S_o \quad (\text{ for inflow side }) \qquad\qquad (5.2.22)$$

Where, A_i [L] is amplitude of each tidal element, δ [L] is angle of delay, T_o, Cl_o [-] are the prescribed values of temperature and salinity at open sea boundary.

As for the boundary condition at (3) sea surface (z = ζ), stress caused by the wind, heat flux, and the free surface condition are given as follows.

$$\rho N_z \frac{\partial u}{\partial z} = \tau_x^s \tag{5.2.23}$$

$$\rho N_z \frac{\partial v}{\partial z} = \tau_y^s \tag{5.2.24}$$

$$-K_z \frac{\partial T}{\partial z} = Q \tag{5.2.25}$$

$$\frac{D\zeta}{Dt} = 0 \tag{5.2.26}$$

Where, τ_x^s and τ_y^s [Mass/ LT^2] are the wind stress, Q is the downward heat flux.

The final boundary condition at (4) seabottom (z = $-H_b$) is the slip condition for the bottom stress and the dynamic boundary condition as follows.

$$\rho N_z \frac{\partial u}{\partial z} = \tau_x^b \tag{5.2.27}$$

$$\rho N_z \frac{\partial v}{\partial z} = \tau_y^b \tag{5.2.28}$$

$$\frac{D H_b}{Dt} = 0 \tag{5.2.29}$$

Where, τ_x^b, τ_y^b are friction stresses at the seabottom.

5.3. Ecological Model

The ecological model is an equation system which set the significant components for hypoxia or for its related environmental phenomena as the model variables. The model describes the interaction among model variables through the biochemical and physical processes in terms of Oxygen (O), Carbon (C), Nitrogen (N), and Phosphorus (P) cycling. Namely, the ecological model describes the changes of various forms (i.e. model variables) which each element (O, C, N, and P) takes due to biochemical processes, while considering the physical transport. The dynamics/spatial distribution of the model variables are described by the partial differential equations (model equations). The equations satisfy the mass conservation of O, C, N and P, and are comprised of the production/consumption terms due to biochemical processes and transport terms due to physical processes. Each path of O-C-N-P coupled cycle caused by biochemical processes is derived from the empirical/experimental formulation. The formulations of each biochemical reaction are based on a first kinetic reaction and include (a) several model variables, (b) environmental variables obtained from prescribed functions/data (i.e. temperature, light intensity, etc.), (c) biochemical parameters, and (d) universal constants. The values of the biochemical processes are calculated and changes at each time step. Changes in the biochemical processes affect the dynamics/spatial distribution of the model variables and vice versa. In this way, the ecological model enables its demonstration of the ecosystem dynamics (O-C-N-P coupled cycle dynamics) as the result of entanglement of various interactions. The details of the ecological model are explained below.

5.3.1. Ecological Diagram (Model Variables and Biochemical Processes)

The pelagic and benthic ecological diagrams treated in each grid of the model are shown in Figures 4 and 5. The model variables in the ecological model are phytoplankton, zooplankton, detritus, dissolved organic matter, NH_4-N, NO_3-N, PO_4-P, benthic algae, suspension feeders, deposit feeders (benthic fauna), dissolved oxygen and ODU (oxygen demand units, representing stoichiometric substitute expression of oxygen demands of Mn^{2+}, Fe^{2+}, S^{2-} (Soetaert et al., 1996a)). In the diagrams, the model variables are illustrated as boxes and biochemical processes as arrows. The model variables and their connecting biochemical processes were selected to describe the hypoxic mechanisms/oxygen dynamics based on the low trophic level.

The oxygen consuming/producing biochemical processes described in the ecological model are (1) photosynthesis of phytoplankton and benthic algae, (b) excretion of phytoplankton, zooplankton and benthic fauna (suspension feeders and deposit feeders), (c) oxic mineralization, (d) nitrification, and (e) oxidization of the total reduced substances, Mn^{2+}, Fe^{2+} and S^{2-} (Figure 6).

The characteristic point of the model is to divide the bacterial mineralization processes into three ways, viz. oxic, suboxic, and anoxic mineralization. All these mineralization processes are formulated as first order kinetics of detritus. Oxic mineralization is limited by oxygen (Michaelis-Menten type kinetics). Suboxic mineralization based on nitrate is inhibited by oxygen (one minus Michaelis-Menten type kinetics) and limited by nitrate (Michaelis-Menten type kinetics). The consumption of oxygen and nitrate as terminal electron acceptors is explicitly modeled. Mineralization processes using other oxidants (manganese oxides, iron oxides, sulphate) are lumped into one process, where this mineralization is inhibited by oxygen and nigrate (one minus Michaelis-Menten type kinetics). Anoxic mineralization produces reduced substances called oxygen demand units, ODU. Re-oxidation of one mole of ODU requires one mole of oxygen (Soetaert et al., 1996a). ODU re-oxdization and nitrification are formulated to be limited by oxygen. Idealized stoichiometric relationships of each bacterial mineralization processes that are used in our model are shown in Table 1 (Sohma et al, 2001). Modeling the mineralization processes into three ways and the stoichiometric relationship in Table 1 enables analysis of the mechanisms underlying the relationship between oxygen consumption, nitrate reduction, de-nitrification and ODU production that are caused by mineralization. As shown in Table 1, the stoichiometric relationships of the consumption/production of living organisms due to respiration and photosynthesis are described in the same way as oxic mineralization due to bacteria.

Detritus is divided into three fractions: fast-labile, slow-labile and refractory organic matter. This means that the model is a Multi-G model (Jørgensen, 1978). Accordingly, the cells of living organisms (phytoplankton, zooplankton, benthic algae and benthic fauna) are defined as being composed of these three fractions (Sohma et al, 2004). Hence, fluxes between living organisms and detritus (fluxes of uptake, extra-release, feces, and mortality, etc.) can be modeled while considering these fractions.

In addition, based on the feeding habitat, benthic fauna is divided into suspension feeders and deposit feeders in order to describe their functional difference which have the potential to affect the hypoxic dynamics, e.g. the transportation of particulate organic matters (phytoplankton, zooplankton, and detritus) from pelagic to benthic due to suspension feeders feeding and mineralization of benthic organic matter due to deposit feeder excretion.

Table 1 Stoichiometric relationships associated with biochemical processes treated in the ecological model

Photosynthesis using NH_4-N

$$m (CO_2) + n (NH_3) + (H_3PO_4) + m (H_2O) \rightarrow (CH_2O)_m(NH_3)_n(H_3PO_4) + m (O_2) \tag{T1.1}$$

Photosynthesis using NO_3-N

$$m(CO_2) + n(NO_3^-) + (H_3PO_4) + (m+n) (H_2O) + nH^+ \rightarrow (CH_2O)_m(NH_3)_n(H_3PO_4) + (2n+m) (O_2) \tag{T1.2}$$

Oxic mineralization, Excretion, Respiration

$$(CH_2O)_m(NH_3)_n(H_3PO_4) + m (O_2) \rightarrow m(CO_2) + n(NH_3) + (H_3PO_4) + m(H_2O) \tag{T1.3}$$

Suboxic mineralization

$$(CH_2O)_m(NH_3)_n(H_3PO_4) + a (HNO_3) \rightarrow m (CO_2) + a (x/2) (N_2) + n (NH_3) + a (1-x) (NH_3) + (H_3PO_4) + b (H_2O) \tag{T1.4}$$

where, $a = -4m/(3x-8)$, $b = m (3x-4) / (3x-8)$, $0 \leq x \leq 1$, These condition satisfy $a \geq 0$ and $b \geq 0$ at anytime

Anoxic mineralization

$$(CH_2O)_m(NH_3)_n(H_3PO_4) + m (TEA) \rightarrow m (CO_2) + n (NH_3) + (H_3PO_4) + m (ODU) + Q (H_2O) \tag{T1.5}$$

where, $ODU = 2Mn^{2+}$, $4Fe^{2+}$ and $1/2S^{2-}$, $TEA = 2 MnO_2$, $2 Fe_2O_3$ and $1/2 SO_4^{2-}$

Nitrification

$$NH_3 + H_2O + 2 O_2 \rightarrow NO_3^- + 2 H_2O + H^+ \tag{T1.6}$$

ODU (Oxygen Demand Unit) oxidization

$$ODU + O_2 \rightarrow TEA \tag{T1.7}$$

where, $ODU = 2 Mn^{2+}$, $4 Fe^{2+}$ and $1/2 S^{2-}$, $TEA = 2 MnO_2$, $2 Fe_2O_3$ and $1/2 SO_4^{2-}$

(1) m, n denote C, N, P ratio of created or mineralized organic matter, i.e., C:N:P=m:n:1. (2) x denotes ratio of nitrogen reducing to nitrogen gas (N_2) and reducing to ammonium from nitrate by suboxic mineralization. i.e., N_2:NH_3= x:(1-x). (3) a, b are coefficients determined from stoichiometric relation. (4) Nitrogen of N_2 and NH_3 shading and written by italic in the right-hand side are derived from HNO_3 in the left-hand side in the equation "T1.4".

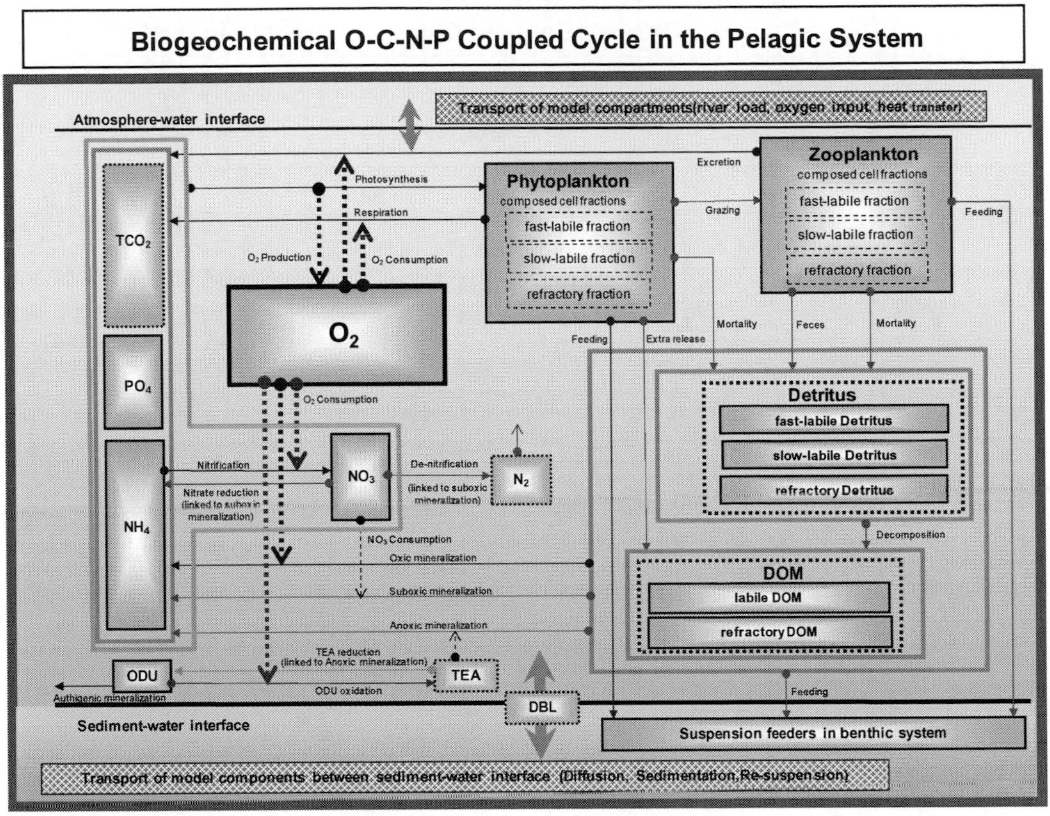

Model variable	Unit	Representation in the above diagram	Notation and [No.] in Tables
Phytoplankton	mgC/l (μgC/ml)	Phytoplankton	PP [01]
Zooplankton	mgC/l (μgC/ml)	Zooplankton	ZP [02]
Fast labile detritus	mgC/l (μgC/ml)	Fast-labile Detritus	WDE_1 [03,1]
Slow labile detritus	mgC/l (μgC/ml)	Slow-labile Detritus	WDE_2 [03,2]
Refractory detritus	mgC/l (μgC/ml)	Refractory Detritus	WDE_3 [03,3]
Labile dissolved organic matter	mgC/l (μgC/ml)	Labile DOM	WDM_1 [04,1]
Refractory dissolved organic matter	mgC/l (μgC/ml)	Refractory DOM	WDM_2 [04,2]
Ammonium	mgN/l (μgN/ml)	NH_4	WNX [05]
Nitrate	mgN/l (μgN/ml)	NO_3	WNY [06]
Phosphate	mgP/l (μgP/ml)	PO_4	WDP [07]
Reduced substances (Fe^{2+}, Mn^{2+}, S^{2-})	mg/l (μg/ml)	ODU	WOU [08]
Dissolved oxygen	mg/l (μg/ml)	O_2	WDO [09]

Figure 4. Model variables and biochemical processes in the pelagic system of the ecological model. Model variables are described by boxes with solid lines while biochemical processes are indicated by arrows with both solid and dotted lines. This ecological diagram is produced from the Oxygen-Carbon-Nitrogen-Phosphorus-coupled cycle. ODU: oxygen demand unit, DOM: dissolved organic matter.

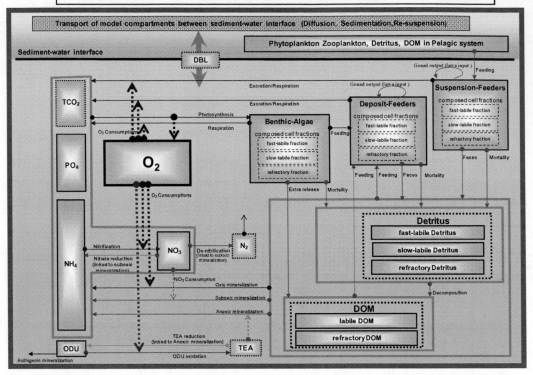

Figure 5. Model variables and biochemical processes in the benthic system of the ecological model. Model variables are described by boxes with solid lines and biochemical processes are indicated by arrows with both solid and dotted lines. This ecological diagram is produced from the Oxygen-Carbon-Nitrogen-Phosphorus-coupled cycle.

Model variable	Unit	Representation in the above diagram	Notation and [No.] in Tables
Suspension feeders	$\mu gC/cm^2$ sediment	Suspension feeders	SFB [51]
Deposit feeders	$\mu gC/cm^2$ sediment	Deposit feeders	DFB [52]
Fast labile detritus	$\mu gC/cm^3$ solid	Fast-labile Detritus	DET_1 [53,1]
Slow labile detritus	$\mu gC/cm^3$ solid	Slow-labile Detritus	DET_2 [53,2]
Refractory detritus	$\mu gC/cm^3$ solid	Refractory Detritus	DET_3 [53,3]
Labile dissolved organic matter	mgC/l ($\mu gC/ml$)	Labile DOM	DOM_1 [54,1]
Refractory dissolved organic matter	mgC/l ($\mu gC/ml$)	Refractory DOM	DOM_2 [54,2]
Ammonium	mgN/l ($\mu gN/ml$)	NH_4	HNX [55]
Nitrate	mgN/l ($\mu gN/ml$)	NO_3	HNY [56]
Phosphate	mgP/l ($\mu gP/ml$)	PO_4	DIP [57]
Reduced substances (Fe^{2+}, Mn^{2+}, S^{2-})	mg/l ($\mu g/ml$)	ODU	ODU [58]
Dissolved oxygen	mg/l ($\mu g/ml$)	O_2	DOO [59]
Benthic algae	$\mu gC/cm^2$ sediment	Benthic-Algae	BAL [60]

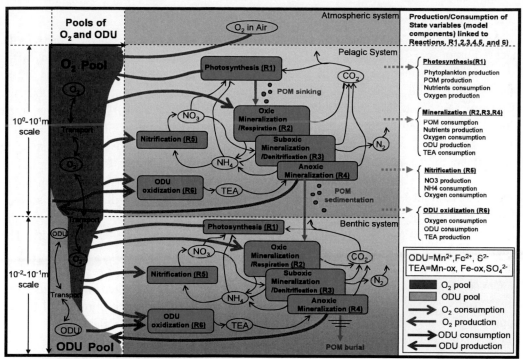

Oxygen/ODU consumption/production mechanisms in coastal marine ecosystem

Figure 6. Modeled oxygen production and consumption mechanisms and associated biochemical processes in the Ecological Connectivity Hypoxia Model, ECOHYM.

Table 1. Stoichiometric relationships associated with biochemical processes treated in the ecological model

Oxic mineralization, Excretion, Respiration

$(CH_2O)_m(NH_3)_n(H_3PO_4) + m(O_2) \rightarrow m(CO_2) + n(NH_3) + (H_3PO_4) + m(H_2O)$ (T1.3)

Suboxic mineralization

$(CH_2O)_m(NH_3)_n(H_3PO_4) + a(HNO_3) \rightarrow m(CO_2) + a(x/2)(N_2) + n(NH_3) + a(1-x)(NH_3) + (H_3PO_4) + b(H_2O)$ (T1.4)

where, a = -4m/(3x-8), b = m (3x-4) / (3x-8), $0 \leq x \leq 1$, These condition satisfy $a \geq 0$ and $b \geq 0$ at anytime

Anoxic mineralization

$(CH_2O)_m(NH_3)_n(H_3PO_4) + m(TEA) \rightarrow m(CO_2) + n(NH_3) + (H_3PO_4) + m(ODU) + Q(H_2O)$ (T1)
where, ODU = $2Mn^{2+}$, $4Fe^{2+}$ and $1/2S^{2-}$, TEA = $2 MnO_2$, $2 Fe_2O_3$ and $1/2 SO_4^{2-}$

Nitrification

$NH_3 + H_2O + 2 O_2 \rightarrow NO_3^- + 2 H_2O + H^+$ (T1.6)

ODU (Oxygen Demand Unit) oxidization

$ODU + O_2 \rightarrow TEA$ (T1.7)

where, ODU = $2 Mn^{2+}$, $4 Fe^{2+}$ and $1/2 S^{2-}$, TEA = $2 MnO_2$, $2 Fe_2O_3$ and $1/2 SO_4^{2-}$

(1) m, n denote C, N, P ratio of created or mineralized organic matter, i.e., C:N:P=m:n:1. (2) x denotes ratio of nitrogen reducing to nitrogen gas (N_2) and reducing to ammonium from nitrate by suboxic mineralization. i.e., $N_2:NH_3$= x:(1-x). (3) a, b are coefficients determined from stoichiometric relation. (4) Nitrogen of N_2 and NH_3 shading and written by italic in the right-hand side are derived from HNO_3 in the left-hand side in the equation "T1.4".

5.3.2. Physical Processes

As for the physical processes, (a) water current (flow velocity) and eddy viscosity in the water column, and (b) molecular diffusion, irrigation, bioturbation and deposition in the sediment or sediment-water interface are described in the ecological model equations. In the water column, model variables are transported by water flow (advection) and eddy viscosity (diffusion). As mentioned before, water current and eddy viscosity/diffusion in the water column is calculated by the hydrodynamics model based on the well established fluid dynamics theory. The time series data of the flow velocity and eddy viscosity/diffusion are implemented to the pelagic equation in the ecological model at each time step. In the sediment or at the sediment-water interface, the transportation of model variables are different from dissolved substances to particulate substances. The model variables of dissolved substances in the benthic system are transported by molecular diffusion, bioturbation and irrigation due to the activity of benthic fauna and advection from pore water velocity. The model variables of particulate substances in the benthic system are transported by advection (sediment deposition) and bioturbation.

Concerning molecular diffusion, the model uses the well established theory based on statistical physics and the value of the molecular diffusion coefficient is formulated by the temperature. Here, the molecular diffusion coefficient in the sediment reflects the effect of tortuosity (Berner 1980) as an empirical function of porosity. The molecular diffusive fluxes of dissolved substances at the sediment-water interface are obtained from the theory of the diffusive boundary layer (DBL) (Boudreau and Jølgensen, 2001). Other physical processes in the sediment or sediment-water interface are influenced by the biomass/species composition of benthic fauna and granularity composition of sediment, but no general theory exits for those processes. There are many variations on how to model or decide on values. In our model, bioturbation is modeled as a diffusion-like process. Irrigation is modeled as a diffusion-like process and as a proportional process of the differences in concentration between surface water and each depth of sediment (Berner, 1980; Berg et al., 1998). The diffusive coefficients of bioturbation/irrigation are formulated as a hyperbolic function of suspension feeders and deposit feeders (Sohma et al., 2004) and the vertical dependence of bioturbation/irrigation are formulated as exponential shape (Soetaert et al. 1996a).

The advection of solid substances and dissolved substances is obtained from the porosity and sedimentation rate to conserve the volume of liquid phase and solid phase.

During the period of emersion of the tidal flat area, the fluxes across the sediment-water interface i.e., diffusion of dissolved substances, sediment deposition and feeding of suspension feeders, are set at zero. The oxygen flux from the atmosphere to pore water due to aeration is solved based on the DBL theory between atmosphere and water.

The details of the formulation of the coefficient of bioturbation/irrigation and the advection of solid and liquid phase are described below.

(1) Coefficient of Bioturbation/Irrigation

$$D = D_{max} \frac{(SFB + DFB)}{Hf + (SFB + DFB)} \times f_B(z)$$

$$f_B(z) = \begin{cases} 1 & for : z < z_B \\ \exp(-(z - z_B) / c_B) & for : z > z_B \end{cases} \qquad (5.3.1)$$

Where, D is bioturbation/irrigation coefficient denoted as D_B, D_B', D_I, or α in the governing equations mentioned in the next section. Namely, D_B = solid biodiffusion coefficient (intraphase mixing expression) [L^2-sediment/T], D_B' = solid biodiffusion coefficient (interphase mixing expression) [L^2-sediment/T], D_I = irrigation coefficient (diffusion-like expression) [L^2-sediment/T], α = irrigation coefficient2 [1/T]. *SFB* and *DFB* are suspension feeder [mass/L^2] and deposit feeder [Mass/L^2] that are treated as the model variables in the model (refer to the table described under Figure 5), z_B [L] and c_B [-] are the parameters that decide the vertical distribution of bioturbation/irrigation.

(2) Advection of Solid and Liquid Phase in the Benthic System
Equations of mass transportation for liquid phase and for solid phase are described as follows:

$$\frac{\partial(\rho_l \phi)}{\partial t} + \frac{\partial(v \cdot \rho_l \phi)}{\partial z} = \frac{\partial}{\partial z}\left(D_B \frac{\partial(\rho_l \phi)}{\partial z} \right) \text{ (for liquied phase)} \qquad (5.3.2)$$

$$\frac{\partial(\rho_s(1-\phi))}{\partial t} + \frac{\partial(\omega \cdot \rho_s(1-\phi))}{\partial z} = \frac{\partial}{\partial z}\left(D_B \frac{\partial(\rho_s(1-\phi))}{\partial z} \right) \text{ (for solid phase)} \qquad (5.3.3)$$

Where, $\phi =$ porosity [-], $\rho_l =$ density of liquid phase [mass-liquid/ L^3-liquid], $\overline{\rho}_s =$ density of solid phase [mass-solid/ L^3-solid], v = velocity of burial of water below the sediment-water interface [L-sediment/T] and w = rate of depositional burial of solids [L-sediment/T].

The following assumptions are set for equations (5.3.2), and (5.3.3).

(i) Density of liquid phase does not change with space or time: ρ_l is constant

(ii) Density of solid phase does not change with space or time: $\overline{\rho}_s$ is constant

As a result, equations (5.3.2), and (5.3.3) describing mass transportation are transformed to the following equations describing volume transportation.

$$\frac{\partial \phi}{\partial t} + \frac{\partial(v \cdot \phi)}{\partial z} = \frac{\partial}{\partial z}\left(D_B \frac{\partial \phi}{\partial z} \right) \qquad (5.3.4)$$

$$\frac{\partial(1-\phi)}{\partial t}+\frac{\partial\{w\cdot(1-\phi)\}}{\partial z}=\frac{\partial}{\partial z}\left\{D_B\frac{\partial(1-\phi)}{\partial z}\right\} \tag{5.3.5}$$

(5.3.4) + (5.3.5) lead the equation of volume conservation.

$$\frac{\partial(v\cdot\phi)}{\partial z}+\frac{\partial\{w\cdot(1-\phi)\}}{\partial z}=0 \tag{5.3.6}$$

The advection of solid substances and dissolved substances are derived from equations (5.3.4) and (5.3.5) with the information of data: ϕ, D_B, v_0 and w_0. Where, v_0, w_0 are v, w at the sediment-water interface.

5.3.3. Governing/General Equations and Assumptions

The following seven equations and four assumptions are applied in the ecological model. These equations, especially for the benthic system, are mostly based on the equations propounded by Berner (1980).

(1) Equation for the Pelagic Substances
The general equation for pelagic system is as follows:

$$\frac{\partial C_w}{\partial t}=-\left(\mathbf{v}_w\bullet\nabla\right)C_w+\nabla\bullet\left(\mathbf{K}\bullet\nabla C_w\right)+\sum R$$

$$=-u_w\frac{\partial C_w}{\partial x}-v_w\frac{\partial C_w}{\partial y}-w_w\frac{\partial C_w}{\partial z}+\frac{\partial}{\partial x}\left(K_x\frac{\partial C_w}{\partial x}\right)+\frac{\partial}{\partial y}\left(K_y\frac{\partial C_w}{\partial y}\right)+\frac{\partial}{\partial z}\left(K_z\frac{\partial C_w}{\partial z}\right)+\sum R \tag{5.3.7}$$

Where, C_w = concentration of pelagic substances i.e., phytoplankton, zooplankton, detritus (fast labile detritus detritus, slow labile detritus, refractory detritus), dissolved organic matter (labile DOM, refractory DOM), NH_4-N, NO_3-N, PO_4-P, dissolved oxygen (DO), and ODU [mass/L^3-liquid], $\mathbf{v}_w=\left(u_w,v_w,w_w\right)$ = flow velocity that already has been calculated by the hydrodynamics model [L/T], t = time [T], x, y, z = space coordinates [L], $\sum R$ = biochemical reactions and, fluxes from outside the system [mass/L^3-liquid/T], K = eddy diffusion (viscosity) tensor [L^2-liquid/T].

(2) Diagenetic Equation for Benthic Dissolved Substances

$$\frac{\partial(\phi C)}{\partial t}=\frac{\partial\left\{D_B\frac{\partial(\phi\,C)}{\partial z}+\phi\left(D_S+D_I+D_B'\right)\frac{\partial C}{\partial z}\right\}}{\partial z}+\phi\alpha\left(C_0-C\right)$$

$$-\frac{\partial\left(\phi vC\right)}{\partial z}+\phi R_{ads}+\phi \sum R'$$

$$(5.3.8)$$

Where, $\phi =$ porosity [-], C = concentration dissolved substances i.e. dissolved organic matter (labile DOM, refractory DOM), NH_4-N, NO_3-N, PO_4-P, DO, and ODU [mass/L^3-liquid], C_0 = concentration of dissolved substances at sediment-water interface [mass/L^3-liquid], D_S = molecular diffusion coefficient in sediment including the effects of tortuosity [L^2-sediment/T], D_B = solid biodiffusion coefficient (intraphase mixing expression) [L^2-sediment/T], D'_B = solid biodiffusion coefficient (interphase mixing expression) [L^2-sediment/T], D_I = irrigation coefficient (diffusion-like expression) [L^2-sediment/T], $\alpha =$ irrigation coefficient2 [1/T], v = velocity of burial of water below the sediment-water interface [L-sediment/T], R_{ads} = reactions of dissolved materials due to equilibrium adsorption or desorption [mass/L^3-liquid/T]. $\sum R'$ = all other slow (irreversible) biochemical reactions [mass/L^3-liquid/T].

(3) Diagenetic Equation for Benthic Particulate Substances

$$\frac{\partial\left\{(1-\phi)\overline{\rho}_s\,\overline{C}\right\}}{\partial t}=\frac{\partial\left[D_B\dfrac{\partial\left\{(1-\phi)\overline{\rho}_s\,\overline{C}\right\}}{\partial z}\right]}{\partial z}+\frac{\partial\left\{D'_B(1-\phi)\overline{\rho}_s\dfrac{\partial \overline{C}}{\partial z}\right\}}{\partial z}$$

$$-\frac{\partial\left\{(1-\phi)\overline{\rho}_s\,w\overline{C}\right\}}{\partial z}+(1-\phi)\overline{\rho}_s\,\overline{R}_{ads}+(1-\phi)\overline{\rho}_s\sum \overline{R'}$$

$$(5.3.9)$$

Where, $\overline{C}$ = concentration of a particulate substance in terms of mass per unit mass of total solids i.e., detritus (fast labile detritus, slow labile detritus, refractory detritus), absorbed DOM, absorbed NH_4-N, $\overline{\rho}_s$ = density of total solid phase [mass-solid/ L^3-solid], w = rate of depositional burial of solids [L-sediment/T], $\overline{R}_{ads}$ = reactions of dissolved materials due to equilibrium adsorption or desorption [mass/mass-solid/T], $\sum \overline{R'}$ = all non-equilibrium slow biochemical reactions [mass/mass-solid/T].

(4) Equations for suspension feeders, deposit feeders and benthic algae

$$\frac{\partial B}{\partial t}=\sum R_B$$

$$(5.3.10)$$

Where, B = biomass, expressed per square of sediment [mass/L^2-sediment], R_B = biochemical reactions [mass/L^2-sediment].

(5) Equation for the relation of adsorption-desorption reaction

$$\overline{R}_{ads} = \frac{-\phi}{(1-\phi)\;\overline{\rho}_s}\, R_{ads}$$

(5.3.11)

(6) Equation for mass/volume conservation of benthic solid phase

$$\frac{\partial \phi}{\partial t} + \frac{\partial (v \cdot \phi)}{\partial z} = \frac{\partial}{\partial z}\left(D_B\, \frac{\partial \phi}{\partial z} \right)$$

(5.3.12) = (5.3.4)

(7) Equation for mass/volume conservation of benthic liquid phase

$$\frac{\partial (1-\phi)}{\partial t} + \frac{\partial \{w \cdot (1-\phi)\}}{\partial z} = \frac{\partial}{\partial z}\left\{ D_B\, \frac{\partial (1-\phi)}{\partial z} \right\}$$

(5.3.13) = (5.3.5)

Note that w and v in the benthic system are calculated to meet the relationship among equations (5.3.12) and (5.3.13) as mentioned in section 5.3.3.

The following assumptions are imposed on the equations described above (Berner, 1980).

(i)   Seawater is treated as incompressible liquid: $div\ v_w = 0$

(ii) Density of solid does not change with space or time: $\overline{\rho}_s$ is constant

(iii) The equilibrium expression for simple linear adsorption: $\overline{C} = K'C$, K' = adsorption coefficient.

(iv) Adsorptive property does not change with space or time, in other words, K' is constant.

(v) If $\overline{C}$ is adsorbed substances, then there are no slow diagenetic reactions, hence, $\sum \overline{R'} = 0$ in equation (5.3.9).

5.3.4. Model Equations and Computational Algorithms

The following three model equations and computational algorithms are used to calculate the temporal/spatial distribution of model variables. Here, the model equations mean the equations transferred from the general equations and assumption in section 5.3.3 for their application to the computational algorithm named Finite Volume Method (FVM). The merit of FVM conserves mass of model variables among the state of each time step.

(1) Model equation for the pelagic substances

$$\frac{\partial C_w}{\partial t} = -\frac{\partial(uC_w)}{\partial x} - \frac{\partial(vC_w)}{\partial y} - \frac{\partial(wC_w)}{\partial z} + \frac{\partial}{\partial x}\left(K_x \frac{\partial C_w}{\partial x}\right) + \frac{\partial}{\partial y}\left(K_y \frac{\partial C_w}{\partial y}\right) + \frac{\partial}{\partial z}\left(K_z \frac{\partial C_w}{\partial z}\right) + \sum R$$

(5.3.14)

(2) Model equation for the benthic dissolved substances

$$\frac{\partial}{\partial t}\left[\{K'\overline{\rho}_s(1-\phi)+\phi\}C\right] = \frac{\partial}{\partial z}\left[\{K'\overline{\rho}_s(D_B+D'_B)\cdot(1-\phi)+\phi(D_S+D_I+D_B+D'_B)\}\frac{\partial C}{\partial z}\right]$$

$$+ \frac{\partial}{\partial z}\left[\left\{(1-K'\overline{\rho}_s)D_B\frac{\partial\phi}{\partial z} - K'\overline{\rho}_s(1-\phi)w - \phi v\right\}C\right]$$

$$+ \phi\alpha(C_0 - C) + \phi\sum R'$$

(5.3.15)

(3) Model equation for the particulate substances in the benthic system

$$\frac{\partial\{(1-\phi)S\}}{\partial t} = \frac{\partial\left\{(D_B+D'_B)\cdot(1-\phi)\dfrac{\partial S}{\partial x}\right\}}{\partial z}$$

$$+ \frac{\partial\left[\left\{D_B\dfrac{\partial(1-\phi)}{\partial x} - (1-\phi)w\right\}S\right]}{\partial z} + (1-\phi)\overline{\rho}_s\sum\overline{R'}$$

(5.3.16)

Where, $\overline{\rho}_s\overline{C} = S$ = concentration of a particulate or adsorbed substances in terms of mass per unit volume of total solids (mass/ L^3-solid).

From the spatial integration of equations (5.3.14), (5.3.15) and (5.3.16), they can be transferred to use FVM. The details are shown in Figures 7 and 8 and Tables 2, 3, and 4.

The algorithm shown in Table 2 is used for the model variables in the pelagic system, i.e., phytoplankton, zooplankton, detritus (fast labile POM, slow labile POM, refractory POM), dissolved organic matter (labile DOM, refractory DOM), NH_4-N, $NO_{2,3}$-N, PO_4-P, DO, and ODU. The algorithm in Table 3 is used for dissolved organic matter (labile DOM, refractory DOM), NH_4-N, $NO_{2,3}$-N, PO_4-P, DO and ODU in the benthic system, and the algorithm in Table 4 is used for the benthic detritus (fast labile POM, slow labile POM, refractory POM).

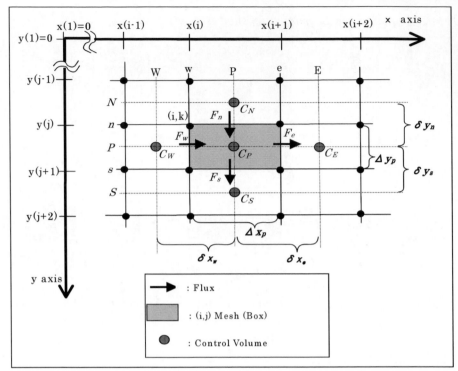

In the horizontal direction

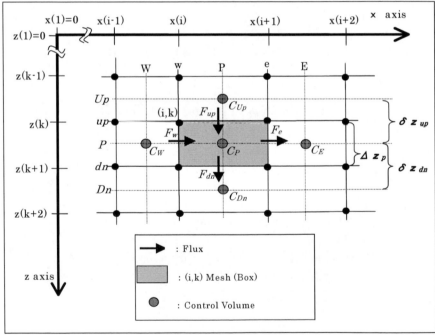

In the vertical direction

Figure 7. Model coordinates in the horizontal direction (upper) and the vertical direction (lower) in the ecological model.

Table 2 Model equation and algorithm in the pelagic ecosystem (refer to Figure 7)

Surface layer in the pelagic system=free surface: (i,j,1) Mesh/Box
From the spatial integration of (i,j,1) mesh/box

$$\iiint_{\zeta(t)}^{z(2)} \frac{\partial}{\partial t} C \, dxdydz + \iiint_{\zeta(t)}^{z(2)} \frac{\partial}{\partial x}\left(uC - K_x \frac{\partial C}{\partial x}\right) dxdydz$$

$$+ \iiint_{\zeta(t)}^{z(2)} \frac{\partial}{\partial y}\left(vC - K_y \frac{\partial C}{\partial y}\right) dxdydz + \iiint_{\zeta(t)}^{z(2)} \frac{\partial}{\partial z}\left(wC - K_z \frac{\partial C}{\partial z}\right) dxdydz = \iiint_{\zeta(t)}^{z(2)} \sum R \, dxdydz \quad (1)$$

$$\Leftrightarrow$$

$$\frac{\partial}{\partial t}\left(\Delta x \cdot \Delta y \cdot \int_{\zeta(t)}^{z(2)} C_P \cdot dz\right) + \Delta x \cdot \Delta y \cdot \frac{\partial \zeta}{\partial t} C\big|_{z=\zeta(t)} + \Delta y \cdot \left(\int_{\zeta(t)}^{z(2)} F_e \, dz - \int_{\zeta(t)}^{z(2)} F_w \, dz\right) + \Delta x \cdot \left(\int_{\zeta(t)}^{z(2)} F_s \, dz - \int_{\zeta(t)}^{z(2)} F_n \, dz\right)$$

$$+ \Delta x \cdot \Delta y \cdot \left(F_{dn} - F_{up}\big|_{z=\zeta(t)}\right) = \Delta x \cdot \Delta y \cdot \Delta z \cdot \sum R_P \quad (2)$$

$$\Leftrightarrow$$

$$\frac{\Delta x \cdot \Delta y \cdot (\Delta z - \zeta(t+\Delta t)) \cdot C_P^{t+\Delta} - \Delta x \cdot \Delta y \cdot (\Delta z - \zeta(t)) \cdot C_P^t}{\Delta t} + \Delta y \cdot \left\{(\Delta z - \zeta(t)) \cdot F_e^t - (\Delta z - \zeta(t)) \cdot F_w^t\right\}$$

$$+ \Delta x \cdot \left\{(\Delta z - \zeta(t)) \cdot F_s^t - (\Delta z - \zeta(t)) \cdot F_n^t\right\} - \Delta x \cdot \Delta y \cdot F_{dn}^{t+\Delta} = \Delta x \cdot \Delta y \cdot \Delta z \cdot \sum R_P^t \quad (3)$$

Any layer without surface in the pelagic system: (i,j,k) Mesh/Box
from the spatial integration of (i,j,k) mesh/box

$$\iiint \frac{\partial}{\partial t} C \, dxdydz + \iiint \frac{\partial}{\partial x}\left(uC - K_x \frac{\partial C}{\partial x}\right) dxdydz$$

$$+ \iiint \frac{\partial}{\partial y}\left(vC - K_y \frac{\partial C}{\partial y}\right) dxdydz + \iiint \frac{\partial}{\partial z}\left(wC - K_z \frac{\partial C}{\partial z}\right) dxdydz = \iiint \sum R \, dxdydz \quad (4)$$

$$\Leftrightarrow$$

$$\frac{\partial}{\partial t}\left(\Delta x \cdot \Delta y \cdot \int_{z(k)}^{z(k+1)} C_P \cdot dz\right) + \Delta y \cdot \left(\int_{z(k)}^{z(k+1)} F_e \, dz - \int_{z(k)}^{z(k+1)} F_w \, dz\right) + \Delta x \cdot \left(\int_{z(k)}^{z(k+1)} F_s \, dz - \int_{z(k)}^{z(k+1)} F_n \, dz\right)$$

$$+ \Delta x \cdot \Delta y \cdot \left(F_{dn} - F_{up}\right) = \Delta x \cdot \Delta y \cdot \Delta z \cdot \sum R_P \quad (5)$$

$$\Leftrightarrow$$

$$\frac{\Delta x \cdot \Delta y \cdot \Delta z \cdot C_P^{t+\Delta t} - \Delta x \cdot \Delta y \cdot \Delta z \cdot C_P^t}{\Delta t} + \Delta y \cdot \Delta z \cdot \left(F_e^t - F_w^t\right)$$

$$+ \Delta x \cdot \Delta z \cdot \left(F_s^t - F_n^t\right) - \Delta x \cdot \Delta y \cdot \left(F_{dn}^{t+\Delta t} - F_{up}^{t+\Delta t}\right) = \Delta x \cdot \Delta y \cdot \Delta z \cdot \sum R_P^t \quad (6)$$

$$where: \quad F_s = \left(vC - k_y \frac{\partial C}{\partial y}\right)\bigg|_{y=s}, \quad F_e = \left(uC - k_x \frac{\partial C}{\partial x}\right)\bigg|_{x=e}, \quad F_{up} = \left(wC - k_z \frac{\partial C}{\partial z}\right)\bigg|_{z=up},$$

$$\varsigma(t) \leq z(2) \quad for \quad \forall t$$

Where the model used the implicit scheme in the vertical direction, and applied the high-lateral flux modification (Spalding, 1972; Patankar, 1980) both in the vertical and horizontal fluxes. The model solves this equation by Tri-Diagonal Matrix Algorithm (TDMA).

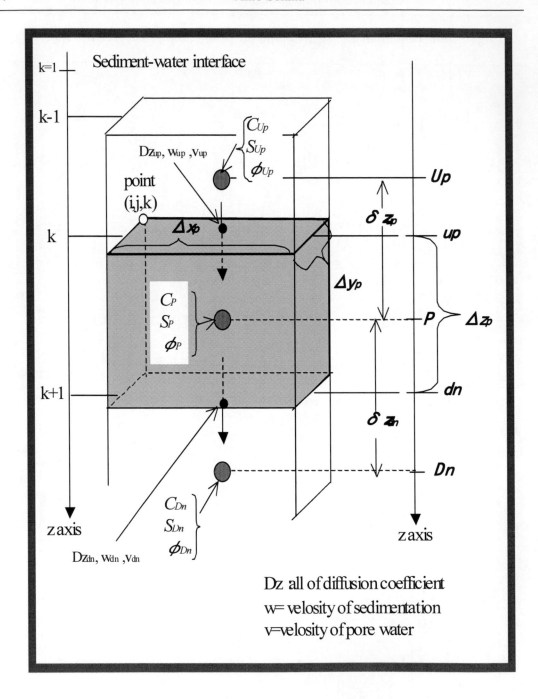

Figure 8. Defined points of variables and coefficients in (i, j, k) box/mesh of benthic system in the ecological model.

Table 3 Model equation and algorithm for dissolved substances in the benthic ecosystem (refer to Figure 8)

Dissolved materials: (i,j,k) Mesh/Box

from the spatial integration of (i,j,k) mesh /box,

$$\Delta x \cdot \Delta y \cdot \frac{\partial}{\partial t} \int_{up}^{dn} \left[\{ K' \overline{\rho}_s (1-\phi) + \phi \} C \right] dz = \Delta x \cdot \Delta y \cdot \left[\{ K' \overline{\rho}_s (D_B + D_B') \cdot (1-\phi) + \phi (D_S + D_I + D_B + D_B') \} \frac{\partial C}{\partial z} \right]_{up}^{dn}$$

$$- \Delta x \cdot \Delta y \cdot \left[\left\{ -(1 - K' \overline{\rho}_s) D_B \frac{\partial \phi}{\partial z} + K' \overline{\rho}_s (1-\phi) w + \phi v \right\} C \right]_{up}^{dn}$$

$$+ \Delta x \cdot \Delta y \cdot \int_{up}^{dn} \alpha (C_0 - C) dz + \Delta x \cdot \Delta y \cdot \int_{up}^{dn} (\phi \sum R') dz \tag{1}$$

$$\Longleftrightarrow$$

$$\frac{\Delta x \cdot \Delta y \cdot \Delta z \cdot \left(h_P^{t+\Delta t} C_P^{t+\Delta t} - h_P^t C_P^t \right)}{\Delta t} =$$

$$\Delta x \cdot \Delta y \cdot \left(F_{up}^{t+\Delta t} - F_{dn}^{t+\Delta t} \right) + \Delta x \cdot \Delta y \cdot \Delta z \cdot \alpha \left(C_0 - C_P^t \right) + \Delta x \cdot \Delta y \cdot \Delta z \cdot \phi_P^t \sum R'^t_P \tag{2}$$

$$\Longleftrightarrow$$

$$\frac{\Delta x \Delta y \Delta z \left(h_P^{t+\Delta t} C_P^{t+\Delta t} - h_P^t C_P^t \right)}{\Delta t} = \Delta x \Delta y \cdot \left(a_{Dn}^{t+\Delta t} C_{Dn}^{t+\Delta t} + a_{Up}^{t+\Delta t} C_{Up}^{t+\Delta t} - a_P^{t+\Delta t} C_P^{t+\Delta t} \right)$$

$$+ \Delta x \Delta y \Delta z \cdot \alpha \left(C_0 - C_P^t \right) + \Delta x \Delta y \Delta z \cdot \phi_P^t \sum R'^t_P \tag{3}$$

$$\Longleftrightarrow$$

$$A_P^{t+\Delta t} C_P^{t+\Delta t} = a_{Dn}^{t+\Delta t} C_{Dn}^{t+\Delta t} + a_{Up}^{t+\Delta t} C_{Up}^{t+\Delta t} + b_P^t \tag{4}$$

Where, $\quad F_x^t = f_x^t C_x^t - g_x^t \left[\frac{\partial C}{\partial z} \right]_x^t, \quad h_z^t = K' \overline{\rho}_s (1 - \phi_z^t) + \phi_z^t,$

$$g_z^t = K' \overline{\rho}_s \left(D_{Bz}^t + D_{Bz}'^t \right) \cdot (1 - \phi_x^t) + \phi_{xz}^t \left(D_{Sz}^t + D_{Iz}^t + D_{Bz}^t + D_{Bz}'^t \right),$$

$$f_z^t = -(1 - K' \overline{\rho}_s) D_{Bz}^t \left[\frac{\partial \phi}{\partial z} \right]_z^t + K' \overline{\rho}_s (1 - \phi_x^t) w_z^{t - \frac{\Delta t}{2}} + \phi_x^t v_z^{t - \frac{\Delta t}{2}}$$

$$a_{Dn}^t = \max \left[-f_{dn}^t, \frac{g_{dn}^t}{\delta z_{dn}} - \frac{f_{dn}^t}{2}, 0 \right], \quad a_{Up}^t = \max \left[f_{up}^t, \frac{g_{up}^t}{\delta z_{up}} + \frac{f_{up}^t}{2}, 0 \right], \quad a_P^t = a_{Up}^t + a_{Dn}^t + \left(f_{dn}^t - f_{up}^t \right)$$

$$A_P^t = a_P^t + \frac{\Delta z}{\Delta t} h_P^t, \quad b_P^t = \frac{\Delta x}{\Delta t} h_P^t C_P^t + \Delta z \cdot \phi_P^t \alpha \left(C_0 - C_P^t \right) + \Delta z \cdot \phi_P^t \sum R'^t_P$$

Where the model used the implicit scheme in the vertical direction, and applied the high-lateral flux modification (Spalding 1972, Patankar 1980) both in the vertical and horizontal fluxes. The model solves this equation by Tri-Diagonal Matrix Algorithm (TDMA).

Table 4 Model equation and algorithm for particulate substances in the benthic system (refer to Figure 9)

Particulate materials: (i,j,k) Mesh/Box
From the spatial integration of (i,j,k) mesh /box,

$$\Delta x \Delta y \frac{\partial}{\partial t} \int_{up}^{dn} \{(1-\phi)S\}dz = \Delta x \Delta y \left\{ (D_B + D_B') \cdot (1-\phi)\frac{\partial S}{\partial z} \right\} \Bigg|_{up}^{dn}$$

$$- \Delta x \Delta y \left[\left\{ -D_B \frac{\partial(1-\phi)}{\partial z} + (1-\phi)w \right\} S \right] \Bigg|_{up}^{dn} + \Delta x \Delta y \int_{up}^{dn} (1-\phi)\overline{\rho}_s \sum \overline{R'}\ dz \qquad (1)$$

$$\Leftrightarrow$$

$$\frac{\Delta x \Delta y \Delta z \left(hs_P^{t+\Delta t} S_P^{t+\Delta t} - hs_P^t S_P^t \right)}{\Delta t} = \Delta x \Delta y \left(Fs_{up}^{t+\Delta t} - Fs_{dn}^{t+\Delta t} \right) + \Delta x \Delta y \Delta z \cdot \left(1-\phi_P^t \right)\overline{\rho}_s \sum \overline{R'}_P^t \qquad (2)$$

$$\Leftrightarrow$$

$$\frac{\Delta x \Delta y \Delta z \left(hs_P^{t+\Delta t} S_P^{t+\Delta t} - hs_P^t S_P^t \right)}{\Delta t} = \Delta x \Delta y \left(as_{Dn}^{t+\Delta t} S_{Dn}^{t+\Delta t} + as_{Up}^{t+\Delta t} S_{Up}^{t+\Delta t} - as_P^{t+\Delta t} S_P^{t+\Delta t} \right)$$

$$+ \Delta x \Delta y \Delta z \cdot \left(1-\phi_P^t \right)\overline{\rho}_s \sum \overline{R'}_P^t \qquad (3)$$

$$\Leftrightarrow$$

$$As_P^{t+\Delta t} S_P^{t+\Delta t} = as_{Dn}^{t+\Delta t} S_{Dn}^{t+\Delta t} + as_{Up}^{t+\Delta t} S_{Up}^{t+\Delta t} + bs_P^t \qquad (4)$$

Where,

$$Fs_z^t = fs_z^t S_z^t - gs_z^t \left[\frac{\partial S}{\partial z} \right]_z^t, \ \ hs_z^t = 1-\phi_z^t, \ gs_z^t = \left(D_{Bz}^t + D_{Bz}'^t \right)\cdot \left(1-\phi_z^t \right),$$

$$fs_z^t = -D_{Bz}' \left[\frac{\partial(1-\phi)}{\partial z} \right]_z^t + \left(1-\phi_z^t \right)w_z^{t-\frac{1}{2}\Delta t}\ z$$

$$as_{Dn}^t = \max\left[-fs_{dn}^t, \frac{gs_{dn}^t}{\delta z_{dn}} - \frac{fs_{dn}^t}{2}, 0 \right], as_{Up}^t = \max\left[fs_{up}^t, \frac{gs_{up}^t}{\delta z_{up}} + \frac{fs_{up}^t}{2}, 0 \right]$$

$$as_P^t = as_{Up}^t + as_{Dn}^t + \left(fs_{dn}^t - fs_{up}^t \right), \ As_P^t = as_P^t + \frac{\Delta z}{\Delta t} hs_P^t$$

$$bs_P^t = \frac{\Delta z}{\Delta t} hs_P^t S_P^t + \Delta z \cdot \left(1-\phi_P^t \right)\overline{\rho}_s \sum \overline{R'}_P^t$$

Where the model used the implicit scheme in the vertical direction, and applied the high-lateral flux modification (Spalding 1972, Patankar 1980) both in the vertical and horizontal fluxes. The model solves this equation by Tri-Diagonal Matrix Algorithm (TDMA).

---

6. IMPLEMENTATION

In this section, the application methodology for ECOHYM on Tokyo Bay is discussed.

6.1. Spatial Resolution

6.1.1. Vertical Spatial Resolution

The vertical and horizontal spatial resolution setting in the ecological model is shown in Figure 9. In the benthic system, as for the biochemical processes, the vertical spatial distributions of processes related to hypoxia – namely, oxic, suboxic and anoxic mineralization, denitrification, nitrification, ODU oxidation and photosynthesis of benthic algae – have a steep gradient within a scale of 1 mm to 10 mm (Kuwae et al. 2005; Sayama 2005). As for the transport processes, the advection flow in the sediment (deposition of sediment and advection of pore water) is 0.3cm~2cm/year of deposition rate in Tokyo Bay (Matsumoto 1983), and the diffusion in the sediment is dominated by molecular diffusion (i.e. small-scale diffusion) as compared to the pelagic system. Thus, the vertical grid interval in the benthic system was set to 0.1mm–1.2cm pitch.

In the pelagic system, because the mass of transportation due to water current is larger than that in the benthic system, the vertical profile of pelagic metabolism, which is significant to the hypoxic mechanism, is more moderate than in the benthic profile. In this implementation, the vertical spatial resolution in the pelagic system is set at 1~2m which is enough to demonstrate accurately the following hypoxia-related processes: vertical mixing/stratification, sedimentation of particulate substances, oxygen consumption and nutrients flux from the benthic system.

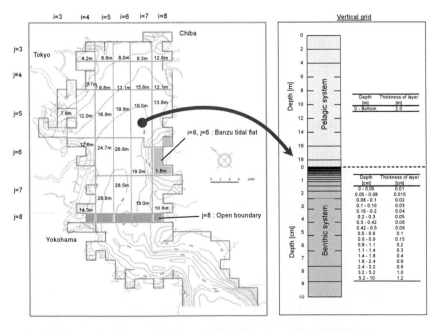

Figure 9. Geographical description of Tokyo Bay (Calculated area, spatial resolution both in the vertical and horizontal directions, and coordinates (i, j) on the ecological model application).

6.1.2. Horizontal Spatial Resolution

Horizontal spatial resolution is different for the hydrodynamics and ecological models. In the hydrodynamics model, a 2km x 2km grid was implemented, whereas in the ecological model, the Tokyo Bay area under consideration was divided into 26 zones (boxes). The input data of the ecological model from the hydrodynamics model, i.e. flow velocity, eddy viscosity and temperature etc, were averaged spatially and adjusted to the 26 boxes of the ecological model, while respecting the flow continuity equation of water volume. The reasons for the low horizontal spatial resolution used in the ecological model are as follows: (a) a shorter calculation time is advantageous to duplicate model tuning/calibration in the first stage of model development, (b) our focus is on vertical micro-mechanisms, particularly for the benthic and short timescale dynamics in the tidal flat and (c) setting the ideal vertical and temporal resolution lies at the limit of computational performance, while maintaining a reasonable calculation time. Although the resolution of the 26 boxes is not sufficient to demonstrate details of the horizontal spatial distribution, it still enables us to describe the difference in ecological mechanisms from the central bay to tidal flat area, the difference in benthic-pelagic mutual interaction depending on the water depth (i.e. the achievable mass of pelagic particulate substances to the seafloor), and the difference in the central bay-tidal flat area interaction depending on the distance between each area.

6.2. Simulation Period and Time Step

6.2.1. Simulation Period

A simulation was carried out to demonstrate the daily and seasonal dynamics of an average year. The prescribed functions (forcing function) of the hydrodynamics and ecological models, i.e. freshwater and nutrient input from rivers, meteorological conditions (light intensity, wind, etc.), and open boundary condition of model variables were set as one-year periodical functions. These functions were created based on the observed data from 1998 to 2002. The convergence state of this simulation describes the dynamics with a one-year period. This state hereafter is called the "annual periodical steady state of the existing Tokyo Bay". The convergence time to achieve the annual periodical steady state was different for (a) the fluid dynamics of pelagic system, (b) ecological dynamics of the pelagic system in the central bay, (c) the ecological dynamics of the benthic system in the central bay and (d) the ecological dynamics of tidal flat. The differences mainly result from the differences in the physical or biochemical turn-over periods of the modeled substances (model variables) among (a), (b), (c) and (d). Here, the turn-over period is defined as the inverse number of turn-over rate, and the turn-over rate is defined as shown in Figure 10. As far as the turn-over period due to physical processes is concerned, the value of the pelagic system, 30-100 days, is the time scale for all existing water in Tokyo Bay to be exchanged once. The value of the benthic system, 30-100 years, is the time scale for the benthic solid phase at the sediment-water interface to achieve the modeled benthic bottom layer depth (10cm). Meanwhile, in the simulation, the hydrodynamics model achieved the annual periodical steady state after a one year calculation. The ecological model achieved the annual periodical steady state after a 100 year calculation. These results suggest that the convergence time of both the hydrodynamics and ecological models are controlled by their turn-over periods due to physical processes.

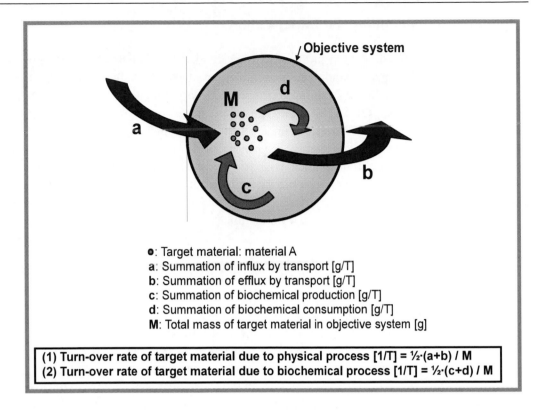

Figure 10. Concept and formulations of turn-over rate due to physical and biochemical processes.

6.2.2. Time Step

The significant time scale of ecosystem dynamics differs between the benthic and pelagic systems, or between the tidal flat and central bay areas. In the tidal flat, tidal or daily scale dynamics of biochemical processes are outstanding because of the shallow depth (e.g. changes in feeding behavior of suspension feeders due to submersion/emersion and changes in the level of photosynthesis of benthic algae due to light intensity at the sediment surface, etc.). These dynamics have the potential to operate dominantly the interaction between the tidal flat and the central bay area. In addition, seasonal dynamics of light and temperature also affect the metabolisms in the tidal flat. In the central bay area, photosynthesis by phytoplankton, and the transport of substances in the pelagic system change on a tidal and daily scale. Seasonal dynamics of light and temperature also affect metabolisms in the central bay area. However, for the benthic system in the central bay where light does not reach the seafloor and there is no submersion/emersion phenomenon, it is suggested that the daily/tidal scale dynamics are not as dominant as in the tidal flat area.

The notable time scale of dynamics in calculations made by ECOHYM ranges from a daily scale to a seasonal time scale. The time step was thus set at 0.2 hrs, a value which enables us to demonstrate both the daily/tidal and seasonal dynamics.

6.3. Boundary Conditions and Initial Values

The boundary conditions used in the model are divided into values set at the open sea boundary (open sea boundary conditions) and values set at the atmosphere/inflow points of river (other boundary conditions). For the "open sea boundary conditions," tidal level of four component tides (M2, S2, O1 and K1), the concentrations of model variables, salinity, and water temperature were set. For "other boundary conditions," air temperature, surface wind, light, and nutrients/detritus/fresh water discharged from rivers were set. These were prescribed based on the field measurements and were interpolated by the spline method or linear interpolation. Initial values of the model variables in the ecological model were set at the yearly averaged value on April 1, from 1998 to 2002. The dynamics of these boundary conditions become the driving force of the simulation.

6.4. PARAMETER TUNING

6.4.1. Values of the Parameters
Many biochemical parameters are included in the formulation of biochemical processes. We investigated a range of values for these parameters that have been observed or used in other models, and set values within the range or at the same order as the investigated value. Relevant biochemical parameters set in this study and references of the sources of the parameters are listed in Tables 5 and 6. On setting the parameter values, the parameters, either analyzed or calculated values based on observations in Tokyo Bay, were initially set within the known ranges in consideration of regionality. Such parameters are indicated as "OD" under "Source" in Tables 5 and 6. Second, the parameters whose values were estimated from data in other coastal areas were set and are indicated as "OE" under "Source". Finally, the values of unknown parameters were given to tune the simulation.

6.4.2. Tuning Scheme
It is very difficult to control ECOHYM due to the inclusion of both benthic and pelagic systems as well as the central bay and tidal flat area. Therefore, in the first stage of the development process, each ecological sub-model, namely, the sub-model of the pelagic system of the central bay area, the sub-model of the benthic system of the central bay area, and the sub-model of the benthic-pelagic-coupled system of the tidal flat area were developed and calculated separately for verification, calibration and validation. In these calculations, the values at the boundary of the focused system of each sub-model (e.g. the benthic boundary condition of pelagic sub-models (the model variables/fluxes from benthic system), or the pelagic boundary condition of the benthic sub-model (the model variables of bottom water of the pelagic system, etc.)) were given by prescribed functions based on the observed data. In the next stage, mutual interactions between sub-models were incorporated by (a) unsetting the prescribed functions at the boundary between sub-models, (b) assigning the value of the boundary from the calculation of each sub-model at each time step, and (c) making the mutual linkage (dependence) work among each sub-model. The tuning of the mutually interacting model was implemented/tuned based on the decided value of the parameters used in the first stage.

Table 5. Values of major biochemical parameters in the pelagic system (refer to Appendix)

Name	Unit	Value	Description	Major Source [a]
[Phytoplankton : 01]				
α_{pho01}, β_{pho01}	1/h, 1/°C	0.0625, 0.0693	Maximum growth rate	1,2, $Q_{10}=2$
$Hf_{n,pho01}$	mgN/l	0.1	Half saturation constant for nitrogen limitation	3
$Hf_{p,pho01}$	mgP/l	0.05	Half saturation constant for phosphorus limitation	4
I_{min01}	$\mu E/m^2/s$	7.1	Minimum light intensity for photosynthesis	5
I_{hf01}	$\mu E/m^2/s$	56.5	Half saturation constant for light	5
R_{chl}	gChl-a/gC	0.0333	Chl-a:C ratio	6
α_{res01}, β_{res01}	1/h, 1/°C	0.00125, 0.0693	Rate of respiration	7, $Q_{10}=2$
α_{mor01}, β_{mor01}	1/h, 1/°C	0.00042, 0.0693	Rate of mortality	Tu, $Q_{10}=2$
R_{nc01}	gN/gC	0.2094	Ratio of nitrogen to carbon	8
R_{pc01}	gP/gC	0.0338	Ratio of phosphate to carbon	8
$R_{PP,1}$, $R_{PP,2}$, $R_{PP,3}$	-	0.85, 0.10, 0.05	Composition ratio (ratio of fast-labile, slow-labile and refractory / very slow-labile part)	Tu
$R_{ext01,1}$, $R_{ext01,2}$	-	0.2, 0.8	Ratio of labile and refractory DOM to extra-release	Tu
[Zooplankton : 02]				
α_{gra02}, β_{gra02}	1/h, 1/°C	0.0150, 0.0693	Maximum ration	9, $Q_{10}=2$
A_{ivl02}	l/mgC	6.3	Ivlev's constant	10, 11
A_{kai02}	$\mu gC/cm^3$	0.1	Feeding threshold	Tu
R_{ege02}	-	0.7	Assimilation efficiency	12, 13
R_{grt02}	-	0.3	Growth efficiency	11
α_{mor02}, β_{mor02}	1/h	0.0021, 0.0693	Rate of mortality	Tu
$R_{ZP,1}$, $R_{ZP,2}$, $R_{ZP,3}$	-	0.80, 0.15, 0.05	Composition ratio (ratio of fast-labile, slow-labile and refractory/very slow-labile part)	Tu
[Detritus, Dissolved organic matter : 03,i (i=1,2,3), 04,j (j=1,2)]				
$\alpha_{mi03,1}$, $\beta_{mi03,1}$	1/h, 1/°C	5.0×10^{-4}, 0.0693	Mineralization rate of fast-labile detritus	14,15,16,17, $Q_{10}=2$
$\alpha_{mi03,2}$, $\beta_{mi03,2}$	1/h, 1/°C	5.0×10^{-5}, 0.0693	Mineralization rate of slow-labile detritus	14,15,16,17, $Q_{10}=2$
$\alpha_{mi03,3}$, $\beta_{mi03,3}$	1/h, 1/°C	5.0×10^{-7}, 0.0693	Mineralization rate of refractory/very slow-labile detritus	14,15,16,17, $Q_{10}=2$
$\alpha_{mi04,1}$, $\beta_{mi04,1}$	1/h, 1/°C	1.0×10^{-3}, 0.0693	Mineralization rate of fast-labile DOM	17, 18, $Q_{10}=2$
$\alpha_{mi04,2}$, $\beta_{mi04,2}$	1/h, 1/°C	0.0, 0.0693	Mineralization rate of refractory DOM	17, 18, $Q_{10}=2$
$Hf_{o2w,omi}$	mgO_2/l	0.096	Half saturation constant for O_2 limitation in oxic mineralization	19
$Hf_{o2w,smi}$	mgO_2/l	0.32	Half saturation constant for O_2 inhibition in suboxic mineralization	19

Table 5. (Continued)

$Hf_{no3w,smi}$	mgN/l	1.86	Half saturation constant for NO_3 limitation in suboxic mineralization	19(OE)
$Hf_{o2w,ami}$	mgO_2/l	0.16	Half saturation constant for O_2 inhibition in anoxic mineralization	19
$Hf_{no3w,ami}$	mgN/l	0.50	Half saturation constant for NO_3 inhibition in anoxic mineralization	19(OE)
$R_{dec03,1}$	-	0.25	Ratio of decomposition to mineralization (fast-labile detritus)	15
$R_{dec03,2}$	-	0.25	Ratio of decomposition to mineralization (slow-labile detritus)	15
$R_{dec03,3}$	-	0.25	Ratio of decomposition to mineralization (refractory/very slow-labile detritus)	15
$R_{nc03,1}$	gN/gC	0.2235	Ratio of nitrogen to carbon (fast-labile detritus)	20(OD), 8(OE)
$R_{nc03,2}$	gN/gC	0.1570	Ratio of nitrogen to carbon (slow-labile detritus)	20(OD), 8(OE)
$R_{nc03,3}$	gN/gC	0.0739	Ratio of nitrogen to carbon (refractory/very slow-labile detritus)	20(OD), 8(OE)
$R_{nc04,1}$	gN/gC	0.2401	Ratio of nitrogen to carbon (labile DOM)	20(OD), 8(OE)
$R_{nc04,2}$	gN/gC	0.0739	Ratio of nitrogen to carbon (refractory DOM)	20(OD), 8(OE)
$R_{pc03,1}$	gP/gC	0.0362	Ratio of phosphate to carbon (fast-labile detritus)	20(OD), 8(OE)
$R_{pc03,2}$	gP/gC	0.0210	Ratio of phosphate to carbon (slow-labile detritus)	20(OD), 8(OE)
$R_{pc03,3}$	gP/gC	0.0020	Ratio of phosphate to carbon (refractory/very slow-labile detritus)	20(OD), 8(OE)
$R_{pc04,1}$	gP/gC	0.0400	Ratio of phosphate to carbon (labile DOM)	20(OD), 8(OE)
$R_{pc04,2}$	gP/gC	0.0020	Ratio of phosphate to carbon (refractory DOM)	20(OD), 8(OE)
$R_{DOM,11}$, $R_{DOM,21}$	-	0.9, 0.1	Ratio of labile and refractory DOM to decomposition (fast-labile detritus)	Tu
$R_{DOM,12}$, $R_{DOM,22}$	-	0.5, 0.5	Ratio of labile and refractory DOM to decomposition (slow-labile detritus)	Tu
$R_{DOM,13}$, $R_{DOM,23}$	-	0.0, 1.0	Ratio of labile and refractory DOM to decomposition (refractory /very slow-labile detritus)	Tu
[NH_4-N, NO_3-N : 05, 06]				
α_{nit05}, β_{nit05}	1/h, 1/°C	0.001, 0.0693	Rate of nitrification	21 (OE), Q_{10}=2
$Hf_{o2,nit05}$	mgO_2/l	0.032	Half saturation constant for O_2 limitation in nitrification	22(OE)
R_{den06}	-	0.75	Denitrification ratio to suboxic mineralization	Tu
[ODU : 08]				
α_{oxi08}, β_{oxi08}	1/h, 1/°C	1.0, 0.0693	Rate of ODU oxidation	19(OE), Q_{10}=2
$Hf_{o2,oxi08}$	mgO_2/l	0.032	Half saturation constant for O_2 limitation in ODU oxidation	19
R_{oxi08}	-	0.0	Ratio of ODU oxidation to anoxic mineralization	Tu
α_{aut08}, β_{aut08}	1/h, 1/°C	0.0, 0.0	Rate of authigenic mineralization	Tu
R_{aut08a}	-	0.0	Ratio of authigenic mineralization to ODU oxidation	Tu
R_{aut08b}	-	0.0	Ratio of authigenic mineralization to ODU production	Tu

Source : 1. Macedo et al. (2001); 2. Horiguchi (2001); 3. Epply et al.(1969); 4. Fuhs et al.(1972); 5. Nishikawa et al. (2002); 6. Strickland (1965); 7. Jørgensen et al. (1991); 8. Jørgensen (1979); 9. Baretta and Ruardij (1988); 10. Zilloux (1970); 11. Suschenya (1970); 12. Marshall and Orr (1955a); 13. Marshall and Orr (1955b); 14. Matsunaga (1981); 15. Ishikawa and Nishimura (1983); 16. Ogura (1972); 17. Emerson and Hedges (1988); 18. Ogura (1975); 19. Soetaert et al. (1996a); 20. National Institute for Land and Infrastructure Management (2006); 21. Jørgensen and Bendoricchio (2001); 22. Oguz (2002) Tu. Tuning. (OD: Observed data in Tokyo Bay; OE. Order estimated value referred to the literature)
[a] Some of the values of these sources were not used directly, but used after analysis to convert the format of model parameters.

Table 6. Values of major biochemical parameters in the benthic system (refer to Appendix)

Name	Unit	Value	Description	Major Source [a]
[Suspension feeders : 51]				
A_{Wwet51}	gw/ind	0.436	Basal weight	1, 2 (OD), 3
A_{wd51}	gd/gw	0.05	Dry weight : wet weight ratio	1(OD)
A_{cd51}	gC/gdry	0.328	Carbon : dry weight ratio	3
A_{Depq}	Cm	100	Double filtering effective depth	Tu
$R_{O2mor51}$	-	0.15	Oxygen saturation rate that starts suspension feeder's mortality from oxygen deficiency	4,5 (OE)
α_{grt51}	1/h	0.002208	Efficient growth rate	6
α_{bas51}	1/h	0.0001256	Basal metabolism	7
R_{ege51}	-	0.6	Digestivity efficiency	8,9
R_{exc51}	-	0.9	Activity metabolism	7
β_{mor51}	1/°C	0.0693	Temperature dependency of mortality	$Q_{10}=2$
α_{mor51a}	1/h	0.00017	Natural mortality rate at 0 °C	Tu
α_{mor51b}	1/h	0.017	Rate of mortality from oxygen deficiency	Tu
R_{lar51}	-	0.01	Ratio of larva input to feeding	10
$R_{SFB,1}, R_{SFB,2}, R_{SFB,3}$	-	0.98, 0.01, 0.01	Composition ratio (ratio of fast-labile, slow-labile and refractory/very slow-labile part)	Tu
$R_{ZWfec51}$	-	0.5	Ratio of feces to pelagic system	Tu
[Deposit feeders : 52]				
$\alpha_{fee52}, \beta_{fee52}$	1/h, 1/°C	0.0058, 0.0693	Maximum ration	11(OE),$Q_{10}=2$
$A_{temp,fee52}$	degree	10	Maximum saturation temperature about eating	12
A_{ivl52}	cm³/μgC	0.005	Ivlev's constant	Tu
A_{kai52}	μgC/cm³	200	Feeding threshold	7(OE)
$Hf_{dfb,fee52}$	μgC/cm²	150	Half saturation constant for cannibalism efficiency	7(OE)
$R_{O2mor52}$	-	0.1	Oxygen saturation rate that starts deposit feeder's mortality from oxygen deficiency	Tu
$Hf_{fod52,fee52}$	μgC/cm²	2500.0	Half saturation constant for digestive efficiency	7(OE)
R_{undg52}	-	0.4	Minimum undigestive efficiency	13
R_{exc52}	-	0.61	Ratio of excretion to assimilated food	7(OE)
β_{mor52}	1/°C	0.0693	Temperature dependency of mortality	$Q_{10}=2$
α_{mor52a}	1/h	0.00025	Natural mortality rate at 0 °C	Tu
α_{mor52b}	1/h	0.0025	Rate of mortality from oxygen deficiency	Tu
R_{lar52}	-	0.01	Ratio of larva input to feeding	10
$R_{DFB,1}, R_{DFB,2}, R_{DFB,3}$	-	0.55, 0.40, 0.05	Composition ratio (ratio of fast-labile, slow-labile and refractory/very slow-labile part)	Tu
[Detritus (POM), Dissolved organic matter (DOM) : 53,i (i=1,2,3), 54,j (j=1,2)]				
$Hf_{o2b,omi}$	mgO₂/l	0.001	Half saturation constant for O_2 limitation in oxic mineralization	Tu
$Hf_{o2b,smi}$	mgO₂/l	1.00	Half saturation constant for O_2 inhibition in suboxic mineralization	14(OE)
$Hf_{o2b,ami}$	mgO₂/l	0.03	Half saturation constant for O_2 inhibition in anoxic mineralization	Tu
$Hf_{no3b,smi}$	mgN/l	0.16	Half saturation constant for NO_3 limitation in suboxic mineralization at oxic layer	14(OE)
$Hf_{no3b,ami}$	mgN/l	0.32	Half saturation constant for NO_3 inhibition in anoxic mineralization at oxic layer	14(OE)
$\alpha_{mi53,1}, \beta_{mi53,1}$	1/h, 1/°C	5.0×10^{-4}, 0.0693	Maximum mineralizatoin rate of fast labile detritus	
$\alpha_{mi53,2}, \beta_{mi53,2}$	1/h, 1/°C	5.0×10^{-5}, 0.0693	Maximum mineralization rate of slow labile detritus	15,16,17,18,$Q_{10}=2$
$\alpha_{mi53,3}, \beta_{mi53,3}$	1/h, 1/°C	5.0×10^{-7}, 0.0693	Maximum mineralization rate of refractory/very slow labile detritus	

Table 6. (Continued)

$\alpha_{mi54,1}$, $\beta_{mi54,1}$	1/h, 1/°C	1.0×10^{-3}, 0.0693	Maximum mineralization rate of labile DOM	18, 19,Q_{10}=2
$\alpha_{mi54,2}$, $\beta_{mi54,2}$	1/h, 1/°C	0, 0.0693	Maximum mineralization rate of refractory DOM	
$R_{dec53,1}$	-	0.1	Ratio of decomposition to mineralization (fast-labile detritus)	Tu
$R_{dec53,2}$	-	0.25	Ratio of decomposition to mineralization (slow-labile detritus)	20
$R_{dec53,3}$	-	0.25	Ratio of decomposition to mineralization (refractory/very slow-labile detritus)	20
$K_{ads54,1}$, $K_{ads54,2}$	ml/g	25.03, 0.69	Adsorption coefficient for labile DOM and refractory DOM	Tu
[NH_4-N, NO_3-N : 55, 56]				
α_{nit55}, β_{nit55}	1/h, 1/°C	0.3, 0.0693	Rate of nitrification	14(OE), Q_{10}=2
$Hf_{o2,nit55}$	mgO_2/l	0.032	Half saturation constant for O_2 limitation in nitrification	14
R_{den56}	-	0.75	Denitrification ratio to suboxic mineralization	Tu
K_{ads55}	ml/g	1.58	Adsorption coefficient	21
[ODU : No. 58]				
α_{oxi58}, β_{oxi58}	1/h, 1/°C	5, 0.0693	Rate of oxidation	14(OE),Q_{10}=2
$Hf_{o2,oxi58}$	mgO_2/l	0.032	Half saturation constant for O_2 limitation in oxidation	14
R_{oxi58}	-	0.0	Ratio of ODU oxidation to anoxic mineralization	Tu
α_{aut58}, β_{aut58}	1/h, 1/°C	0.0, 0.0	Rate of authigenic mineralization	Tu
R_{aut58a}	-	0.0	Ratio of authigenic mineralization to ODU oxidation	Tu
R_{aut58b}	-	0.0	Ratio of authigenic mineralization to ODU production	Tu
[Benthic algae : 60]				
α_{pho60}, β_{pho60}	1/h, 1/°C	0.045, 0.0693	Maximum growth rate	7, 22(OE), Q_{10}=2
$Hf_{n,pho60}$	mgN/l	0.018	Half saturation constant for nitrogen limitation	23
$Hf_{p,pho60}$	mgP/l	0.001	Half saturation constant for phosphorus limitation	24
$Hf_{o2,res60}$	mg/l	0.03	Half saturation constant for oxygen limitation to respiration	Tu
I_{opt60}	$\mu E/m^2/s$	30	Half saturation constant for light limitation	25
k_b	1/cm	59.0	Light attenuation coefficient in sediment	26(OE)
$R_{pho60,55}$	-	10	Ammonium intake ratio coefficient	Tu
R_{ext60}	-	0.122	Ratio of extra-release to photosynthesis	Tu
R_{res60a}	-	0.0002	Ratio of rest respiration rate to maximum growth rate	7
R_{res60b}	-	0.1	Ratio of activity respiration to photosynthesis	7(OE)
α_{mor60}, β_{mor60}	1/h, 1/°C	0.00019, 0.0693	Rate of natural mortality	7
R_{nc60}	gN/gC	0.2148	Ratio of nitrogen to carbon	24
R_{pc60}	gP/gC	0.0342	Ratio of phosphate to carbon	24
$R_{BAL,1}$, $R_{BAL,2}$, $R_{BAL,3}$	-	0.900, 0.075, 0.025	Composition ratio (ratio of fast-labile, slow-labile and refractory/very slow-labile part)	Tu
$R_{ext60,1}$, $R_{ext60,2}$	-	0.1, 0.9	Ratio of labile and refractory DOM to extra-release	Tu

Source : 1. Kuwae (2001); 2. Kuwae et al. (2005); 3. Japan Environmental Management Association for Industry (1998); 4. Kurashige (1942); 5. Kakino (1982); 6. Isono et al. (1998); 7. Barreta and Ruardij (1988); 8. Yamamuro and Koike (1993); 9. Hiwatai et al. (2002); 10. Conover (1978); 11. Cammen (1980); 12.Kremer and Nixon (1978), 13. Valiela (1984); 14. Soetaert et al. (1996a); 15. Matsunaga (1981); 16. Ishikawa et al. (1983); 17. Ogura (1972); 18. Emerson and Hedges (1988); 19. Ogura (1975); 20. Ishikawa et al. (1983); 21. Rosenfeld (1979) 22. Admiraal et al. (1982); 23. Epply et al. (1969); 24. Jørgensen (1979); 25. Hiroshima Environment & Health Association (2002); 26. Kamio et al. (2004); Tu. Tuning. (OD: Observed data in Tokyo Bay; OE: Order estimated value referred to the literature)

[a] Some of the values of these sources were not used directly, but used after analysis to convert the format of model parameters.

6.4.3. Focus Point on Tuning

In the early stage of tuning, we initially focused on getting the model to reproduce the observed values of particulate substances (phytoplankton, zooplankton, and detritus in both the sediment and water column). Second, we checked the model's capacity to reproduce the inorganic substance values (dissolved oxygen, NH_4-N, NO_3-N, PO_4-P). This method is based on the assumption that the dynamics of inorganic matters are controlled by the organic matters, given the higher turn-over rate of the former, especially in the benthic system. In the case where the model reproduced the observed values of organic matter well but did not reproduce the values of inorganic matter, the fraction rate of fast-labile detritus, slow-labile detritus and refractory detritus, and physical parameters such as bioturbation and irrigation were tuned to adjust the observed data.

It is relatively difficult to control the dynamics of benthic fauna (suspension feeders and deposit feeders), which are higher organisms compared to plankton and bacteria, by formulating their metabolism based on statistical data. Therefore, during the early stage of tuning, the temporal/spatial distribution of benthic fauna was given by the prescribed functions of the observed value, while during the final stage of tuning, the biomass of the benthic fauna was treated as a model variable and distributions were calculated autonomously. In the tidal flat area, the biomass of benthic fauna is particularly affected by processes such as digging, sowing transported clams (artificial activities) and bird feeding. Therefore, although the dynamics of benthic fauna on the tidal flat were calculated as a model variable, the value was compared to the observed data at each time step and the gaps between the observed value and the calculated value were compensated using the nagging technique in the simulation.

7. Validation - Comparison between Model Output and Observed Data

Calculated model variables and fluxes were compared with the time series of measurements recorded at monitoring stations in Tokyo Bay, including the tidal flat located at Banzu. The compared model outputs here are the results of calculations in which the values of the parameters are set within the known values, and the mutual interactions between benthic-pelagic or the central bay-tidal flat area are functional (outputs from the final stage calculation mentioned in "Tuning scheme" discussed in section 6.4.2). Because the temporal/spatial distribution of model variables depend on each other through the ecological network described in the model, it is necessary for the validation to be performed by checking all model variables. Therefore, in this section, we demonstrated the simultaneous reproduction of the observed temporal/spatial dependence of all major model variables.

7.1. Pelagic Ecosystem in the Central Bay Area

The seasonal dynamics of the model variables were compared to the observed data. The results at four boxes, i.e. (i, j) = (5, 6), (6, 4), and (7, 5), are shown in Figure 11, and the corresponding relationship between the described areas and the coordinates (i, j) are shown in

Figure 9. The compared model variables at each area are of a similar magnitude as for the observed results. The seasonal variation of the observed items, total organic matter (TOM), dissolved organic matter (DOM), and dissolved oxygen (DO), in which conspicuous seasonal periodical dynamics appeared, are also reproduced by the model. In particular, hypoxia in the bottom layer during summer (from the end of July to mid-September) is reproduced by the model. The calculated DO concentration during hypoxia gives a value under 2.0 mgO_2/l, which is almost the same as the observed value. This result denotes that the model fills the requirement for providing precise descriptions of the intensity of vertical mixing processes determined by turbulent flow modeling, and the flux of oxygen consumption around sediment-water interface determined by the sum of each oxygen consuming/producing biochemical process described in the model.

The observed data show that the relationships between the surface and the bottom concentrations of NH_4-N, PO_4-P, and NO_3-N change seasonally and spatially. For example, the data of NH_4-N and NO_3-N in area $(i, j) = (5, 6)$ show a greater difference between the surface and the bottom compared to that of areas $(i, j) = (6, 4)$ and $(7, 5)$ throughout one year, while in area $(i, j) = (5, 6)$, NH_4-N and NO_3-N at the surface is higher than at the bottom. The model's output reproduces the trend of this spatial distribution of NH_4-N and NO_3-N concentrations. As for PO_4-P, the observed value for all boxes in Figure 8 shows that the bottom concentration tends to be higher than that of the surface during the period from June to September. In contrast, from November to March, PO_4-P at the bottom is either similar to that at the surface or lower. In particular, in area $(i, j) = (5, 6)$, the trend for a lower value at the bottom compared to the surface is observed from November to March. The model reproduces this temporal/spatial distribution of PO_4-P concentration. A good agreement of the model results with observed data for the temporal/spatial distribution of NH_4-N, PO_4-P and NO_3-N mentioned above is a required condition for the model to describe precisely and mechanically the dynamics or spatial distribution of nutrients due to photosynthesis, and nutrient fluxes from the benthic to pelagic systems.

7.2. Benthic Ecosystem in Central Bay Areas

The seasonal dynamics and vertical profiles of the model variables are compared with observed data in Figures 12 and 13. A comparison of the seasonal changes of oxygen and NH_4-N fluxes at the sediment-water interface is shown in Figure 14. The areas (boxes) in these figures were selected based on differences/variations in topographic characteristics among selected areas in the vertical and horizontal plane (depth, distance from open sea, from the land, or from the tidal flat, etc.) with due consideration of the observed conditions.

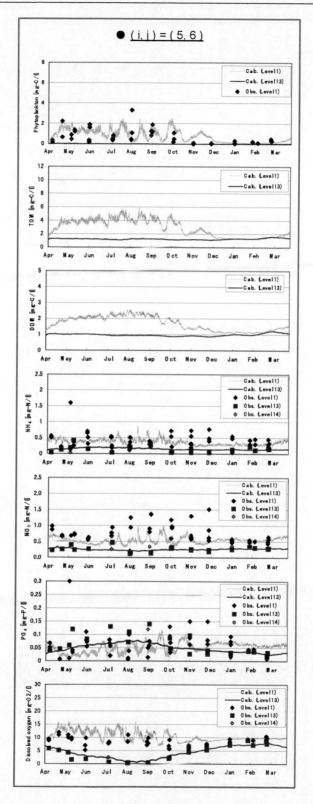

Figure 11. (Continued)

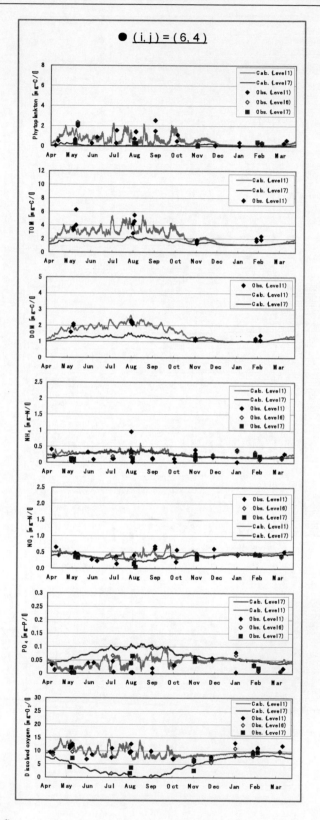

Figure 11. (Continued)

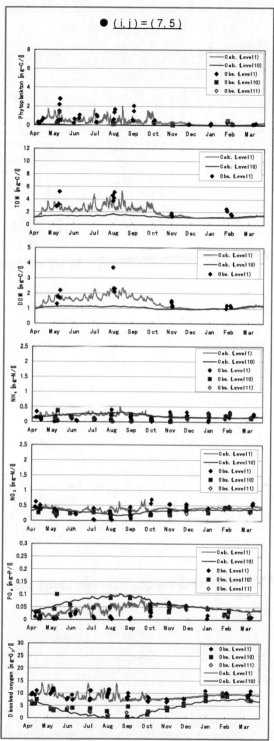

* Observed data are drawn from Chiba Prefecture (1998-2002), Kanagawa Prefecture (1998-2002), Tokyo Metropolitan (1998-2002), Ministry of the Environment (1998-2002)

Figure 11. Seasonal variation of the pelagic system in the central bay area - Comparison between model output (lines) and observed value (dots). Grid number (i, j) = (5, 6), (6, 4), (7, 5).

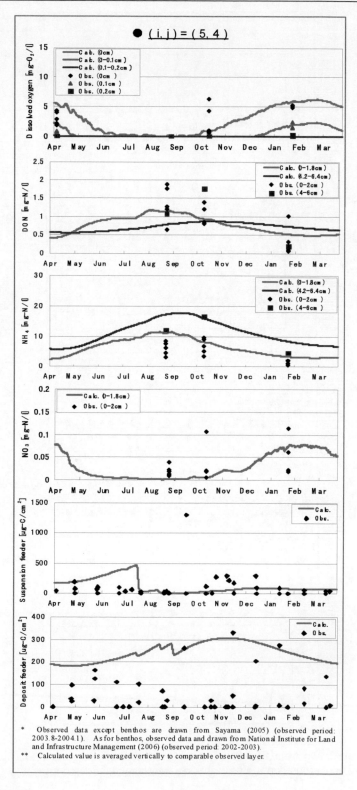

* Observed data except benthos are drawn from Sayama (2005) (observed period: 2003.8-2004.1). As for benthos, observed data and drawn from National Institute for Land and Infrastructure Management (2006) (observed period: 2002-2003).
** Calculated value is averaged vertically to comparable observed layer.

Figure 12. (Continued)

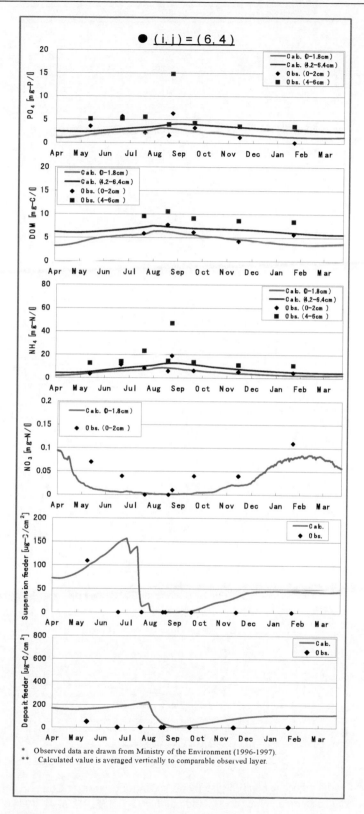

Figure 12. (Continued)

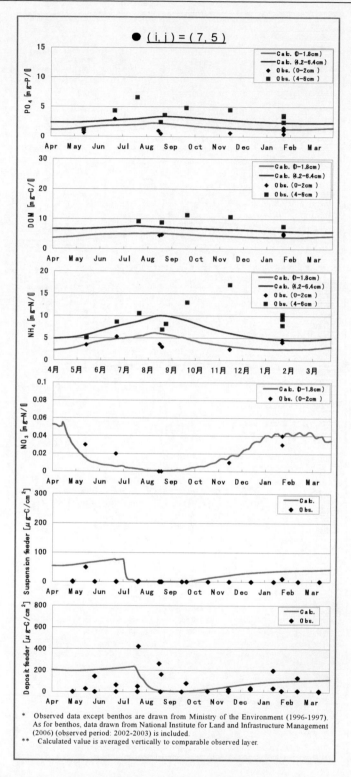

Figure 12. Seasonal variation of the benthic system in the central bay area - Comparison between model output (lines) and observed value (dots). Grid number (i, j) = (5, 4), (6, 4) , (7, 5).

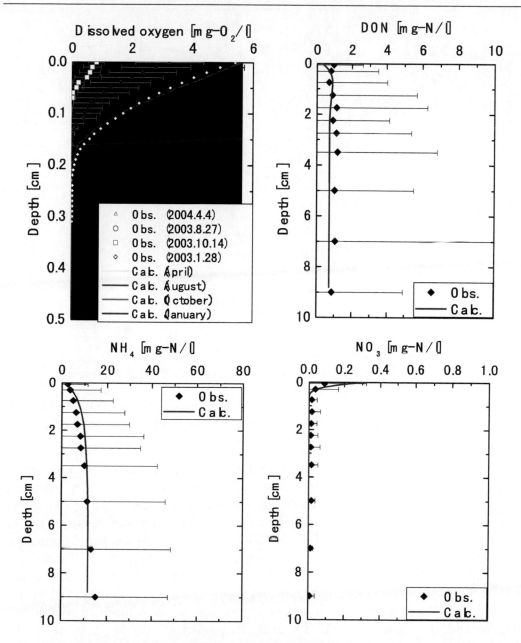

*   Observed data are drawn from Sayama (2005) (observed period: 2003.8 - 2004.1).
**   Except for dissolved oxygen, annual averaged data are compared.   As for dissolved oxygen, monthly averaged data are compared.
***   Error bars indicate 99% confidence interval.

Figure 13. Vertical micro-profiles of the benthic system in the central bay area - Comparison between model output (lines) and observed value (dots). Grid number (i, j) = (5, 4).

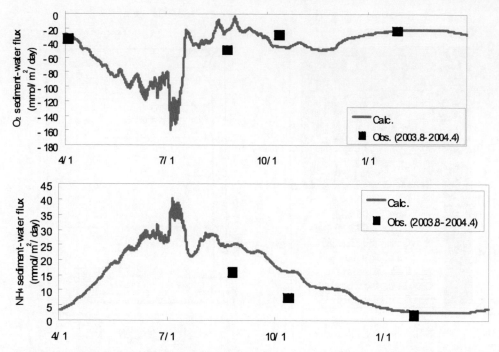

* Positive value indicates efflux from sediment to bottom water, negative value indicates uptake from bottom water to sediment
** Observed data are drawn from Sayama (2005).

Figure 14. Oxygen (left) and NH₄-N (right) fluxes at the sediment-water interface in the central bay area - Comparison between model output (lines) and observed value (squares). Grid number (i, j) = (5, 4).

As for the benthic system, validation of the model based on a comparison of the model output and observed data is more difficult compared to the pelagic system due to much less benthic observation than the pelagic observation, and large differences in the benthic aspect despite only small differences in sampling/observation points. Seasonal dynamics shown in Figures 12 and 13, for example, reveal the difficulty to find a trend in seasonal variation due to the limited amount of observed data. It is therefore impossible to determine whether the model reproduces the seasonal dynamics or not in this situation. However, the model variables and fluxes of the model output, are of a similar magnitude as the observed values. Furthermore, for the model's results related to oxygen, our emphasized/focused point in this study, namely (a) the microstructure of vertical DO profiles and (b) oxygen fluxes at the sediment-water interface, reproduces the observed data (Figures 13 and 14). The model outputs shown in Figures 13 and 14 result from the mutual interaction between the benthic and pelagic systems, and are compared with observed data taken *in situ* (Sayama, 2005). Therefore, the calculated situation is closer to the observed situation compared to the non-mutual interaction model, and the experimental situation is closer to the calculated situation compared to experiments in the laboratory. Describing the vertical micro-spatial scale and the mutual interaction between the benthic and the pelagic systems, the model could compare the model outputs to a micro-scale observed data *in situ* with more realistic situations.

The rapid decrease in suspension feeders and deposit feeders during summer shown in Figure 9 is attributed to the mortality caused by oxygen depletion of the bottom water. The causes of the lower mortality rate of deposit feeders compared to suspension feeders in area (i, j) = (5, 4) are that the parameter of mortality due to oxygen depletion was set for the deposit feeders at a higher tolerance to hypoxia, and that the decrease in DO concentration during summer in area (i, j) = (5, 4) was not so much as to lead to the mortality of deposit feeders compared to other areas.

7.3. Benthic and Pelagic Ecosystems in the Tidal Flat Area

As for the tidal flat area, the seasonal dynamics of the model variables for benthic and pelagic systems are shown in Figures 15 and 16, and the vertical profiles of the benthic system are shown in Figure 17. The plotted data in Figures 15 and 16 were observed at six stations located on the Banzu tidal flat with 30-60 min intervals from low to high tide at each month/season. From the observed data in Figures 15 and 16, seasonal trends are not found for any model variable, thus making it difficult to comment on the model's capacity to reproduce any aspect of seasonal variation. However, the results for every model variable were within the range of observed values. Concerning the benthic vertical profiles shown in Figure 17, the observed POM data decrease exponentially from the surface to the bottom layer. The model reproduces this exponential decrease, although its output values are higher than the observed data. As for the observed data of vertical profiles of NH_4-N, NO_3-N and PO_4-P, the shapes of the profiles vary from station to station and season to season. The cause of these variations is unclear; however, the model output values are within the ranges of the observed values. The model's output for vertical micro-profiles of dissolved oxygen is in good agreement with the observed data during the months of June, August and February.

● (i, j) = (8, 6)

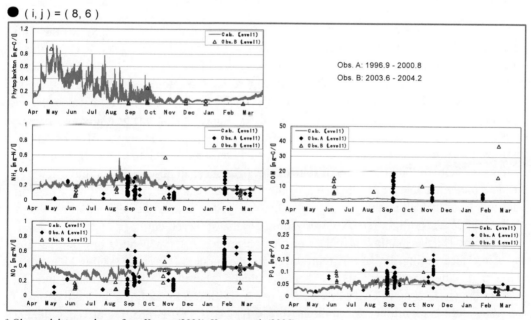

* Observed data are drawn from Kuwae (2001), Kuwae et al. (2005).

Figure 15. Seasonal variation of the pelagic system in the tidal flat area - Comparison between model output (lines) and observed value (symbols). Grid number (i, j) = (8, 6).

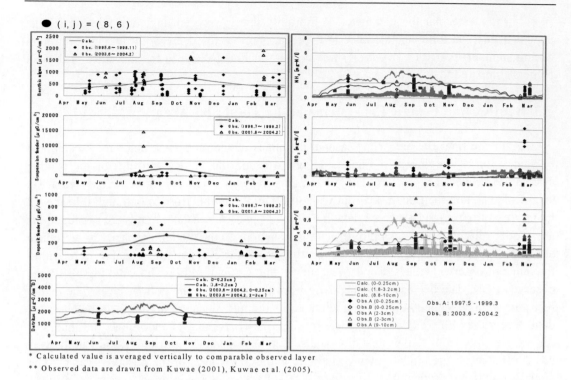

* Calculated value is averaged vertically to comparable observed layer
** Observed data are drawn from Kuwae (2001), Kuwae et al. (2005).

Figure 16. Seasonal variation of the benthic system in the tidal flat area - Comparison between model output (lines) and observed value (symbols). Grid number (i, j) = (8, 6).

In order to demonstrate the short time scale dynamics of the tidal flat area, (a) the daily changes of DO vertical micro-profiles in the sediment, (b) the daily changes of oxygen fluxes at sediment-water interface, and (c) the daily changes of DO, DIN, and Chl-a in the water column are shown in Figures 18, 19 and 20. In Figure 18, the calculated value of the vertical profiles of dissolved substances in pore water, i.e. ODU, NH_4-N and PO_4-P, are also shown. In addition, in Figure 20, the time series of tidal level and light intensity, the major driving forces attributed to daily dynamics (Sohma et al 2000), are provided. Figure18, 19 and 20 demonstrate that the model reproduces the observed data well in relation to daily time scale dynamics of the tidal flat ecosystem in spite of the differences in observed year and calculated year (The model calculates the situation for the averaged year from 1998 to 2002. Thus, the model output used for the comparison with the observed data was the selected value where both the tidal level and light intensity were similar to the observed day).

In Figure 18, the maximum of DO profiles at around 0.1 cm depth from the benthic-pelagic interface during daytime is reproduced by the model. The cause of this DO profile maximum is the photosynthesis of benthic algae. Therefore, the calculated nutrient concentration (NH_4-N and PO_4-P) at the same depth for the DO maximum during daytime is lower than during nighttime. In addition, the NH_4-N and PO_4-P profiles at the sediment surface (0 cm depth) during submersion are different from the emersion period, and their concentrations during emersion are higher than during submersion. This phenomenon results from the model's description of ceasing the nutrient fluxes from the benthic to pelagic systems during emersion.

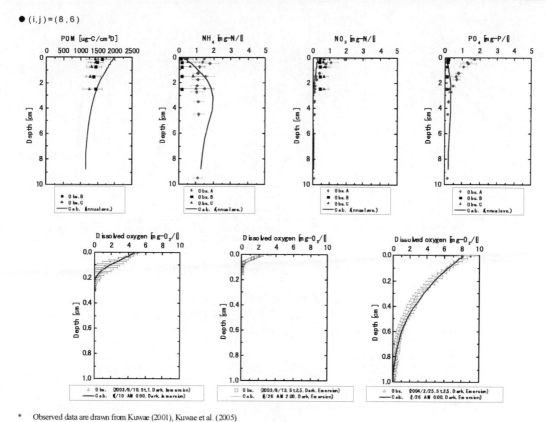

Figure 17. Vertical micro profiles of the benthic system in the tidal flat area - Comparison between model output (lines) and observed value (symbols). Grid number (i, j) = (8, 6).

Concerning oxygen flux at the sediment-water interface, the analyzed value of *in situ* data using the eddy correlation method (Berg et al, 2003; Kuwae et al, 2006) is compared to the model output in Figure 19. The model result indicates values of the same order as observed values. The temporal trend of higher oxygen flux during daytime than nighttime is also reproduced by the model. In our parameter setting, the major physical processes controlling the oxygen flux at the sediment-water interface are irrigation and molecular diffusion. The value of the irrigation coefficient is uncertain in general; however, the fact that oxygen flux of a similar magnitude was reproduced by the model indicates that the requirement for the appropriate setting of the irrigation coefficient is at least satisfied. The higher oxygen flux during daytime is attributed to the higher concentration of DO due to photosynthesis of benthic algae.

Akio Sohma

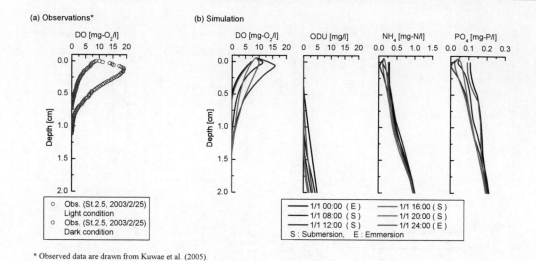

Figure 18. Daily variation of vertical micro-profiles of benthic system in the tidal flat area – (a) Observed dissolved oxygen micro-profile (left graph, Kuwae et al. (2005)) and (b) calculated micro-profiles (all four graphs on right). Grid number (i, j) = (8, 6).

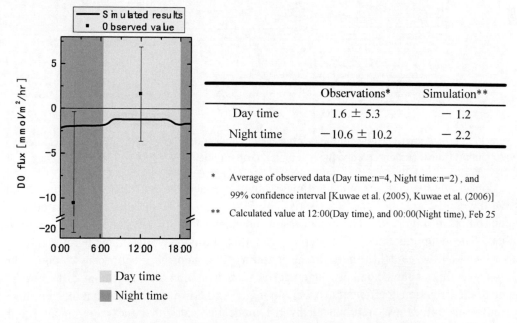

Figure 19. Oxygen fluxes at the sediment-water interface in the tidal flat area - Comparison between model output (lines) and observed value (symbols). Grid number (i, j) = (8, 6).

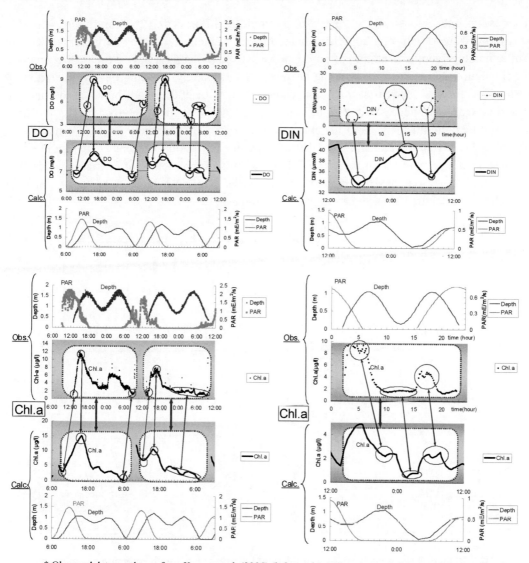

* Observed data are drawn from Kuwae et al. (2005) (left panels), Nakamura et al. (2004) (right panels).

Figure 20. Daily variation of the pelagic system in the tidal flat area - Comparison between model output and observed value - DO (upper left), Chl-a (lower left), DIN (upper right), and Chl-a (lower right). The data in graphs on the left side (DO, Chl-a) were observed from 11 to 13 August, 2003. The data in graphs (DIN, Chl-a) on the right side were observed from 8 to 9 August, 2001. Calculated values were selected from the period when both the light and tidal conditions were similar to the observed period. PAR is Photosynthetically Active Radiation.

Figure 20 demonstrates that if the actual temporal trend of both tidal level and light intensity are the same as in the calculation, the time series of DO, Chl-a and DIN calculated in the model are similar to those of the observed data (Nakamura et al, 2004, Kuwae et al, 2005). For example, the first maximum value of DO is explained by the balance between the increase factor, namely "the transportation of surface water including high DO due to photosynthesis of phytoplankton from the central bay to the tidal flat (the transportation is due to tidal current)" and the decrease factor of DO, namely "weakened photosynthesis due to a decrease in light intensity". As for Chl-a, the trend is different for 2003 and 2001. The first maximum value of Chl-a in 2003 is explained by the balance caused by an increase factor in Chl-a, namely "supplementation of phytoplankton from the central bay to the tidal flat due to the tidal current", and a corresponding decrease factor, namely "feeding of phytoplankton by suspension feeders". The first minimum value of Chl-a in 2001 is attributed to the balance resulting from the decrease factor in Chl-a, namely "weakened photosynthesis due to a decrease in light intensity" and the increase factor in Chl-a, namely "supplementation of phytoplankton from the central bay to the tidal flat due to the tidal current". The reproduction of these phenomena results from (a) modeling the mutual interaction of central bay-tidal flat ecosystems, (b) modeling the mutual interaction of benthic-pelagic ecosystems, and (c) describing the daytime scale dynamics, simultaneously.

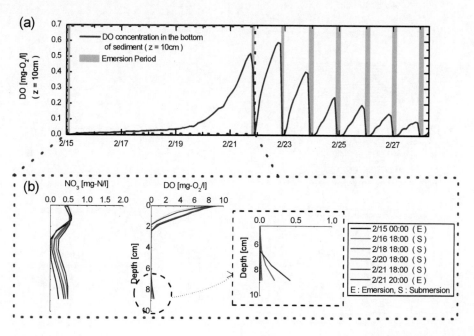

Figure 21. Variation in sediment properties of the tidal flat area related with emersion and submersion. (a) Oxygen concentration in the sediment at 10cm depth from the sediment-water interface (model output). (b) Vertical profiles of oxygen and nitrate (model output).

Another interesting phenomenon at the tidal flat is that the observed data revealed a finite value of NO_3-N at an increased depth of the sediment (under 5cm depth from sediment-water interface) (Kuwae, 2001). NO_3-N is produced by nitrification only, and this process requires oxygen. Therefore, oxygen ought to be supplied at an increased depth by physical transportation (there is no oxygen produced due to biochemical processes at deep locations). However, the observed thickness of the oxic layer was 2 cm even at the deepest condition. Meanwhile, for the oxygen transport processes, irrigation is considered in the model. The irrigation coefficient in the model was set at the optimized value through tuning by the comparison of model variables and fluxes at the sediment-water interface. As a result, the model demonstrated the existence of a period when oxygen takes a finite value under 5cm depth from the sediment-water interface during the submerged period in winter (Figure 21). It is impossible to validate this result because of lack of DO observations during submersion at under 5 cm depth from the sediment-water interface on the Banzu tidal flat. However, DO profiles, which have a zero value at a shallow sediment depth and a finite value at a greater depth, were observed in tidal flat areas in Germany (Gundersen et al., 1995, de Beer et al., 2005).

8. WHAT WAS REVEALED FROM THE COLLABORATION OF "PHILOSOPHY" AND "TECHNOLOGY"?

Several analysis for revealing ecosystem mechanisms and for predicting ecosystem responses have been performed by using validated ECOHYM applied to Tokyo Bay. The results help to inspire our interest and idea of how we should understand the environmental problem and the ecosystem in the eutrophic and hypoxic estuary. Some results are introduced in this section.

8.1. Benthic Oxygen Consumption during Summer Is Low - The Motivation why Hypoxia Potential Is Focused on

In Figure 22, the vertical profiles of oxygen producing/consuming fluxes and model variables at the benthic system in the central bay area are shown. The values in Figure 22 are picked up from the model output at 12:00 one day of each spring (May), summer (August), autumn (November) and winter (January) in area $(i, j) = (5, 4)$. At the benthic system in the central bay, light does not reach the sea floor. Thus, there is no photosynthesis and no oxygen producing processes. As for the oxygen consuming processes, oxic mineralization, nitrification and ODU oxidization exist. As for the excretion of the benthic faunas, they use oxygen around the benthic-pelagic boundary. Therefore, it is possible for us to assume benthic fauna using oxygen at the benthic, at the pelagic, or both at the benthic and at the pelagic systems. In this study, since the model was set for the benthic fauna using oxygen at the pelagic system (bottom layer of pelagic systems), the oxygen consumption due to excretion of benthic fauna was not included in the benthic system. What were revealed from the net biochemical (oxygen) consumption in the benthic system are that (a) ODU oxidization is the largest among all biochemical processes during any season and that (b) oxygen consumption in summer is the smallest among all seasons in terms of vertical integrated value. The result of high contribution of ODU oxidization to oxygen consumption is conjectured to be the same characteristic as most of the hypoxic estuaries in Japan. The small oxygen consumption during summer is attributed to no oxygen for the biology to use in their metabolisms at the seafloor because of hypoxia. In contrast to oxygen consumption, ODU concentration during summer is largest among all seasons. This phenomenon results from the higher anoxic mineralization rate in summer than other seasons. The causes of higher anoxic mineralization of summer are (a) larger mass of detritus due to high production of phytoplankton, (b) higher rate of anoxic mineralization than oxic mineralization due to no oxygen and (c) high temperature during summer.

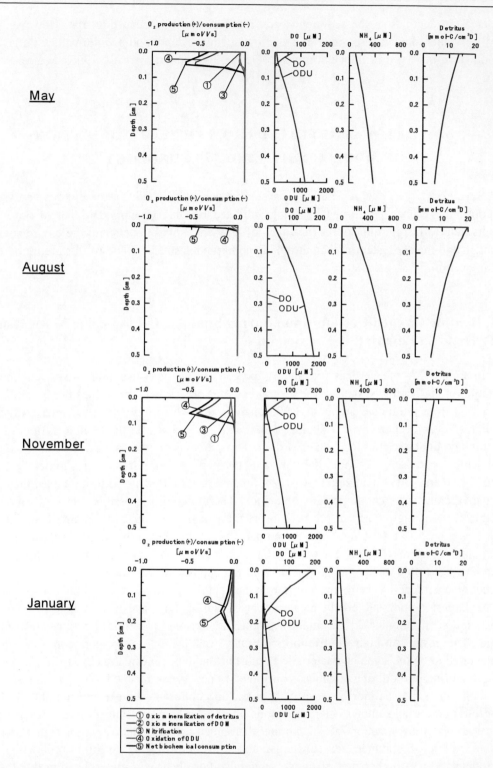

Figure 22. Vertical profiles of oxygen consumed biochemical processes and model variables (dissolved oxygen, ODU, NH₄-N and detritus) at the benthic system in the central bay area. Grid number (i, j) = (5, 4): spring (May), summer (August), autumn (November) and winter (January).

Actually, from not only the model results but also from the observation, under hypoxia during summer, higher level of detritus, ODU and NH_4-N are known to be accumulated in the benthic system. Under this situation, even if the oxygen is supplied to the seafloor due to some kind of vertical disturbance, all of the supplied oxygen is consumed immediately due to the higher rate of oxic mineralization, oxidization of ODU, and nitrification. However, whether oxygen consumption substrates are highly accumulated or not, oxygen consumption at seafloor is few or zero under hypoxic circumstances because there is no oxygen to consume. Therefore, the oxygen consumption rate under a hypoxic situation is unsuitable to estimate the tenacity of hypoxia or the tendency to be hypoxic. The difference in the oxygen consumption rate from the state of high accumulation of oxygen consumption substrates to the state of low accumulation is revealed only when oxygen is supplied to water at the sea-floor. Based on these backgrounds, I have defined the oxygen consumption rate under the situation of the defined standard concentration of dissolved oxygen level at sea-floor as "hypoxia potential" (Sohma et al, 2005a, Sohma, 2005b), and its value is used to the estimation of the tenacity of hypoxia and the tendency to become hypoxic. The concept of hypoxia potential can be used for both experiments (i.e., measurement of oxygen consumption rate by the experiment of the core sample of sediments soaked by the sea-water which oxygen-level is controlled at a defined standard concentration) and numerical model analysis.

On the estimation of hypoxia potential by using ECOHYM, hypoxia potential was calculated as the oxygen consumption rate both in the pore water of benthic system (at the range from sediment-water interface to α mm) and in the bottom water of the pelagic system (at the range from sediment-water interface to β cm) under the defined standard level (γ mgO_2/l) of dissolved oxygen concentration. Here, the counting oxygen consumption processes are oxic mineralization ($R_{oxic-min}$), nitrification ($R_{nitrification}$), ODU oxidization ($R_{ODU-oxidization}$), phytoplankton respiration ($R_{PP-respiration}$), zooplankton respiration ($R_{ZP-respiration}$), and benthic algae respiration ($R_{DIA-respiration}$). In addition, in order to estimate the hypoxia potential at the shallow water areas where light can reach the sea-bottom on the same standard, the oxygen production processes, phytoplankton photosynthesis ($R_{PP-photosynthesis}$) and benthic algae photosynthesis ($R_{DIA-photosynthesis}$) are counted as the decrease effects of the hypoxia potential. Each oxygen consumption/production process is calculated by the formulation described in Tables A5 and A6 in Appendix. On the calculation, the values of the model variables in the formulation are, for the dissolved oxygen set at the value of γ mgO_2/l and for others set at the values of pre-calculated results (i.e. the values of the results of regular simulation without fixing the level of dissolved oxygen at γ mgO_2/l) at each time step. The calculation method and the conceptual figure of the hypoxia potential: HP (α, β, γ) is described in Figure 23. In this study, the values of α, β, γ and   were set as   = 1mm, β = 10cm and γ = 2 mgO_2/l. Here, each values of α, β,   and γ are assumed the mean values of thickness of the benthic oxic layer during hypoxia, the ranges/thickness of the hypoxic water column in the pelagic system, and the lowest dissolved oxygen level for mortality not to occur for benthic fauna due to oxygen depletion.

8.2. A Clean Ocean Is Different from A Bountiful Ocean

Tokyo Bay on the scenario of reproduction of the early tidal flat reclaimed in the past (tidal flat reproduction case) was compared with Tokyo Bay on the scenario of 50% reduction of the nutrient load from rivers (nutrient load reduction case). The geographical condition of tidal flat reproduction case were shown in Figure 24. These two scenarios were calculated as the sensitive analysis based on the model implementation in section 6, "the annual periodical steady state of the existing Tokyo Bay (control case)". Both scenarios were started from the ecological state of the control case on 1st April. Namely, the initial values of model variables on the two scenarios were set at the state on 1st April of the control case. On the tidal flat reproduction case, the reproduced tidal flat ecosystems were set geographically in the simulation and the model variables of the reproduced tidal flats were initially set as the same values as the existing tidal flat (Banzu tidal flat in Figure 9 or 24). Then their states were calculated autonomously in the model subsequently. On the calculation of the nutrient load reduction case, 50% reduction of the nutrient load from rivers was set in the simulation. Both cases achieved a new annual periodical steady state after a 15 years' run.

The comparison between the new annual periodical steady state of the two cases, the tidal flat reproduction case and the nutrient load reduction case, are shown in Figure 25. Both cases decrease in the level of particulate organic matter (Detritus) compared to the control case (existing system). In addition, both cases demonstrate the hypoxic improvement in the same level compared to the control case. In other words, measures of tidal flat reproduction and nutrient load reduction lead to the "clean ocean" in terms of water quality. However, concerning the benthic fauna, the tidal flat reproduction system had a different response from the 50% nutrient load reduction system. Namely, the benthic fauna in the tidal flat reproduction system increased compared to the existing system, while benthic fauna in the nutrient load reduction system decreased compared to the existing system. It means that although 50% decrease of nutrient load reduction leads to the recovery of eutrophication, it does not lead to the rich ecosystem with biodiversity/many living organisms but leads to the poor ecosystem with little living organisms. In contrast, tidal flat reproduction promotes the effective utilization of excess nutrients/organic matter by the assimilation of nutrients/organic matter into benthic fauna, and it leads to the rich ecosystem/bountiful ocean. Although the model output shown here includes several assumptions and hypothesis, the output reveals the essential differences from tidal flat reproduction to nutrient load reduction, namely: the tidal flat reproduction recovering the rich ecosystem is the measure to assimilate nutrients into many biology and to raise the effective utilized nutrient potential, while nutrient load reduction is not the positive measure to raise the effective utilized nutrient potential by increasing the biology, although it may prevent the mortality of biology due to hypoxia[2].

---

[2] The result demonstrated here does not mean that the reduction of nutrient loads from river is not necessary. Excess nutrients should be carried away from the ocean. However, if recovering the bountiful ocean existing various/huge living organisms is succeeded by the reproduction of tidal flats, the amount of the utilized nutrients by living organisms increase. As a result, the amount of nutrients taken as excess decreases and the required amount of reduced nutrients also decreases.

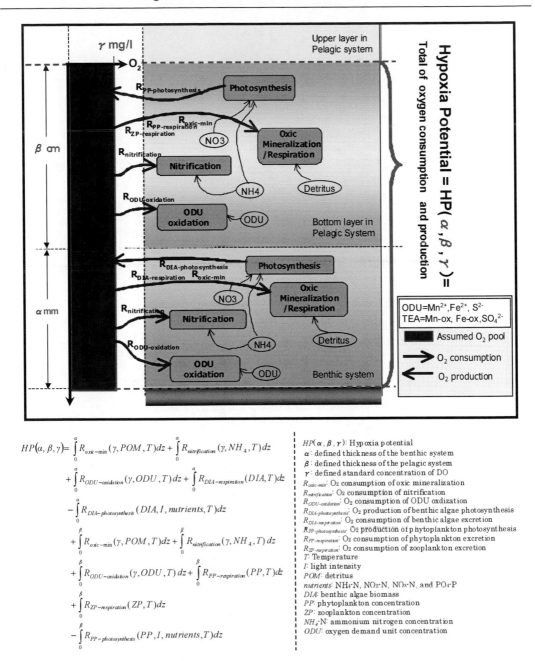

$$HP(\alpha,\beta,\gamma)= \int_0^a R_{oxic-min}(\gamma,POM,T)dz + \int_0^a R_{nitrification}(\gamma,NH_4,T)dz$$

$$+ \int_0^a R_{ODU-oxidation}(\gamma,ODU,T)dz + \int_0^a R_{DIA-respiration}(DIA,T)dz$$

$$- \int_0^a R_{DIA-photosynthesis}(DIA,I,nutrients,T)dz$$

$$+ \int_0^\beta R_{oxic-min}(\gamma,POM,T)dz + \int_0^\beta R_{nitrification}(\gamma,NH_4,T)dz$$

$$+ \int_0^\beta R_{ODU-oxidation}(\gamma,ODU,T)dz + \int_0^\beta R_{PP-respiration}(PP,T)dz$$

$$+ \int_0^\beta R_{ZP-respiration}(ZP,T)dz$$

$$- \int_0^\beta R_{PP-photosynthesis}(PP,I,nutrients,T)dz$$

$HP(\alpha,\beta,\gamma)$: Hypoxia potential
α: defined thickness of the benthic system
β: defined thickness of the pelagic system
γ: defined standard concentration of DO
$R_{oxic-min}$: O_2 consumption of oxic mineralization
$R_{nitrification}$: O_2 consumption of nitrification
$R_{ODU-oxidation}$: O_2 consumption of ODU oxidization
$R_{DIA-photosynthesis}$: O_2 production of benthic algae photosynthesis
$R_{DIA-respiration}$: O_2 consumption of benthic algae excretion
$R_{PP-photosynthesis}$: O_2 production of phytoplankton photosynthesis
$R_{PP-respiration}$: O_2 consumption of phytoplankton excretion
$R_{ZP-respiration}$: O_2 consumption of zooplankton excretion
T: Temperature
I: light intensity
POM: detritus
$nutrients$: NH_4-N, NO_2-N, NO_3-N, and PO_4-P
DIA: benthic algae biomass
PP: phytoplankton concentration
ZP: zooplankton concentration
NH_4-N: ammonium nitrogen concentration
ODU: oxygen demand unit concentration

Figure 23. The concept and formulation of hypoxia potential (HP). Hypoxia potential is the oxygen consumption rate under the defined standard concentration of dissolved oxygen. Right side describes the formulation of hypoxia potential in ECOHYM. 1st to 5th terms of the right hand side are oxygen consumption/production flux in the benthic system. 6th to 11th terms are oxygen consumption/production flux in the pelagic system. Each term is calculated by the biochemical formulation in Tables A5 and A6.

116   Akio Sohma

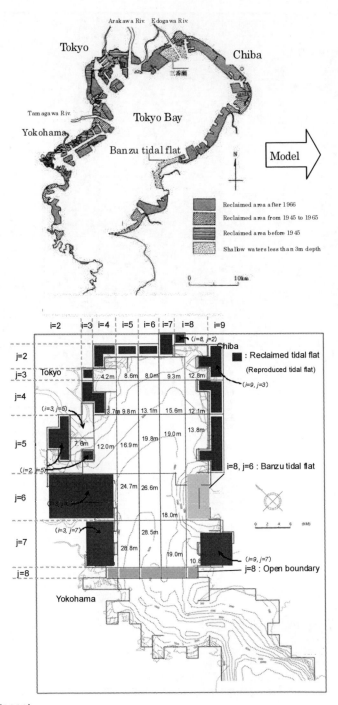

Reference: Koike (2000)

Figure 24. Left figure; the early Tokyo Bay before reclamation (tidal flat reproduction system) and the existing Tokyo bay after reclamation (existing system/tidal flat disappearance system). Right figure; the boxes/axis set on the simulation using ECOHYM.

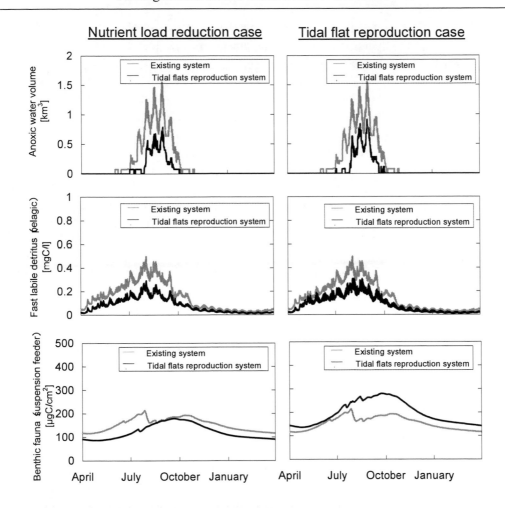

Figure 25. The seasonal variation of Tokyo Bay on the scenario of 50% nutrient load reduction from rivers (nutrient load reudction system) and the scenario of reproduction of early tidal flats reclaimed in the past (tidal flats reproduction system), compared to the existing Tokyo Bay (existing system). (a) anoxic water volume (lower DO level than 2 mgO$_2$ l^{-1}) in Tokyo Bay, (b) concentration of detritus (fast labile detritus) in the pelagic system, and (c) biomass of benthic fauna (suspension feeders) in the benthic system. The values in figures are mean values of the existing Tokyo Bay areas.

8.3. Robust/healthy Balance of the Ecosystem - A Motivation why Lower Trophic Production and Quasi-higher Trophic Production are the Focus.

The vision of a smooth transition from a lower trophic level to a higher trophic level is significant. Coastal areas where eutrophication is proceeding and hypoxia is a serious problem has slight/poor nutrient paths/channels from the lower trophic level to the higher trophic level. As a result, ecosystems being out of balance such as "red tide (rapidly growth of phytoplankton)" have the possibility to be formed easily[3]. In contrast, by accelerating the

---

[3] The discussion here is based on the concept that robust/healthy environment of estuaries is the state of well-balanced material cycling. Although it is difficult to define the well-balanced or unbalanced material cycling directly, the overview explanation from the vision of "stocks" and "flows" is as follows. For example, carbon

smooth transition from a lower to higher trophic level, a robust ecosystem balance and bountiful ocean with biodiversity/many living organisms may be led. The state of ecosystem should be estimated from such a vision/perspective.

Based on the above background, by demonstrating the fluxes of lower and higher trophic productions quantitatively, (1) the smooth transition from a lower trophic level to a higher trophic level, (2) effective utilization of excess nutrients by higher level living organisms and (3) the recovery of a robust ecosystem balance derived from (1) and (2) were estimated. Lower trophic production were defined as the total flux of phytoplankton production due to photosynthesis and zooplankton production due to grazing in Figure 4. Concerning the definition of higher trophic production, fish representing high trophic level, is not modeled in the ECOHYM. Therefore, the transition from phytoplankton, zooplankton and detritus to benthic fauna, relatively higher trophic level than planktons, was regarded as "quasi-higher trophic production". Specifically, "quasi-higher trophic production" was defined as the total flux of feeding of benthic faunas (suspension feeder and deposit feeder) in Figure 5. All required fluxes on the calculation of lower trophic production and quasi-higher trophic production are calculated in the simulation of ECOHYM.

8.4. It's Possible that the Environmental Recovery Spiral Could Occur

The differences of ecosystem response to the increase and decrease of load from rivers, was estimated in terms of the following indexes, (1) hypoxia potential, (2) lower trophic production and (3) quasi-higher trophic production, between the early Tokyo Bay before reclamation of the tidal flats (tidal flats reproduction system) and the present Tokyo Bay after reclamation of most of tidal flats (existing system/tidal flat disappearance system). Figure 26 shows the dependences of hypoxia potential, lower trophic production and quasi-higher trophic production on increase/decrease of load from rivers. The values of the three indexes shown here are annually averaged and integrated values of all areas of the existing Tokyo Bay (the indexes evaluated for the tidal flat reproduction system does not include the value of the reproduced tidal flat areas[4]) and are analyzed on the periodical seasonal steady state which were achieved from the control case after reducing/increasing load from livers. As shown in Figure 26, level of hypoxia potential and lower trophic production indicate lower and the level of quasi-higher trophic potential indicate higher on the tidal flat reproduction system

---

exists as various forms (i.e. "stocks") such as plankton, cell of fish, detritus, and CO_2 in the estuary. The form varies through biological, chemical, and physical processes (i.e. "flows") and these processes make the paths of carbon cycling (material cycling). Red tide is the state of most of the carbon/nutrients forming in the phytoplankton and it is the unbalance state of "stocks" of carbon/nutrients/materials. The biological, chemical and physical processes are the driven forces of carbon/nutrients/material cycling. To recover the ecosystem inhabiting various living organisms means to make the many and various "stocks" and "flows" of material cycling.

[4] The area of Tokyo Bay's tidal flat reproduction system has expanded compared to the tidal flat disappearance system due to the additional areas of reproduced tidal flats (Figure 24). However, the results shown in Figures 25, 26, 27, 28 and 29 are estimated excluding the reproduced tidal flats area for the estimation of the tidal flat reproduction and disappearance systems to be compared in the same areas (i.e. the integrated zone on calculating three indexes, i.e., the hypoxia potential, lower trophic production and quasi-higher trophic production does not include the reproduced tidal flats areas). In the case of the estimation including the reproduced tidal flat areas, the differences between tidal flat reproduction system and tidal flat disappearance system in the three indexes are more obvious/higher than the results shown in Figures 25, 26, 27, 28 and 29.

compared to the tidal flat disappearance system for any state of load input. In addition, the levels of hypoxia potential and lower trophic production on tidal flat reproduction and disappearance systems both increase in accordance with load increase. However, the increase rates of hypoxia potential and lower trophic production on the tidal flat disappearance system are higher than on the tidal flat reproduction system. Meanwhile, the level of quasi-higher trophic production increases in accordance with load increase in the range from 0.5 to 1.5 times the current load. In the range of over 1.5 times the current load, the level of quasi-higher trophic production of the tidal flat disappearance system goes into decline with increase of load. The decline results from the mortality of benthic faunas due to oxygen depletion. These ecosystem responses demonstrated by ECOHYM suggests the possibility of significant effects derived from the tidal flat reproduction. The effects are (1) preventing hypoxia (hypoxia potential), (2) lifting up the nutrients from lower trophic to higher trophic level and (3) recovering a bountiful ocean with high biological production, i.e. driving the "environmental improvement spiral".

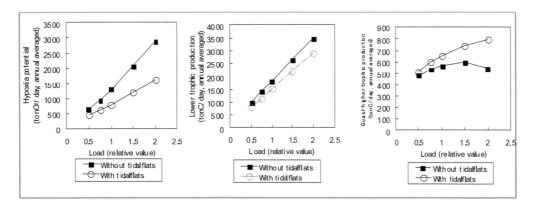

Figure 26. Ecological response of the early Tokyo Bay (tidal flats reproduction system; with tidal flats) and of the existing Tokyo Bay (without tidal flats) to the increase and decrease of nutrient load from rivers. The values in figures are integrated values of the existing Tokyo Bay areas.

8.5. A Bountiful Ocean Has a High Tolerability to the Imbalanced ecosyStem State

In order to investigate the differences from the tidal flat reproduction system to the tidal flat disappearance system on the ecosystem tolerability to environmental stress impact collapsing the ecosystem balance, a red tide pulse (temporal high level of phytoplankton) was forcefully set to both the tidal flat reproduction system and the tidal flat disappearance system. The responses of each system to the red tide were evaluated in terms of hypoxia potential, lower trophic production, and quasi-higher trophic production.

Firstly, the red tide pulse impact of 3 mgC/l in the level of phytoplankton, starting from 0:00 on 15th August and continuing 24 hrs, was forced to be set on the tidal flat reproduction system and the tidal flat disappearance system. With this, the time series of ecosystem response was investigated. Here, the red tide pulse impact was set at the surface layer of the entire Tokyo Bay. The results are shown in Figure 27. Longitudinal axis in Figure 27

120 Akio Sohma

represents the differences between (1) the case with the red tide impact and (2) the case without the red tide impact, i.e. (1) minus (2) in the level of hypoxia potential, lower trophic production, and quasi-higher trophic production. For hypoxia potential, the maximum peak of the tidal flat reproduction system is not higher than the tidal flat disappearance system, and increase of hypoxia potential of the tidal flat reproduction system is alleviated promptly compared to the tidal flat disappearance system. For lower trophic production, its levels of both tidal flat reproduction and disappearance systems reach the maximum value immediately after the red tide impact. Then, 5-6 days later, the levels are alleviated to the same value as the case without the red tide impact (zero of longitudinal axis in Figure 27). For the maximum value of lower trophic production on 16th August, the tidal flat reproduction system results in a higher value than the tidal flat disappearance system. For quasi-higher trophic production, at the moment the red tide impact arises, the values of both tidal flat reproduction and disappearance systems increase. However, 2-3 days later, for the case of tidal flat disappearance system, the value of quasi-higher trophic production with the red tide impact results in a lower value compared to the case without the red tide impact (negative value of longitudinal axis in Figure 27). In contrast, for the case of tidal flat reproduction system, the value of quasi-higher trophic production with red tide impact returns immediately to the same value as without the red tide impact (zero value of longitudinal axis in Figure 27). The model results shown here represent that the tidal flat reproduction system absorbs red tide immediately, prevents the process/paths from red tide to hypoxic generation, and alleviates the decrease of the quasi-higher trophic production due to the red tide impact.

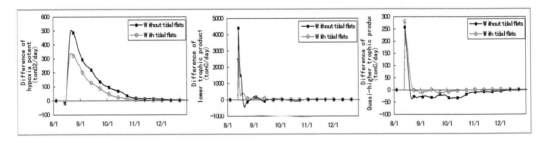

Figure 27. Ecological response of the early Tokyo Bay (tidal flats reproduction system; with tidal flats) and of the existing Tokyo Bay (without tidal flats) to the red tide pulse impact (temporal dynamics after the red tide impact). The values in figures are integrated values of the existing Tokyo Bay areas.

Secondly, the dependence of ecosystem tolerability on the concentration/level of the red tide impact was evaluated for the tidal flat reproduction system and the tidal flat disappearance system. Figure 28 shows hypoxia potential, lower trophic production and quasi-higher trophic production as the results from the ecosystem response to the red tide impact of various concentration in phytoplankton. The red tide was set to start from 0:00 on 15th August and to continue for 24hrs at the surface layer of the entire Tokyo Bay. Longitudinal axis in Figure 28 represents, as well as Figure 27, the differences between (1) the case with the red tide impact and (2) the case without the red tide impact, i.e. (1) minus (2) in the level of hypoxia potential, lower trophic production, and quasi-higher trophic production. The result shown in Figure 28 represents that as the level in the concentration of red tide (phytoplankton) increase, hypoxia potential and lower trophic level production increase on both systems of the tidal flat reproduction and disappearance. For quasi-higher level production, the value on the tidal flat disappearance system decreases as the concentration of red tide increase. However, on the tidal flat reproduction system, quasi-higher trophic production has a higher activity (positive value in Figure 28) due to red tide at the range of less than 4mgC/l level in the concentration of red tide. The activity turns weaker than the case without red tide (negative value in Figure 28) when the level of red tide is higher than 4mgC/l.

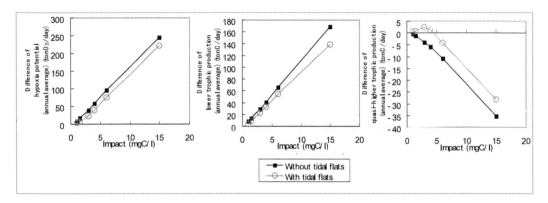

Figure 28. Ecological response of the early Tokyo Bay (tidal flats reproduction system; with tidal flats) and of the existing Tokyo Bay (without tidal flats) to the red tide pulse impact (dependence on the concentration of phytoplankton). The values in figures are temporal-spatial integrated values. Integrated period is from 15th August to 1st April the following year and the integrated area is the existing Tokyo Bay areas.

Finally, the dependence of ecosystem tolerability on the running days of the red tide was evaluated for the tidal flat reproduction system and the tidal flat disappearance system. Figure 29 shows the results of ecosystem response to the red tide in terms of hypoxia potential, lower trophic production and quasi-higher trophic production. In this study, the forced red tide started from 0:00 on 15th August and periods were set from half day to one week and the occurring area of red tide was the entire surface layer of Tokyo Bay. In all cases, the level of red tide (phytoplankton) was fixed at 3 mgC/l. Longitudinal axis in Figure 29 represents, as well as Figures 27 and 28, the differences from (1) the case with the red tide impact to (2) the case without the red tide impact, i.e. (1) minus (2). From the results of this analysis, hypoxia potential and lower trophic production increase as the running days of the red tide are longer both in tidal flat disappearance and reproduction systems. However, for the tidal flat reproduction systems, the quasi-higher trophic production is more activating rather than decreasing due to the red tide up to running 4 days compared to the non-red tide situation. If the red tide continues for more than 4 days, the quasi-higher trophic production decreases. In addition, in the case red tide runs for over 4 days of more than 4 days, the increase rate of hypoxia potential and lower trophic production have differences from the tidal flat reproduction system to the tidal flat disappearance system, and their increase rates in the tidal flat disappearance system are higher. As for the decrease rate of the quasi-higher trophic production, the tidal flat disappearance system indicates a higher decrease rate compared to the tidal flat reproduction system in the ranges of red tide running for over 4 days.

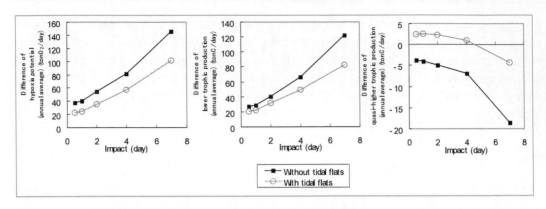

Figure 29. Ecological response of the early Tokyo Bay (tidal flats reproduction system; with tidal flats) and of the existing Tokyo Bay (without tidal flats) to the red tide pulse impact (dependence on the running days of red tide). The values in figures are temporal-spatial integrated values. Integrated period is from 15th August to 1st April the following year and the integrated area is the existing Tokyo Bay areas.

The results described above reveal the higher differences in the tolerability to red tide from the tidal flat reproduction system to the tidal flat disappearance system. In summary, for the tidal flat reproduction system, the increase of hypoxia potential with red tide generation rapidly reduces after red tide annihilation and the quasi-higher trophic production does not decrease except in the case where a strong red tide impact occurs. Red tide running for short days has the potential to lead to a higher activity of quasi-higher trophic production rather than lower it. In contrast, for the tidal flat disappearance system, hypoxia potential does not reduce rapidly after red tide annihilation and it decreases even if a weak red tide impact occurs. If a red tide running for more than several days occurs, the ecosystem gets caught in an environmental deterioration spiral (negative spiral), in which hypoxia generation and lower trophic production are accelerated and quasi-higher trophic production is decreased.

9. CONCLUSION AND REMARKS - ECOSYSTEM MODEL AS A COMMUNICATION PLATFORM

Development of ECOHYM contributed to establish the bases of the conversion of philosophy such as "bountiful ocean", "Robust/healthy balance of the ecosystem" and "environmental deterioration/improvement spirals" into a tangible form, and to establish the technology of prediction, explanation, and evaluation of the philosophy mentioned above quantitatively. However, the analyzed scenarios/cases demonstrated in this manuscript are bold scenarios (i.e., reproduction of all reclaimed tidal flats, and 50% reduction of nutrient loads from rivers) and they may be not realistic to actualize. However, bold scenarios/exciting hypothesis demonstrated here are meaningful to understand/discuss the direction of the environmental restoration of the estuary. In addition, even if feasible measures of tidal flat reproduction/creation or nutrient reduction from rivers is small in scale, the ecosystem response at the surrounding area where measures are applied will be the same trend as the ecosystem response presented in this manuscript with a high possibility. The detailed analysis of the ecosystem response to small scale measures is now proceeding by using the advanced

ECOHYM application having higher vertical spatial resolution. If the small areas of the ecosystem response will be revealed to have the trend as mentioned above, from the results of this detailed analysis, one of the ideas of a more realistic method to manage the eutrophic estuaries may be (1) to select the environmental model (narrow) area in accordance with the expected ecosystem recovery or with the intended use from the perspectives of not only the environmental conservation, but also the development of industry and national land/ocean using plan, (2) to make/decide a goal/direction for each environmental model area and (3) to make an actual strategy of applying measures to achieve them. An ecosystem model is a useful tool to perform items (1), (2) and (3) described above.

Based on the ECOHYM, the availability and extendability of ecosystem models brought out hereafter, will be expected as follows.

(1) contribution on natural science
   (i) understanding the autonomus response of ecosystems (chain response of ecosystems)
   (ii) revealing the unknown biological and chemical processes.
   (iii) predicting the fact overlooked by observation/monitoring
   (iv) making an efficient and effective strategy on observation/monitoring
(2) contribution on policy science
   (i) utilization as the environmental management tool
      - estimating the cost-performance on environmental improvement measures (including the effect of autonomous ecological responses)
      - predicting the effect of environmental development and improvement measures (including the effect of autonomous ecological responses)
   (ii) utilization as the communication tool on consensus-building
      - revealing the discussed points on consensus-building through a quantitative explanation about the contribution rate of each process on the focusing problem.

The availability and extendability of ecosystem models mentioned above set my expectation that ecosystem models become a powerful tool as the communication platform on the collaboration among investigation of physics, biology and chemistry or on the field of policymaking.

ACKNOWLEDGEMENT

Part of this research was supported by the Program for Promoting Fundamental Transport Technology Research from the Japan Railway Construction, Transport and Technology Agency (JRTT).

APPENDIX. FORMULATION OF MAJOR BIOCHEMICAL PROCESSES

The formulations of biochemical processes in ECOHYM are shown in Tables A1-A6.

Table A1. Notation of model variables (pelagic system)

Notation and No. of Model Variables	Unit	Description
PP : [01]	mgC/l	Phytoplankton
ZP : [02]	mgC/l	Zooplankton
WDE_1, WDE_2, WDE_3 : [03,1], [03,2], [03,3]	mgC/l	Detritus (fast-labile, slow-labile, and refractory/very slow-labile detritus)
WDM_1, WDM_2 : [04,1], [04,2]	mgC/l	Dissolved organic matter (labile and refractory DOM)
WNX : [05]	mgN/l	Ammonium
WNY : [06]	mgN/l	Nitrate
WDP : [07]	mgP/l	Phosphate
WOU : [08]	mg/l	Oxygen demand unit (ODU)
WDO : [09]	mg/l	Dissolved oxygen

Table A2. Notation of model variables (benthic system)

Notation of Model Variables	Unit	Description
SFB : [51]	$\mu gC/cm^2$	Suspension feeders
DFB : [52]	$\mu gC/cm^2$	Deposit feeders
DET_1, DET_2, DET_3 : [53,1], [53,2], [53,3]	$\mu gC/cm^3$ solid	Detritus (fast-labile, slow-labile, and refractory/very slow-labile detritus)
DOM_1, DOM_2 : [54,1], [54,2]	$\mu gC/ml$	Dissolved organic matter (labile and refractory DOM)
HNX : [55]	$\mu gN/ml$	Ammonium
HNY : [56]	$\mu gN/ml$	Nitrate
DIP : [57]	$\mu gP/ml$	Phosphate
ODU : [58]	$\mu g/ml$	Oxygen demand unit (ODU)
DOO : [59]	$\mu g/ml$	Dissolved oxygen
BAL : [60]	$\mu gC/cm^2$	Benthic algae

Table A3. General variables and prescribed functions in the model

General Variables and Prescribed Functions	Unit	Description
z	cm	Water depth or sediment depth
Δz	cm	Thickness of layer
dt	h	Calculation time step
ϕ	-	Porosity
$\bar{\rho}_s$	g/cm^3	Density of sediment
TmpW, TmpB	°C	Temperature of sea water and sediment
I_0 , I_B	$\mu E/m^2/s$	Light intensity on sea surface and sediment surface

Table A4. Ratio and distribution functions
(prescribed function or calculated in the model)

Functions	Description
$R_{FOD51,1}$, $R_{FOD51,2}$, $R_{FOD51,3}$	Composition ratio (ratio of fast-labile, slow-labile and refractory/very slow-labile part) of prey of suspension feeders
$R_{ncFOD51}$, $R_{pcFOD51}$	N/C, P/C ratio of prey of suspension feeders
R_{Zfec51}, R_{Zmor51}	Vertical distribution of feces and mortality of suspension feeders
$R_{FOD52,1}$, $R_{FOD52,2}$, $R_{FOD52,3}$, $R_{FOD52,541}$, $R_{FOD52,542}$	Composition ratio (ratio of fast-labile, slow-labile and refractory/very slow-labile part) of prey of deposit feeders
$R_{ncFOD52}$, $R_{pcFOD52}$	N/C, P/C ratio of prey of deposit feeders
R_{Zfee52}, R_{Zfec52}, R_{Zexc52}, R_{Zmor52}	Vertical distribution of feeding, feces, excretion and mortality of deposit feeders
R_{Zpho60}, R_{Zres60}, R_{Zmor60}	Vertical distribution of photosynthesis, base respiration and mortality of benthic algae
$R_{FOD52,60}$	Ratio of benthic algae to prey of deposit feedeers

Table A5. Formulation of essential biochemical processes (pelagic system)

Biochemical Processes	Formulation [min(a, b) = a (a > b), or b (a < b) ; g(X, a_{half}) = X / (X + a_{half})]	Unit	Parameters [see Table 2, Table 3]
[Phytoplankton : 01]			
Biochemical net production / consumption	$\dot{C}_{pp} = Dpp_{pho} - Dpp_{ext} - Dpp_{res} - Dpp_{mor} - Dzp_{gra} - Dpp_{sff}$	mgC/l/h	
Photosynthesis	$Dpp_{pho} = v_{pho01} \cdot u_{pho01a} \cdot u_{pho01b} \cdot PP$	mgC/l/h	
Maximum growth rate	$v_{pho01} = \alpha_{pho01} \cdot exp(\beta_{pho01} \cdot TmpW)$	1/h	α_{pho01}, β_{pho01}
Nutrient limitation	$u_{pho01a} = min[\ g(WNX+WNY, Hf_{n,pho01}), g(WDP, Hf_{p,pho01})\ ]$	-	$Hf_{n,pho01}$, $Hf_{p,pho01}$
Light availability	$u_{pho01b} = (I_0 e^{-kz} - I_{min01}) / ((I_0 e^{-kz} - I_{min01}) + (I_{hf01} - I_{min01}))$	-	I_{hf01}, I_{min01}
Light attenuation	k: Calculated in model depend on PP, ZP, WDE_i	1/cm	
Extra-release	$Dpp_{ext} = Dpp_{pho} \cdot 0.135 \cdot exp(-0.00201 \cdot R_{chl} \cdot PP \cdot 10^3)$	mgC/l/h	R_{chl}
Respiration	$Dpp_{res} = \alpha_{res01} \cdot exp(\beta_{res01} \cdot TmpW) \cdot PP$	mgC/l/h	α_{res01}, β_{res01}
Mortality	$Dpp_{mor} = \alpha_{mor01} \cdot exp(\beta_{mor01} \cdot TmpW) \cdot PP$	mgC/l/h	α_{mor01}, β_{mor01}
Suspension feeder feeding (bottom layer only)	$Dpp_{sff} = \dfrac{PP}{PP + ZP + \sum_{i=1}^{3} WDE_i} \cdot \dfrac{1}{A_{Depq}} \cdot Dsfb_{fee}$	mgC/l/h	A_{Depq}
[Zooplankton : 02]			
Biochemical net production / consumption	$\dot{C}_{zp} = Dzp_{gra} - Dzp_{fee} - Dzp_{exc} - Dzp_{mor} - Dzp_{sff}$	mgC/l/h	
Grazing	$Dzp_{gra} = \alpha_{gra02} \cdot exp(\beta_{gra02} \cdot TmpW) \cdot (1 - exp(A_{ivl02} \cdot (A_{kai02} - PP))) \cdot ZP$	mgC/l/h	α_{gra02}, β_{gra02}, A_{ivl02}, A_{kai02}
Feces	$Dzp_{fee} = (1 - R_{ege02}) \cdot Dzp_{gra}$	mgC/l/h	R_{ege02}
Excretion	$Dzp_{exc} = (R_{ege02} - R_{grt02}) \cdot Dzp_{gra}$	mgC/l/h	R_{ege02}, R_{grt02}
Mortality	$Dzp_{mor} = \alpha_{mor02} \cdot exp(\beta_{mor02} \cdot TmpW) \cdot ZP$	mgC/l/h	α_{mor2}, β_{mor02}
Suspension feeder feeding (bottom layer only)	$Dzp_{sff} = \dfrac{ZP}{PP + ZP + \sum_{i=1}^{3} WDE_i} \cdot \dfrac{1}{A_{Depq}} \cdot Dsfb_{fee}$	mgC/l/h	A_{Depq}
[Detritus : 03,i] *[i = 1 (fast-labile), 2 (slow-labile), 3 (refractory / very slow-labile)]*			
Biochemical net production / consumption	$\dot{C}_{wde,i} = R_{PP,i} \cdot Dpp_{mor} + R_{PP,i} \cdot Dzp_{fee} + R_{ZP,i} \cdot Dzp_{mor}$ $- Dwde_{omi,i} - Dwde_{smi,i} - Dwde_{ami,i} - Dwde_{dec,i}$ $- Dwde_{sff,i} + Dwde_{sfc,i}$	mgC/l/h	$R_{PP,i}$, $R_{ZP,i}$ [i = 1 – 3]
Oxic mineralization	$Dwde_{omi,i} = m_{03,i} \cdot g(WDO, Hf_{o2w,omi}) \cdot WDE_i / G$	mgC/l/h	$Hf_{o2w,omi}$
Suboxic mineralization	$Dwde_{smi,i} = m_{03,i} \cdot g(WNY, Hf_{no3w,smi})$ $\cdot (1 - g(WDO, Hf_{o2w,smi})) \cdot WDE_i / G$	mgC/l/h	$Hf_{o2w,smi}$, $Hf_{no3w,smi}$

Table A5. (Continued)

Biochemical Processes	Formulation [min(a, b) = a (a > b), or b (a < b) ; g(X, a$_{half}$) = X / (X + a$_{half}$)]	Unit	Parameters [see Table 2, Table 3]
Anoxic mineralization	$Dwde_{ami,i} = m_{03,i} \cdot (1-g(WNY, Hf_{no3w,ami}))$ $\cdot (1-g(WDO, Hf_{o2w,ami})) \cdot WDE_i /G$	mgC/l/h	Hf$_{o2w,ami}$, Hf$_{no3w,ami}$
Mineralization rate	$m_{03,i} = \alpha_{mi03,i} \cdot \exp(\beta_{mi03,i} \cdot TmpW)$	1/h	$\alpha_{mi03,i}$, $\beta_{mi03,i}$ [i = 1 – 3]
	$G = g(WDO, Hf_{o2w,omi})$ $+g(WNY, Hf_{no3w,smi}) \cdot (1-g(WDO, Hf_{o2w,smi}))$ $+(1-g(WNY, Hf_{no3w,ami})) \cdot (1-g(WDO, Hf_{o2w,ami}))$	-	Hf$_{o2w,omi}$, Hf$_{o2w,smi}$, Hf$_{no3w,smi}$ Hf$_{o2w,ami}$, Hf$_{no3w,ami}$
Decomposition	$Dwde_{dec,i} = R_{dec03,i} \cdot (Dwde_{om,i} + Dwde_{sm,i} + Dwde_{am,i})$	mgC/l/h	R$_{dec03,i}$ [i = 1 – 3]
Suspension feeder feeding (bottom layer only)	$Dwde_{sff,i} = \dfrac{WDE_i}{PP + ZP + \sum_{i=1}^{3} WDE_i} \cdot \dfrac{1}{A_{Depq}} \cdot Dsfb_{fee}$	mgC/l/h	A$_{Depq}$
Suspension feeder feces (bottom layer only)	$Dwde_{sfc,i} = R_{ZWfee51} \cdot R_{FOD51,i} \cdot \dfrac{1}{\Delta z} \cdot Dsfb_{fee}$	mgC/l/h	R$_{ZWfee51}$

[Dissolved organic matter (DOM) : 04,j] [j = 1 (labile), 2 (refractory)]

Biochemical net production / consumption	$\dot{C}_{wdm,j} = R_{ext01,j} \cdot Dpp_{ext} + \sum_{i=1}^{3} R_{DOM,ji} \cdot Dwde_{dec,i}$ $- Dwdm_{omi,j} - Dwdm_{smi,j} - Dwdm_{ami,j}$	mgC/l/h	R$_{ext01,j}$, R$_{DOM,ji}$ [i = 1 – 3, j = 1 – 2]
Oxic mineralization	$Dwdm_{omi,j} = m_{04,j} \cdot g(WDO, Hf_{o2w,omi}) \cdot WDM_j /G$	mgC/l/h	Hf$_{o2w,omi}$
Suboxic mineralization	$Dwdm_{smi,j} = m_{04,j} \cdot g(WNY, Hf_{no3w,smi})$ $\cdot (1-g(WDO, Hf_{o2w,smi})) \cdot WDM_j /G$	mgC/l/h	Hf$_{o2w,smi}$, Hf$_{no3w,smi}$
Anoxic mineralization	$Dwdm_{ami,j} = m_{04,j} \cdot (1-g(WNY, Hf_{no3w,ami}))$ $\cdot (1-g(WDO, Hf_{o2w,ami})) \cdot WDM_j /G$	mgC/l/h	Hf$_{o2w,ami}$, Hf$_{no3w,ami}$
Mineralization rate	$m_{04,j} = \alpha_{mi04,j} \cdot \exp(\beta_{mi04,j} \cdot TmpW)$	1/h	$\alpha_{mi04,j}$, $\beta_{mi04,j}$ [j = 1 – 2]

[NH$_4$-N : 05]

Biochemical net production / consumption	$\dot{C}_{wnx} = R_{nc01} \cdot Dpp_{res} + R_{nc01} \cdot Dzp_{exc} + Dwnx_{red}$ $+ \sum_{i=1}^{3} R_{nc03,i} \cdot (Dwde_{omi,i} + Dwde_{smi,i} + Dwde_{ami,i})$ $+ \sum_{j=1}^{2} R_{nc04,j} \cdot (Dwdm_{omi,j} + Dwdm_{smi,j} + Dwdm_{ami,j})$ $- R_{nc01} \cdot Dpp_{pho} \cdot \dfrac{WNX}{WNX + WNY} - Dwnx_{nit}$ $+ R_{ncFOD51} \cdot \dfrac{1}{\Delta z} \cdot Dsfb_{exc}$ (bottom layer only)	mgN/l/h	R$_{nc01}$, R$_{nc03,i}$, R$_{nc04,j}$ [i = 1 – 3, j = 1 – 2]
Nitrification	$Dwnx_{nit} = \alpha_{nit05} \cdot \exp(\beta_{nit05} \cdot TmpW) \cdot g(WDO, Hf_{o2,nit05}) \cdot WNX$	mgN/l/h	α_{nit05}, β_{nit05}, Hf$_{o2,nit05}$
Nitrate reduction	$Dwnx_{red} = (14/12) \cdot (4/(8-3 \cdot R_{den06})) \cdot (1-R_{den06})$ $\cdot \left(\sum_{i=1}^{3} Dwde_{smi,i} + \sum_{j=1}^{2} Dwdm_{smi,j} \right)$	mgN/l/h	R$_{den06}$

[NO$_3$-N : 06]

Biochemical net production / consumption	$\dot{C}_{wny} = Dwnx_{nit} - Dwny_{den} - Dwnx_{red}$ $- R_{nc01} \cdot Dpp_{pho} \cdot \dfrac{WNY}{WNX + WNY}$	mgN/l/h	R$_{nc01}$
De-nitrification	$Dwny_{den} = (14/12) \cdot (4/(8-3 \cdot R_{den06})) \cdot R_{den06}$ $\cdot \left(\sum_{i=1}^{3} Dwde_{smi,i} + \sum_{j=1}^{2} Dwdm_{smi,j} \right)$	mgN/l/h	R$_{den06}$

[PO$_4$-P : 07]

Biochemical net production / consumption	$\dot{C}_{wdp} = R_{pc01} \cdot Dpp_{res} + R_{pc01} \cdot Dzp_{exc}$ $+ \sum_{i=1}^{3} R_{pc03,i} \cdot (Dwde_{omi,i} + Dwde_{smi,i} + Dwde_{ami,i})$ $+ \sum_{j=1}^{2} R_{pc04,j} \cdot (Dwdm_{omi,j} + Dwdm_{smi,j} + Dwdm_{ami,j})$ $- R_{pc01} \cdot Dpp_{pho}$ $+ R_{pcFOD51} \cdot \frac{1}{\Delta z} \cdot Dsfb_{exc}$ (bottom layer only)	mgP/l/h	$R_{pc01}, R_{pc03,i}, R_{pc04,j}$ [$i = 1 - 3, j = 1 - 2$]

[ODU : 08]

Biochemical net production / consumption	$\dot{C}_{wou} = (32/12) \cdot \left(\sum_{i=1}^{3} Dwde_{ami,i} + \sum_{j=1}^{2} Dwdm_{ami,j} \right)$ $- Dwou_{aut} - Dwou_{oxi}$	mg/l/h	
Oxidation	$Dwou_{oxi} = \alpha_{oxi08} \cdot exp(\beta_{oxi08} \cdot TmpW) \cdot g(WDO, Hf_{o2,oxi08}) \cdot WOU$ $+ R_{oxi08} \cdot (32/12) \cdot \left(\sum_{i=1}^{3} Dwde_{ami,i} + \sum_{j=1}^{2} Dwdm_{ami,j} \right)$	mg/l/h	$\alpha_{oxi08}, \beta_{oxi08},$ $Hf_{o2,oxi08}, R_{oxi08}$
Authigenic mineralization	$Dwou_{aut} = \alpha_{aut08} \cdot exp(\beta_{aut08} \cdot TmpW) \cdot WOU + R_{aut08a} \cdot Dwou_{oxi}$ $+ R_{aut08b} \cdot (32/12) \cdot \left(\sum_{i=1}^{3} Dwde_{ami,i} + \sum_{j=1}^{2} Dwdm_{ami,j} \right)$	mg/l/h	$\alpha_{aut08}, \beta_{aut08},$ R_{aut08a}, R_{aut08b}

[Dissolved oxygen (DO) : 09]

Biochemical net production / consumption	$\dot{C}_{wdo} = \frac{32}{12} \cdot Dpp_{pho} - \frac{32}{12} \cdot Dpp_{res} - \frac{32}{12} \cdot Dzp_{exc}$ $- \frac{32}{12} \cdot \left(\sum_{i=1}^{3} Dwde_{omi,i} + \sum_{j=1}^{2} Dwdm_{omi,j} \right)$ $- 2 \cdot \frac{32}{14} \cdot Dwnx_{nit} - Dwou_{oxi}$ $- \frac{32}{12} \cdot \frac{1}{\Delta z} \cdot (Dsfb_{exc} + Ddfb_{exc})$ (bottom layer only)	mgO$_2$/l/h	

Table A6. Formulation of essential biochemical processes (benthic system)

Biochemical Processes	Formulation [min(a, b) = a (a > b), or b (a < b) ; g(X,a$_{half}$) = X / (X + a$_{half}$)]	Unit	Parameters [see Table 2, Table 3]
[Suspension feeders : 51]			
Biochemical net production / consumption	$\dot{C}_{sfb} = Dsfb_{fee} - Dsfb_{fec} - Dsfb_{exc} - Dsfb_{mor} + Dsfb_{lar}$	µgC/ cm^2/h	
Feeding	$Dsfb_{fee} = min(Lim_{filter}, Lim_{growth})$	µgC/cm^2/h	
Filter rate limitation	$Lim_{filter} = v_{fee51} \cdot u_{fee51} \cdot (PP + ZP + WDE_1 + WDE_2 + WDE_3)$ $\cdot R_{cor0} \cdot SFB$	µgC/cm^2/h	
Filter rate	$v_{fee51} = 1.2 \times 10^{-5} \cdot TmpB^{1.25} \cdot A_{Wwet51}^{-0.75} / A_{wd51} / A_{cd51}$ (TmpB > 10) $\quad\quad\quad 1.2 \times 10^{-5} \cdot 10^{1.25} \cdot A_{Wwet51}^{-0.75} / A_{wd51} / A_{cd51}$ (TmpB < 10)	ml/h/µgC	$A_{Wwet51}, A_{wd51},$ A_{cd51}
Oxygen saturation limitation	$u_{fee51} = min(1, do_{sat}/R_{O2mor51})$ do_{sat} : Oxygen saturation of bottom water (calculated)	-	$R_{O2mor51}$
Decreasing by double filtering	$R_{cor0} = (1 - exp(-COR0)) / COR0$	-	
Double filtering ratio	$COR0 = v_{fee51} \cdot u_{fee51} \cdot SFB \cdot dt / A_{Depq}$	-	A_{Depq}
Growth rate limitation	$Lim_{growth} = (\alpha_{grt51} + \alpha_{bas51}) / (R_{ege51} \cdot (1 - R_{exc51})) \cdot SFB \cdot F_{temp}$ F_{temp} : function of temperature	µgC/cm^2/h	$\alpha_{grt51}, \alpha_{bas51},$ R_{ege51}, R_{exc51}
Feces	$Dsfb_{fec} = (1 - R_{ege51}) \cdot Dsfb_{fee}$	µgC/cm^2/h	R_{ege51}
Excretion	$Dsfb_{exc} = R_{exc51} \cdot R_{ege51} \cdot Dsfb_{fee} + \alpha_{bas51} \cdot SFB$	µgC/cm^2/h	$R_{exc51}, R_{ege51},$ α_{bas51}
Mortality	$Dsfb_{mor} = u_{mor51} \cdot exp(\beta_{mor51} \cdot TmpB) \cdot SFB$	µgC/cm^2/h	β_{mor51}
Rate of mortality	$u_{mor51} = \alpha_{mor51a} + \alpha_{mor51b} \cdot (1 - u_{fee51})$	µgC/cm^2/h	$\alpha_{mor51a}, \alpha_{mor51b}$
Larva input	$Dsfb_{lar} = R_{lar51} \cdot Dsfb_{fee, av}$	µgC/cm^2/h	R_{lar51}
	$Dsfb_{fee, av}$: spatial and temporal average of feeding (calculated)	µgC/cm^2/h	
[Deposit feeders : 52]			
Biochemical net production / consumption	$\dot{C}_{dfb} = Ddfb_{fee} - Ddfb_{fec} - Ddfb_{exc} - Ddfb_{mor} + Ddfb_{lar}$	µgC/cm^2/h	

Table A6. (Continued)

Biochemical Processes	Formulation [min(a, b) = a (a > b), or b (a < b) ; g(X,a_{half}) = X / (X + a_{half})]	Unit	Parameters [see Table 2, Table 3]
Feeding	$Ddfb_{fee} = \alpha_{fee52} \cdot \exp(\beta_{fee52} \cdot TmpB) \cdot u_{fee52a} \cdot u_{fee52b} \cdot u_{fee52c} \cdot DFB$	$\mu gC/cm^2/h$	$\alpha_{fee52}, \beta_{fee52}$
Food limitation	$u_{fee52a} = 1 - \exp(A_{ivl52} \cdot \min(0, A_{kai52} - Food_{52}))$	-	A_{ivl52}, A_{kai52}
	$Food_{52}$: average concentration of food in mud. (calculated)		
Cannibalism efficiency	$u_{fee52b} = g(DFB, Hf_{dfb,fee52})$	-	$Hf_{dfb,fee52}$
Oxygen saturation limitation	$u_{fee52c} = \min(1, do_{sat} / R_{O2mor52})$	-	$R_{O2mor52}$
Feces	$Ddfb_{fec} = (1 - u_{fee52}) \cdot Ddfb_{fee}$	$\mu gC/cm^2/h$	
Assimilation efficiency	$u_{fee52} = 1 - R_{undg52} \cdot (1 + g(Food_{52}, Hf_{fod52,fee52}))$	-	$R_{undg52}, Hf_{fod52,fee52}$
Excretion	$Ddfb_{exc} = R_{exc52} \cdot u_{fee52} \cdot Ddfb_{fee}$	$\mu gC/cm^2/h$	R_{exc52}
Mortality	$Ddfb_{mor} = v_{mor52} \cdot u_{mor52} \cdot DFB$	$\mu gC/cm^2/h$	
Temperature dependency	$v_{mor52} = \min(\exp(\beta_{mor52} \cdot TmpB), \exp(\beta_{mor52} \cdot A_{temp,fee52}))$	–	$A_{temp,fee52}, \beta_{mor52}$
Rate of mortality	$u_{mor52} = \alpha_{mor52a} + \alpha_{mor52b} \cdot (1 - u_{fee52c})$	1/h	$\alpha_{mor52a}, \alpha_{mor52b}$
Larva input	$Ddfb_{lar} = R_{lar52} \cdot Ddfb_{fee, av}$	$\mu gC/cm^2/h$	R_{lar52}
	$Ddfb_{fee, av}$: spatial and temporal average of feeding (calculated)	$\mu gC/cm^2/h$	

[Detritus : 53,i] *[i = 1 (fast-labile), 2 (slow-labile), 3 (refractory/very slow-labile)]*

Biochemical net production / consumption	$\dot{C}_{det,i} = R_{Zfee51} \cdot R_{FOD51,i} \cdot Dsfb_{fee} + R_{Zmor51} \cdot R_{SFB,i} \cdot Dsfb_{mor}$ $+ R_{Zfee52} \cdot R_{FOD52,i} \cdot Ddfb_{fee} + R_{Zmor52} \cdot R_{DFB,i} \cdot Ddfb_{mor}$ $- R_{Zfee52} \cdot R_{FOD52,i} \cdot Ddfb_{fee} + R_{Zmor60} \cdot R_{BAL,i} \cdot Dbal_{mor}$ $- Ddet_{omi,i} - Ddet_{smi,i} - Ddet_{ami,i} - Ddet_{dec,i}$	$\mu gC/cm^3/h$	$R_{SFB,i}, R_{DFB,i},$ $R_{BAL,i},$ $[i = 1 - 3]$
Oxic mineralization	$Ddet_{omi,i} = m_{53,i} \cdot g(DOO, Hf_{o2b,omi}) \cdot DET_i / G \cdot (1 - \phi)$	$\mu gC/cm^3/h$	$Hf_{o2b,omi}$
Suboxic mineralization	$Ddet_{smi,i} = m_{53,i} \cdot g(HNY, Hf_{no3b,smi})$ $\cdot (1 - g(DOO, Hf_{o2b,smi})) \cdot DET_i / G \cdot (1 - \phi)$	$\mu gC/cm^3/h$	$Hf_{o2b,smi}, Hf_{no3b,smi}$
Anoxic mineralization	$Ddet_{ami,i} = m_{53,i} \cdot (1 - g(HNY, Hf_{no3b,ami}))$ $\cdot (1 - g(DOO, Hf_{o2b,ami})) \cdot DET_i / G \cdot (1 - \phi)$	$\mu gC/cm^3/h$	$Hf_{o2b,ami}, Hf_{no3b,ami}$
Mineralization rate	$m_{53,i} = \alpha_{mi53,i} \cdot \exp(\beta_{mi53,i} \cdot TmpB)$	1/h	$\alpha_{mi53,i}, \beta_{mi53,i} [i = 1 - 3]$
	$G = g(DOO, Hf_{o2b,omi})$ $+ g(HNY, Hf_{no3b,smi}) \cdot (1 - g(DOO, Hf_{o2b,smi}))$ $+ (1 - g(HNY, Hf_{no3b,ami})) \cdot (1 - g(DOO, Hf_{o2b,ami}))$	-	$Hf_{o2b,omi},$ $Hf_{o2b,smi}, Hf_{no3b,smi}$ $Hf_{o2b,ami}, Hf_{no3b,ami}$
Decomposition	$Ddet_{dec,i} = R_{dec53,i} \cdot (Ddet_{omi,i} + Ddet_{smi,i} + Ddet_{ami,i})$	$\mu gC/cm^3/h$	$R_{dec53,i} [i = 1 - 3]$

[Dissolved organic matter (DOM): 54,j] *[j = 1 (labile), 2 (refractory)]*

Biochemical net production / consumption	$\dot{C}_{dom,j} = R_{ext60,j} \cdot Dbal_{ext} + \sum_{i=1} R_{DOM,ji} \cdot Ddet_{dec,i}$ $- R_{Zfee52} \cdot R_{FOD52, 54j} \cdot Ddfb_{fee}$ $+ R_{Zfee52} \cdot R_{FOD52,54j} \cdot Ddfb_{fec}$ $- Ddom_{omi,j} - Ddom_{smi,j} - Ddom_{ami,j}$	$\mu gC/cm^3/h$	$R_{ext60,j}, R_{DOM,ji}$ $[i = 1 - 3, j = 1 - 2]$
Oxic mineralization	$Ddom_{omi,j} = m_{54,j} \cdot g(DOO, Hf_{o2b,omi}) \cdot DOM_j / G \cdot (\phi + \overline{\rho}_s \cdot K_{ads54,j} \cdot (1 - \phi))$	$\mu gC/cm^3/h$	$Hf_{o2b,omi},$ $K_{ads54,j} [j = 1 - 2]$
Suboxic mineralization	$Ddom_{smi,j} = m_{54,j} \cdot g(HNY, Hf_{no3b,smi})$ $\cdot (1 - g(DOO, Hf_{o2b,smi})) \cdot DOM_j / G \cdot (\phi + \overline{\rho}_s \cdot K_{ads54,j} \cdot (1 - \phi))$	$\mu gC/cm^3/h$	$Hf_{o2b,smi}, Hf_{no3b,smi},$ $K_{ads54,j} [j = 1 - 2]$
Anoxic mineralization	$Ddom_{am,j} = m_{54,j} \cdot (1 - g(HNY, Hf_{no3b,ami}))$ $\cdot (1 - g(DOO, Hf_{o2b,ami})) \cdot DOM_j / G \cdot (\phi + \overline{\rho}_s \cdot K_{ads54,j} \cdot (1 - \phi))$	$\mu gC/cm^3/h$	$Hf_{o2b,ami}, Hf_{no3b,ami},$ $K_{ads54,j} [j = 1 - 2]$
Mineralization rate	$m_{54,j} = \alpha_{mi54,j} \cdot \exp(\beta_{mi54,j} \cdot TmpB)$	1/h	$\alpha_{mi54,j}, \beta_{mi54,j} [j = 1 - 2]$

[NH₄-N : 55]

Biochemical net production / consumption	$\dot{C}_{hnx} = R_{nc60} \cdot Dbal_{res} + Dhnx_{red}$ $+ \sum_{i=1}^{3} R_{nc03,i} \cdot (Ddet_{omi,i} + Ddet_{smi,i} + Ddet_{ami,i})$ $+ \sum_{j=1}^{2} R_{nc04,j} \cdot (Ddom_{omi,j} + Ddom_{smi,j} + Ddom_{ami,j})$ $- R_{nc60} \cdot Dbal_{pho} \cdot \dfrac{R_{pho60,55} \cdot HNX}{R_{pho60,55} \cdot HNX + HNY} - Dhnx_{nit}$ $+ R_{Zexc52} \cdot R_{ncFOD52} \cdot Ddfb_{exc}$	$\mu gN/cm^3/h$	$R_{nc60}, R_{nc03,i},$ $R_{nc04,j},$ $R_{pho60,55}$ $[i = 1 - 3, j = 1 - 2]$

Nitrification	$Dhnx_{nit} = \alpha_{nit55} \cdot exp(\beta_{nit55} \cdot TmpB) \cdot g(DOO, Hf_{o2,nit55}) \cdot HNX$ $\cdot (\phi + \bar{\rho}_s \cdot K_{ads55} \cdot (1-\phi))$	$\mu gN/cm^3/h$	$\alpha_{nit55}, \beta_{nit55},$ $Hf_{o2,nit55}, K_{ads55}$
Nitrate reduction	$Dhnx_{red} = (14/12) \cdot (4/(8-3 \cdot R_{den56})) \cdot (1 - R_{den56})$ $\cdot \left(\sum_{i=1}^{3} Ddet_{smi,i} + \sum_{j=1}^{2} Ddom_{smi,j} \right)$	$\mu gN/cm^3/h$	R_{den56}
[NO$_3$-N : 56]			
Biochemical net production / consumption	$\dot{C}_{hny} = Dhnx_{nit} - Dhny_{den} - Dhnx_{red}$ $- R_{nc60} \cdot Dbal_{pho} \cdot \dfrac{HNY}{R_{pho60,55} \cdot HNX + HNY}$	$\mu gN/cm^3/h$	$R_{nc60}, R_{pho60,55}$
De-nitrification	$Dhny_{den} = (14/12) \cdot (4/(8-3 \cdot R_{den56})) \cdot R_{den56} \cdot \left(\sum_{i=1}^{3} Ddet_{smi,i} + \sum_{j=1}^{2} Ddom_{smi,j} \right)$	$\mu gN/cm^3/h$	R_{den56}
[PO$_4$-P : 57]			
Biochemical net production / consumption	$\dot{C}_{dip} = R_{pc60} \cdot Dbal_{res}$ $+ \sum_{i=1}^{3} R_{pc03,i} \cdot (Ddet_{omi,i} + Ddet_{smi,i} + Ddet_{ami,i})$ $+ \sum_{j=1}^{2} R_{pc04,j} \cdot (Ddom_{omi,j} + Ddom_{smi,j} + Ddom_{ami,j})$ $- R_{pc60} \cdot Dbal_{pho} + R_{Zexc52} \cdot R_{pcFOD52} \cdot Ddfb_{exc}$	$\mu gP/cm^3/h$	$R_{pc60}, R_{pc03,i},$ $R_{pc04,j}$ $[i = 1 - 3, j = 1 - 2]$
[ODU : 58]			
Biochemical net production / consumption	$\dot{C}_{odu} = (32/12) \cdot \left(\sum_{i=1}^{3} Ddet_{ami,i} + \sum_{j=1}^{2} Ddom_{ami,j} \right) - Dodu_{aut} - Dodu_{oxi}$	$\mu g/cm^3/h$	
Oxidation	$Dodu_{oxi} = \alpha_{oxi58} \cdot exp(\beta_{oxi58} \cdot TmpB) \cdot g(DOO, Hf_{o2,oxi58}) \cdot ODU \cdot \phi$ $+ R_{oxi58} \cdot (32/12) \cdot \left(\sum_{i=1}^{3} Ddet_{ami,i} + \sum_{j=1}^{2} Ddom_{ami,j} \right)$	$\mu g/cm^3/h$	$\alpha_{oxi58}, \beta_{oxi58},$ $Hf_{o2,oxi58}$ R_{oxi58}
Authigenic mineralization	$Dodu_{aut} = \alpha_{aut58} \cdot exp(\beta_{aut58} \cdot TmpB) \cdot ODU \cdot \phi + R_{aut58a} \cdot Dodu_{oxi}$ $+ R_{aut58b} \cdot (32/12) \cdot \left(\sum_{i=1}^{3} Ddet_{ami,i} + \sum_{j=1}^{2} Ddom_{ami,j} \right)$	$\mu g/cm^3/h$	$\alpha_{aut58}, \beta_{aut58},$ R_{aut58a}, R_{aut58b}
[Dissolved oxygen (DO) : 59]			
Biochemical net production / consumption	$\dot{C}_{doo} = \dfrac{32}{12} \cdot Dbal_{pho} - \dfrac{32}{12} \cdot Dbal_{res}$ $- \dfrac{32}{12} \cdot \left(\sum_{i=1}^{3} Ddet_{omi,i} + \sum_{j=1}^{2} Ddom_{omi,j} \right)$ $- 2 \cdot \dfrac{32}{14} \cdot Dhnx_{nit} - Dodu_{oxi}$	$\mu g/cm^3/h$	
[Benthic algae : 60]			
Biochemical net production / consumption	$\dot{C}_{bal} = \int (Dbal_{pho} - Dbal_{ext} - Dbal_{res} - Dbal_{mor}) dz$ $- R_{FOD52,60} \cdot Ddfb_{fee}$	$\mu gC/cm^2/h$	
Photosynthesis	$Dbal_{pho} = v_{pho60} \cdot u_{pho60a} \cdot u_{pho60b} \cdot BAL \cdot R_{Zpho60}$	$\mu gC/cm^3/h$	
Maximum growth rate	$v_{pho60} = \alpha_{pho60} \cdot exp(\beta_{pho60} \cdot TmpB)$	$1/h$	$\alpha_{pho60}, \beta_{pho60}$
Nutrient limitation	$u_{pho60a} = min[g(HNX+HNY, Hf_{n,pho60}), g(DIP, Hf_{p,pho60})]$	-	$Hf_{n,pho60}, Hf_{p,pho60}$
Light availability	$u_{pho60b} = \dfrac{1}{\Delta z} \int_{z}^{z+\Delta z} \dfrac{I_B}{I_{opt60}} e^{-k_b z} exp \left\{ 1 - \dfrac{I_B}{I_{opt60}} e^{-k_b z} \right\} dz$	-	I_{opt60}, k_b
Extra-release	$Dbal_{ext} = R_{ext60} \cdot Dbal_{pho}$	$\mu gC/cm^3/h$	R_{ext60}
Respiration	$Dbal_{res} = R_{res60a} \cdot exp(\beta_{pho60} \cdot TmpB) \cdot g(DOO, Hf_{o2,res60}) \cdot BAL \cdot R_{Zres60}$ $- R_{res60b} \cdot Dbal_{pho}$	$\mu gC/cm^3/h$	$R_{res60a}, \beta_{pho60},$ $Hf_{o2,res60}, R_{res60b}$
Mortality	$Dbal_{mor} = v_{mor60} \cdot BAL \cdot R_{Zmor60}$	$\mu gC/cm^3/h$	
Rate of mortality	$v_{mor60} = \alpha_{mor60} \cdot exp(\beta_{mor60} \cdot TmpB)$		$\alpha_{mor60}, \beta_{mor60}$

REFERENCES

Admiraal, W., Peletier, H., Zomer, H., 1982. Observations and experiments on the population dynamics of epipelic diatoms from an esturaine mudflat. Estuarine, *Coastal and Shelf Science*, 14, 471-487.

Aoyama, H, Suzuki, T, 1997. In Situ Measurement of Particulate Organic Matter Removal Reates by a Tidal Flat Macrobenthic Community. *Bulletin of the Japanese Society of Fisheries Oceanography,* 61(3), 265-274. (in Japanese with English abstract)

Baretta, J.W., Ruardij, P., 1988. Tidal flat estuaries, simulation and analysis of the Ems Estuary. *Ecological Studies,* 71, Springer-Verlag.

Baretta, J.W., Ebenhöh, W., Ruardij, P., 1995. The European Regional Seas Ecosystem Model, a complex marine ecosystem model. *Neth. J. Sea Res.* 33, 233-246.

Baretta-Bekker, J.G., Baretta, J.W., 1997. Microbial dynamics in the marine ecosystem model ERSEM II with decoupled carbon assimilation and nutrient uptake. *J. Sea Res.,* 38, 195-211.

Berg, P., Risgaard-Petersen, N. and Rysgaard, S., 1998. Interpretation of measured concentration profiles in sediment pore water. *Limnology and Oceanography*, 43, 1500-1510.

Berg, P., Røy, H., Janssen F., Meyer. V., Jørgensen, B.B., Huettel, M. and Beer D., 2003. Oxygen uptake by aquatic sediments measured with a novel non-invasive eddy-correlation technique. *Marine Ecology Progress Series,* 261, 75-83.

Berner, R.A., 1980. *Early diagenesis - A theoretical approach -.* Princeton University Press, New Jersey, 241pp.

Boudreau, B.P., 1996. A method of lines code for carbon and nutrient diagenesis in aquatic sediments. *Computers and Geosciences* 22(5), 479-496.

Boudreau, B.P. and Jørgensen, B.B.(eds.) 2001. The Benthic Boundary Layer. *Transport Processes and Biogeochemistry.* Oxford University Press, Oxford, 440pp.

Blumberg. A.F., G.L. Mellor, 1978. A coastal ocean numerical model. In: J. Sundermann and K.P. Holz (Editors), *Mathematical Modeling of Estuarine Physics, Proceedings of an International Symposium,* Hamburg, August 24 to 26, 1978, Springer-Verlag, Berlin, 203-219.

Cammen, L.M., 1980. *Ingestion rate: an empirical model for aquatic deposit feeders and detritivores. Oecologia,* 44, 303-310.

Canfield, D.E., Jørgensen, B.B., Fossing, H., Glud, R., Gundersen, J., Ramsing, N.B., Thamdrup, B., Hansen, J.W., Nielsen, L.P. and Hall, P.O.J., 1993. Pathways of organic carbon oxidation in three continental margin sediments, *Marine Geology,* 113, 27-40.

Chiba Prefecture, 1998-2002. *Result of the water quality survey of public water areas.* (in Japanese)

Chiba Prefectural Fisheries Research Center, 2001-2006. *Hypoxia quick information.* (web site http://www.awa.or.jp/home/cbsuishi/04tkhinsanso/04tkhinsansoflame.htm) (in Japanese)

Conover, R.J., 1978. Transformation of organic matter. In: Kinne, O. (Editor), *Marine Ecology,* vol. IV. Dynamics, Wiley, New York, pp.221-499.

de Beer, D., Wenzhöfer, F., Ferdelman, T.G., Boehme, S.E., Huettell, M., van Beusekom, J.E.E., Böttcher, M.E., Musat, N., Dubilier, N., 2005. Transport and mineralization rates in North Sea sandy intertidal sediments, Sylt-Rømø Basin, Wadden Sea. *Limnology and Oceanography,* 50, 113-127.

Dedieu, K., Rabouille, G., Gilbert, F., Soetaert, K., Metzger, E., Simonucci, G., Jézéquel, D., Prévot, F., Anschutz, P., Hulth, S.,Ogier, S., Mesnage, V., 2007. Coupling of carbon, nigrogen and oxygen cycles in sediments from a Mediterranian lagoon: a seasonal prespective. *Marine Ecology Progress Series* 346, 45-59.

Emerson, S. and Hedges, J.L., 1988. Processes controlling the organic carbon content of open ocean sediments. *Paleoceanography,* 3, 621-634.

Epply, R.W., Rogers, J.N., McCarthy, J.J., 1969. Half saturation constants for uptake of nitrate and ammonium by marine phytoplankton. *Limnology and Oceanography* 14, 912–920.

Fuhs, W.G., Demmerle, S.D., Canelli, E. and Chen, M., 1972. Characterization of phosphorus-limited plankton algae (with reflections on the limiting nutrient concept). In: Likens, G. E. (Editor), Nutrients and Eutrophication. *Spec. Symp. Vol. 1. Am. Soc. Limnology and Oceanography,* pp.113-133. Allen Press, Lawrence, KS.

Furota, T., 1988. Effects of low-oxygen water on benthic and sessile animal communities in Tokyo Bay. In Symposium: Material cycling and biological environment in Tokyo Bay. *Bulletin on Coastal Oceanography.* 25(2) 104-113. (in Japanese)

Gundersen, J.K., Glud, R.N., Jørgensen, B.B., 1995. Oxygen transformations in the sea floor (in Danish). Marine Research from the Danish Environmental Agency, Vol. 57.

Hata K. and Nakata, K., 1998. Evaluation of eelgrass bed nitrogen cycle using an ecosystem model. *Environmental Modeling & Software.* 13, 491-502.

Hiroshima Environment & Health Association, 2002. *Report about ecological improvement of sediment quality using benthic algae in Seto Inland Sea* (web site http://nippon.zaidan.info/seikabutsu/2001/00614/mokuji.htm) (in Japanese)

Hiwatari, T., Kohata, K. and Iijima, A., 2002. Nitrogen budget of the bivalve Mactra veneriformis, and its significance in benthic-pelagic systems in the Sanbanse area of Tokyo Bay. *Estuar. Coast. Self Sci.,* 55, 299-308.

Horiguchi, F., 2001. Numerical simulations of seasonal cycle of Tokyo Bay using an ecosystem model. *Journal of Advanced Marine Science and Technology Society,* 7(1&2), 1-30. (in Japanese with English abstract)

Imao, K., Suzuki, T., Takabe, T., 2004. New method to predict changes in the structure and function of a macrobenthic community from changes in environmental oxygen concentrations. *Fisheries Engineering,* 41(1), 13-24. (in Japanese with English abstract)

Ishida, M., Hara, T., 1996. Changes in water quality and eutrophication in Ise and Mikawa Bays. *Bulletin of the Aichi Fisheries Research Institute,* 3, 29-41. (in Japanese with English abstract)

Ishikawa, M. and Nishimura, H., 1983. A new method of evaluating the mineralization of particulate and dissolved photoassimilated organic matter. *Journal of the Oceonography Society of Japan,* 39 (2), 29–42.

Isono, R., Kita, J., Kishida, C., 1998. Upper temperature effect on rates of growth and oxygen consumption of the Japanese little neck clam, Ruditapes philippinarum. *Journal of the Oceanography Society of Japan,* 39(2), 29-42.

Japan Environmental Management Association for Industry, 1998. *Survey report of the water quality pollution mechanism in Mikawa Bay.* (in Japanese)

Jørgensen, B.B., 1978. A comparison of methods for the quantification of bacterial sulfate reduction in coastal marine sediments: II Calculations from mathematical models. *Geomicrobiology Journal* 1, 29-47.

Jørgensen, S.E. (Ed.), 1979. Handbook of Environmental Data and Ecological Parameters. International Society for Ecological Modelling, *Pergamon Press, Amsterdam,* 1162 pp.

Jørgensen, S.E., Nielsen, S.N. and Jørgensen, L.A., 1991. *Handbook of Ecological Parameters and Ecotoxicology.* Elsevier Science Publishers, Amsterdam, 1263 pp.

Jørgensen, S.E. and Bendoricchio, G., 2001. *Fundamentals of ecological modelling: Developments in environmental modelling,* 21, 3rd ed. Elsevier, New York, 530pp.

Kakino J., 1982. *Effects of Ao-Shio on the mortality in Manila clams.* Bulletin of Chiba Prefectural Fisheries Research Institute, 40, 1-6. (in Japanese)

Kamio, K., Nomura, M., Nakamura, Y., Kuwae, T., Inoue, T., Konuma, S., 2004. Oxygen variation of the tidal flat overlying water. In: *Proceedings of the 2004 Spring Annual Meeting of the Oceanographic Society of Japan,* p199. (in Japanese)

Kanagawa Prefecture, 1998-2002. *Result of the water quality survey of public water areas.* (in Japanese)

Kanagawa Prefectural Fisheries Research Institute, 2005-2006. *Tokyo Bay Dissolved Oxygen Information.* (web site http://www.agri.pref.kanagawa.jp/suisoken/kankyo/sanso/TokyoBayOxInfo.htm) (in Japanese)

Kikuchi, T., 1993. Ecological characteristics of the tidal flat ecosystem and importance of its conservation. *Japanese Journal of Ecology,* 43, 223-235. (in Japanese)

Koike, K., 2000., *Reclamation of Tokyo Bay and artificial beach in Kanto and Ogasawara areas - Japanese geography -.* University of Tokyo Press. (in Japanese)

Kremer, J.N. and Nixon, S.W., 1978. *A coastal marine ecosystem. simulation and analysis.* Springer-Verlag, Berlin. 217pp.

Kurashige, H., 1942. Resistance of Paphia philippinarum Adams et Reeve to Lack of Oxygen. *Journal of the Oceanographical Society of Japan,* 1 (Nos. 1-2), 123-132. (in Japanese)

Kuwac, T., 2001. Biogeochemical roles of benthic microorganisms in intertidal sandflats. *Ph. D. thesis,* Kyoto University, 93pp.

Kuwae, T., Kibe, E. and Nakamura, Y., 2003. Effect of emersion and immersion on the porewater nutrient dynamics of an intertidal sandflat in Tokyo Bay. *Estuarine, Coastal and Shelf Science,* 57, 929-940.

Kuwae, T., Inoue, T., Miyoshi, E., Konuma, S., Hosokawa, S., Nakamura, Y., 2005. In: Modeling the coastal marine ecosystem coupled with tidal flats based on the study of oxygen cycling in sediments. *Report of Program for Promoting Fundamental Transport Technology Research.* pp.262-423. Japan Railway Construction, Transport and Technology Agency. (in Japanese)

Kuwae, T., Kamio, K., Inoue, T., Miyoshi, E. and Uchiyama, Y., 2006. In situ measurement of oxygen exchange flux between sediment and water of an intertidal sandflat, measured in situ by the eddy-correlation method. *Marine Ecology Progress Series,* 307, 59-68.

Luff, R. and Moll, A., 2004. Seasonal dynamics of the North Sea sediments using a three-dimensional coupled sediment-water model system. *Continental Shelf Research,* 24, 1099-1127.

Marshall, S.M., Orr, A.P., 1955a. Experimental feeding of the copepod Calanus finmarchicus on phytoplankton cultures labeled with radioactive carbon. *Pap. Marine Biology and Oceanography, Deep-Sea Research* 3(Suppl.), 110-114.

Marshall, S.M., Orr, A.P., 1955b, On the biology of Calanus finmarchicus. VIII. Food uptake, assimilation and excretion in adult and stage V. Calanus. *Journal of the Marine Biological Association of United Kindom,* 34, 495-529.

Matsumoto, E., 1983. The sedimentary environment in Tokyo Bay. *Earth Chemistory*, 17, 27-32. (in Japanese)

Matsunaga, K., 1981. Studies on the decomposition processes of phytoplanktonic organic matter. *The Japanese Journal of Limnology,* Vol. 42 (4), 220-229.

Mellor, G.L., Yamada, T., 1982. Development of a turbulent closure model for geophysical fluid problems. *Reviews of Geophysics,* 20, 851-875.

Macedo, M.F., Duarte, P., Mendes, P., Ferreira, J.G., 2001. Annual variation of environmental variables, Phytoplankton species composition and phytosynthetic parameters in a coastal lagoon. *Journal of Plankton Research,* 23(7), 719-732.

Ministry of the Environment, 2006. *Reference data of a basic principle on the regulation of total amount control for chemical oxygen demand, contained amount of nitrogen and phosphorus.* (in Japanese)

Ministry of the Environment, 1998-2002. Comprehensive survey on regional water quality. (in Japanese)

Miyata, M., 2003. Bibliography of Zappai historical sources (*Zappai shiryou kaidai*). Systemized Japanese historical bibliography, *Seisyohdosyoten,* pp.501.

Nakamura, Y., Nomura, M., Kamio, K., 2004. Field observation and analysis of benthic–pelagic coupling in Banzu tidal flat and the adjoincent coastal area of Tokyo Bay. *Report of the Port and Airport Research Institute,* 43(2), 35-71. (in Japanese with English abstract)

Nakata, K., Horiguchi, F., Taguchi, K., Setoguchi, Y., 1983a. Three dimensional simulation of tidal current in Oppa Estuary. *Bulletin of the National Research Institute for Pollution and Resources,* 12(3), 17-36. (in Japanese)

Nakata, K., Horiguchi, F., Taguchi, K., Setoguchi, Y., 1983b. Three dmensional eco-hydrodynamical model in coastal region. *Bulletin of the National Research Institute for Pollution and Resources,* 13(2), 119-134. (in Japanese)

National Institute for Land and Infrastructure Management, 2006. *Integrated Environmental Monitoring at Tokyo Bay* (2002-2003). (web site http://www.nilim.go.jp/) (in Japanese)

Nishikawa, T., Miyahara, K., Nagai S., 2002. The growth response of Coscinodiscus wailesii Gran (Bacillariophyceae) as a function of irradiance isolated from Harima-Nada, Seto Inland Sea, Japan. *Bull. Plankton Soc.* 49(1), 1-8.(in Japanese with English abstract)

Odum, E.P., 1971. *Fundamentals of ecology,* 3rd ed. W.B. Saunders, Philadelphia.

Ogura, N., 1972. Decomposition of dissolved organic matter derived from dead phytoplankton. pp.507-515. In Takenouti, A. Y. (ed.) *Biological Oceanography of the Northern Pacific Ocean.* Idemitsu Shoten, Tokyo.

Ogura, N., 1975. Decomposition of dissolved organic matter in coastal seawater. *Marine Biology* 31, 101–111.

Oguz, T., 2002. The role of physical processes controlling the oxycline and suboxic layer structures in the Black Sea, *Glob. Biogeochem. Cycles,* 16(2), 101029-101042.

Patankar, S.V., 1980. Numerical heat transfer and fluid flow. *Hemisphere Publishing,* USA, pp.1-197.

Revsbech, N. P., Madsen, B. and Jørgensen, B.B., 1986. Oxygen production and consumption in sediments determined at high spatial resolution by computer simulation of oxygen microelectrode data. *Limnology and Oceanography,* 31(2), 293-304.

Rosenfeld, J.K., 1979. Ammonium adsorption in nearshore anoxic sediments. *Limnology and Oceanography,* 24, 356-364.

Rysgaard, S. and Berg, P., 1996. Mineralization in a northeastern Greenland sediment : mathematical modeling, measured sediment pore water profiles and actual activities, *Aquatic Microbial Ecology,* 11, 297-305.

Sayama, M., 2005. In: Modeling the coastal marine ecosystem coupled with tidal flats based on the study of oxygen cycling in sediments. Report of Program for Promoting Fundamental Transport Technology Research. pp.424-456. Japan Railway Construction, *Transport and Technology Agency.* (in Japanese)

Soetaert, K., Herman, P.M.J., Middleburg, J.J., 1996a. A model of early diagenetic processes from the shelf to abyssal depth, *Geochimica et Cosmochimica Acta* 60(6), 1019-1040.

Soetaert, K., Herman, P.M.J., Middelburg, J.J., 1996b. Dynamic response of deep-sea sediments to seasonal variations: *Amodel. Limnol. Oceanogr.,* 41(8), 1651-1668.

Soetaert, K., Middelburg, J.J., Herman, P.M.J., Buis, K., 2000. On the coupling of benthic and pelagic biogeochemical models. *Earth-Science Reviews,* 51, 173-201.

Sohma, A., Sato, T., Nakata, K., 2000. New numerical model study on a tidal flat system - seasonal, daily and tidal variation. *Spill Science Technology Bulletin,* 6, 173-185.

Sohma, A., Sekiguchi, Y., Yamada, H., Sato, T., Nakata, K., 2001. A new coastal marine ecosystem model study coupled with hydrodynamics and tidal flat ecosystem effect, *Marine Pollution Bulletin,* 43, 187-208.

Sohma, A., Sayama, M., 2002. Modeling for coupled cycle of Oxygen, Nitrogen, and Carbon in a coastal marine sediment - A new ecological model for dynamics in the micro profiles -. In: *Proceedings of Coastal Engineering, JSCE,* 49, 1231-1235. (in Japanese)

Sohma, A., Sekiguchi, Y., 2003. Development of a new multiple coastal ecosystem model focused on ecological network and benthic vertical mechanisms in the micro scale - application of a hydrodynamics model and benthic ecosystem model in the central bay area of Tokyo Bay -. *Proceedings of Advanced Marine Science and Technology conference in autumn,* 87-92. (in Japanese)

Sohma, A., Sekiguchi, Y., Nakata, K., 2004. Modeling and evaluating the ecosystem of sea-grass beds, shallow waters without sea-grass, and an oxygen-depleted offshore area. *Journal of Marine Systems,* 45, 105-142.

Sohma, A., Sekiguchi, Y., Kakio, T., 2005a. Development of a new multiple coastal ecosystem Model "ZAPPAI" including benthic, pelagic and tidal flat ecosystems for ecological evaluation in hypoxic estuary. - autonomous response to the tidal flat creation, dredging, sand capping, load reduction and red tide - , *Journal of Advanced Marine Science and Technology Society,* 11, 2, 21-52 (in Japanese with English abstract)

Sohma, A., 2005b. *Development of a multiple coastal ecosystem model including benthic, pelagic and tidal flat ecosystems for ecological evaluation in hypoxic estuary.* Ph. D. thesis, Tokai University, 368pp.

Sohma, A., Sekiguchi, Y., Kuwae, T., Nakamura, Y., 2008. A Benthic-pelagic coupled ecosystem model to estimate the hypoxic estuary including tidal flats - model description and validation of seasonal/daily dynamics -. *Ecological Modeling* 215, 10-39.

Spalding, D.B., 1972. A novel finite-difference formulation for differential expressions involving both first and second derivatives. *International Journal for Numerical Methods in Engineering,* 4, 551-559.

Strickland, J.D.H., 1965. Chemical composition of phytoplankton and method for measuring plant bio-mass, practical considerations composition ratios. *Chemical Oceanography,* 1, 514-518.

Suschenya, L.M., 1970. Food rations, metabolism, and growth of crustaceans. In: Steele, J.H. (Editor), *Marine Food Chains,* University of California Press, Berkeley, CA.

Suzuki, T., Aoyama, H., Kai, M., Imao, K., 1998. Effect of dissolved oxygen deficiency on a shallow benthic community in an embayment. *Oceanography in Japan,* 7(4), 223-236. (in Japanese with English abstract)

Suzumura, M., Kokubun, H., Itoh, M., 2003. Phosphorus cycling at the sediment-water interface in a eutrophic environmnet of Tokyo Bay, Japan. *Oceanography in Japan,* 12(5), 501-516. (in Japanese with English abstract)

Tokyo Metropolitan, 1998-2002. *Result of the water quality survey of public water areas.* (in Japanese)

Valiela, I, 1984. *Marine Ecological Processes,* 1-546pp, Springer, New York.

Yamamuro, M., Koike, I., 1993. Nitrogen metabolism of the filter feeding bivalve Corbicula Japonica and its significance in primary production of a brackish lake in Japan. *Limnol. Oceanogr.,* 38, 997-1007.

Zillioux, E., 1970. Ingestion and assimilation in laboratory cultures of acartia. *Technical Report, the National Marine Water Quality Laboratory,* EPA, Narragansett, RI.

In: Water Purification
Editors: N. Gertsen and L. Sønderby

ISBN 978-1-60741-599-2
© 2009 Nova Science Publishers, Inc.

Chapter 3

BANK FILTRATION OF RIVERS AND LAKES TO IMPROVE THE RAW WATER QUALITY FOR DRINKING WATER SUPPLY

Günter Gunkel[*] and Anja Hoffmann
Berlin University of Technology, Dept. Water Quality Control,
Straße des 17 Juni 135, 10623 Berlin, Germany

ABSTRACT

Bank filtration is a relative low cost system for raw water treatment or pre-treatment for drinking water abstraction, used in European countries since about 140 years with very good experiences, but up to now knowledge of the physical, chemical and biological processes of water purification is still insufficient, especially under consideration of the application of this technology in other countries. Focus of interest are the mechanical, physicochemical, chemical and biological processes during infiltration pathway such as the retention of particulate organic material (POM), especially of algae cells with toxic cell compounds (cyanobacteria), the turnover of natural organic matter (NOM), bacteria and viruses, and the retention of toxicology relevant micro pollutants like cyanotoxins and drugs.

The water infiltration during bank filtration is not only controlled hydraulically but determined by severe clogging processes mainly triggered by accumulation of biological components in the upper sediment layer. Clogging of the interstice is regularly observed in infiltration ponds, and up to now several mechanisms can be distinguished like physical (input of fine sediments, building of gas bubbles), chemical (precipitation mainly of carbonates) and biological processes (excretion of extracellular substances by algae and bacteria). As a consequence water permeability of the interstice will become strictly reduced. The interstice are place of an adapted biocoenosis of bacteria, fungi, algae and meiofauna, which is characterized by the occurrence of extra-cellular polymeric substances (EPS). The meiofauna counteracts the clogging process by detritivorous activity. For many contaminants like DOM, POM, pathogens (*Giardia*,

[*] E-mail: guenter.gunkel@tu-berlin.de

Cryptosporidium) and cyanobacteria as well as cyanotoxins a good removal is given. The active sediment layer is the first meter of infiltration pathway with a decrease in DOC concentration of up to 50 % and high removal rates of $10^2 - 10^4$ for pathogens like bacteria and protozoa.

ABBREVIATIONS

DOC	dissolved organic carbon
DOM	dissolved organic matter
EPS	extra-cellular polymeric substances
FPOM	fine particulate organic matter
GWR	ground water recharge
ind.	individuals
k_f	hydraulic potential
LBF	lake bank filtration
NOM	natural organic matter
POC	particulate organic carbon
POM	particulate organic matter
RBF	river bank filtration
SSF	slow sand filtration

1. INTRODUCTION

On a global point of view excess to clean water is the most limiting factor for human life and prosperity, and many national and international efforts are done to enhance the actual situation. Thus, water treatment systems are in focus of interest, and in general, natural as well as cheap and simple technical systems are favoured in many countries. One simple technology for water treatment is bank filtration at the riverine or next to lake shore as well as natural and artificial ground water recharge systems (Figure 1). Bank filtration has a high potential for further applications in many countries and can serve as low cost technology for water treatment in developing countries as well as pre-treatment of raw water in advanced water treatment systems (Ray, 2008). But before implication of this technology worldwide, capacity for removal, especially for new contaminants such as protozoa or cyanobacteria, as well as the long-time efficiency related to quantity and quality of removal must be proved. Recently several bank filtration research programs such as NASRI (Natural and Artificial Systems for Recharge and Infiltration) with some further research programs were carried out in Berlin, Germany, NASRI II in India (see 'Berlin Centre of Competence for Water'), a 'Verbundprojekt Langsamsandfiltration' investigated slow sand filter systems (DVGW, 2006), and PROSAB (Programa de Pesquisa em Saneamento Basico) in Brazil had focus on water purification processes during infiltration and underground water passage (Pádua, 2006).

Bank filtration is an old water abstraction technology, and first large scale technical systems were installed at the River Clyde, Glasgow, Great Britain, in 1810. Flehe Waterworks, Great Britain, and Düsseldorf Waterworks, Germany, have been using bank filtration since 1870, the Dresden Waterworks since 1876. Due to increased water contamination and advances in hygienic, that means the knowledge of cholera as a sewage transmitted disease, this basic water treatment technology was applied in many European countries such as Czech Republic, Austria (River Danube), Germany (River Rhine, River Havel, River Elbe, River Ruhr) and Netherlands (River Rhine; Ray et al., 2003), but in some tropical countries too, e. g. in Brazil, Acude do Prata, Recife, build up about 1900. Today, these bank filtration systems are still working and are the basis of a water supply with high quality. In terms of yield, most important bank filtration sites are situated along the rivers Danube, Rhine, Elbe and at the lakes Tegel, Wannsee and Müggelsee.

Riverbank filtration (RBF) is the main water abstraction method used in many European regions, but, too, lake bank filtration (LBF) is applied at large scale in a few regions, e. g. in Berlin, Germany, where 2,600,000 habitants are served with LBF water as main water treatment since about 70 years, additionally only oxygenation and iron precipitation by water sprinkling and fast sand filtration for iron elimination are done. The long-time use of bank filtration led to a good practical knowledge in construction of wells and management of infiltration ponds. Many investigations of the RBF describe hydrogeological processes, water purification during infiltration and the occurrence of an interstitial flora and fauna, and verify a biological self-purification process in the sandy infiltration layer (Higgins & Thiel, 1988; Graham & Collins, 1996; Kühn & Müller, 2000; Emtiazi et al., 2004). A good understanding of the geohydrological situation in bank filtration is given, and a huge data collection of water chemistry within raw water and drinking water monitoring exists. But knowledge of physical, chemical and biological processes during bank filtration is still scarce, and only little information is available about turnover and retention of natural organic matter (NOM), that means particulate organic matter (POM) like algae cells and bacteria, dissolved organic matter (DOM) like cell excretion products (exo-enzymes, algae toxins) as well as of the non degradable rests of DOM (Beulker & Gunkel, 1996; Hakenkamp et al., 2002); especially the knowledge of infiltration and water purification at LBF sites is insufficient.

Figure 1. Lake bank filtration site, Lake Tegel, Berlin, Germany, with one well of a well gallery next to the lake shore.

Since many years bank filtration was only regarded as a black box and the good water quality at the abstraction wells did not force more detailed investigations about the processes of water purification, but the implementation of bank filtration to other sites and the occurrence of new contaminants (e. g. *Cryptosporidium parvum*, *Giardia intestinalis*, both protozoa, and toxic cyanobacteria) makes it necessary to investigate the capacity and effectiveness of raw water purification more detailed. Some relevant processes for water purification during bank filtration are

- mechanical retention of POM at the surficial sand layer of the infiltration zone,
- adsorption of NOM and anthropogenic contaminants on soil compounds during underground water passage to the abstraction well,
- biological degradation of NOM and anthropogenic contaminants during sand filter passage,
- decrease of bacteria and encysted bacteria (e. g. *E. coli*) due to limited survival time during the groundwater passage of about 50 days.

These processes are of some significance, but more specific knowledge is needed about the biological self purification and the related organisms, the extension of the zone of increased bioactivity and the structural components within the interstice, the biofilm. Some authors emphasize that most of the biodegradation occurs in the first meters of the bank filtration tranche (Beulker & Gunkel, 1996; Brugger et al., 2001; Hiscock & Grischek, 2002), an observation being of high interest for water treatment by bank filtration.

Since a few decades bank filtration has become more significant for ground water recharge (GWR), and natural as well as artificial systems in ground water recharge have been developed, especially in countries with a deficit in available water (Sens & Dalsasso, 2007; Dash et al., 2008; Ray, 2008). The experience existing in Germany validates RBF and GWR in different environmentally settings, but this can not be transferred to tropical and sub-tropical countries, which are characterized by high temperature of surface waters, extreme and periodically water level changes, and cyanobacteria dominance.

2. METHODS IN BANK FILTRATION INSTALLATION

The geohydrological situation of a bank filtration site is given by aquifers and more impermeable layers with little hydrological potential, and infiltration capacity is determined by surficial sediment characteristic. The theoretical permeability of sediment is estimated by Hazen's method (k_f = permeability in m sec^{-1}; Beyer, 1964) taking into account particle size distribution and water temperature.

The hydraulic potential is measured as k_f values according to Darcy's Law,

$$k_f = \frac{Q}{A} \cdot \frac{dL}{dH} \quad \left[\text{m sec}^{-1} \right]$$

Q = amount of infiltrating water per time (m^3 s^{-1}),
A = area (m^2),

dH = decrease of the hydraulic potential,
dL = length of the sediment segment for dH determination.

Observation wells and physical-chemical characterisation of the bore hole material will give a two or three dimensional mapping of the sediment layers, and a hydro-geological model can be developed to calculate the infiltration capacity. A good approach for determination of infiltration is the temperature oscillation between surface water and water of the observation wells. Determination of water sources and mean transit times can be done by isotopes and chemical tracers, for ground water recharges good application is given by chloride and isotopes like ^{222}Ra, ^{3}H, ^{2}H and ^{18}O (Bertin & Bourg, 1994; Seiler & Gat, 2007). Under natural conditions chloride in water is of atmospheric origin and concentrations differ according to rain intensities, evaporation and geological conditions. Natural tritium is produced in the atmosphere with a half-life of 12.3 years, whereas the open-air nuclear weapon tests tritium emissions are decreased close to natural concentrations and do not influence this method any more. The isotope ratios $^{2}H/^{1}H$ and $^{18}O/^{16}O$ are common parameters to characterize water, given in delta notation[1].

Natural variability of the stable water isotopes ^{2}H and ^{18}O are given by the meteoric water line, and isotope composition is given by altitude, evaporation of surface water and water chemistry interactions. Isotope enrichment or depletion follows straight lines in the $\delta^{2}H/\delta^{18}O$ diagram and also includes water mixing processes. Nowadays, too drugs and other anthropogenic contaminants are used to determine infiltration rates und ground water mixing processes during bank filtration (Heberer et al., 2004).

Direct in situ infiltration rates can be determined by use of plexiglass chambers pressed some centimetres into the sediment and connected pressurelessly to plastic bags, filled with a weighed amount of lake water, which will be reweighed after exposure time (Hoffmann & Gunkel, 2009a).

Positively and negatively charged melamine resin particles of a few micrometres labelled with 7-amino-4-methylcoumarin (AMC) can be used as particulate tracers for infiltration processes, particle size varies over a wide range and is only limited by light microscopy amplification (Gunkel et al., 2008).

Biological self purification is a more small scale process at lake water-sediment-interface during water infiltration, and sediment cores taken with acryl glass tubes of about 0.5 m length, divided in small layers of a few centimetres, allow physical, chemical and biological analyses in vertical extension. Analyses of the interstitial[2] fauna, the meiofauna[3] has to be done by the freeze core technique (Bretschko & Klemens, 1986): A lance, inserted in the sediment, is filled with liquid nitrogen, and a frozen sediment block is withdrawn and divided into small layers. Between driving the lance into the sediment and filling with liquid nitrogen. A new, successful applied method for sediment core analyses is the computer tomography, which gives a three dimensional image with a resolution of a few millimetres (see Chapter 4).

---

[1] $\delta^{2}H = \left(\dfrac{^{2}H_{sample}}{^{2}H_{standard}} - 1 \right) \cdot 1000 \quad \left[\%_{00} \right].$

[2] Sand pores system.
[3] Benthic organisms in the range from 0.03 - 0.1 mm to 0.5 - 1 mm; Higgins & Thiel, (1988).

A novel fluorescein-5-isothiocyanate (FITC) fluorescent labelling method allows the study of vertical particle transfer in sediments by application of FITC-labelled natural organic matter (e. g. algae, leaves) with a large particle size range (Gunkel et al., 2008).

3. BANK FILTRATION SYSTEMS: RIVERS, LAKES, INFILTRATION PONDS AND SLOW SAND FILTERS

Water infiltration as a natural recharge process of groundwater occurs if water infiltration conditions are sufficient, in general with a hydraulic potential k_f of $10^{-3} - 10^{-5}$ m sec^{-1}, that corresponds to middle to coarse sands. Today anthropogenic induced infiltration occurs, too, and different infiltration systems have to be distinguished,

- river bank filtration (RBF) in riverine zone,
- lake bank filtration (LBF) at lake shore,
- ground water recharge (GWR) with water charged artificial ponds, and
- slow sand filtration (SSF) in constructed filters.

RBF at the riverine is the most common system and is characterized by the dynamic of the flow regime of the river and coarse sands or even fine gravels in bank filtration sites. In general an infiltration occurs at the river bed too, but river bank and river bed infiltration is not differentiated. LBF can be done next to lake shore consisting of fine to middle sands, but in lakes sedimentation of organic and inorganic seston[4] occurs, leading to an accumulation of fine material at the lake bottom, and infiltration conditions vary with lake's water depth. Artificial ponds for GWR can be constructed at any sites with middle to coarse sand and good infiltration conditions, they are filled with surface water, and water level determines the hydraulic pressure and thus the infiltration. SSF are small constructed sand filter systems, which are charged with raw water; sand gains and hydraulic pressure determine filter characteristics. Main application of SSF is local water treatment with small quantities.

Water infiltration under natural and induced conditions is not only regulated by the sediment characteristics, but too by clogging of the interstice, a phenomenon being known from ground water recharge ponds, but too from rivers and lakes.

3.1. River Bank Filtration

RBF is a natural process, occurring if the water level of the river is above the ground water table, that means infiltration conditions exists; exfiltration conditions occur during low discharge of the river, that means water level of the river is beneath the ground water table. Thus, river infiltration conditions are dynamic and determined first by the surface water and ground levels, and second by the low and high flood conditions of the river (Figure 2). Formation of unsaturated conditions beneath the river occurs if the river bed is clogged due to accumulation of fine particulate material (most clayey material, in organic polluted rivers

---

[4] Suspended inorganic and organic (dead and alive) matter.

accumulation of organic matter, too), and if RBF water abstraction rates are not adapted to the hydraulic conductivity of the river bed and banks or if the ground water table decreases in a large-scale due to water abstraction at some places. The majority of RBF areas possess in- as well as exfiltration conditions, but the groundwater flow beneath the rivers is mostly neglected (Hiscock & Grischek, 2002).

The infiltration conditions, especially the clogging of the river bed by sedimentation of suspended clayey material, is regulated by floods: The periodically flooding due to snow smelt or rainy season causes a bed load and a resuspension of fine deposited material, thus the interstice is re-opened and the hydraulic conductivity of the river bed increases. Only periodically floods guarantee a long-time infiltration capacity, but flooding affects RBF in many other ways such as shock load of contaminants, high hydraulic pressure with reduced filter contact times, and destruction of the developed filter stratification as a complex biocoenosis within the pore system.

3.2. Lake Bank Filtration

LBF is scarce and only some sites are known such as Naimital, India, Enns Reservoir, Austria, Lagoa do Peri, Brasil, Lake Müggelsee and Lake Tegel, both Berlin, Germany, where most experiences exists (Brugger et al., 2001; Sens et al., 2006, Dash et al., 2008; Massmann et al., 2007). Lake Müggelsee is a shallow lowland lake of the River Spree and Lake Tegel a shallow lowland lake of the River Havel, water residence times are 63 days respectively 70 days. Infiltration conditions in lakes are determined by the colmation layer formed by lake sediments (calcareous mud in oligotrophic lakes, organic mud in eutrophic lakes, with an extension of many meters), and the bottom of the lake is clogged by these lake sediments; strait infiltration zones occur only at the lakes shores (Figure 3). In general the lake littoral zone serves as an infiltration system to ground water if ground water table is lower than lake water level; but LBF water is mixed with groundwater and infiltration water of the opposite lake side. Formation of unsaturated conditions beneath the lake's littoral zone occurs if groundwater abstraction rates are not adapted to the hydraulic conductivity of the littoral zone (= over exploration) or if the hydraulic conductivity of the lake shore is reduced due to clogging (= accumulation of fine particulate material in the littoral sediments). The introduction of air beneath the lake bottom by an unsaturated sand layer causes the aeration of anoxic lake sediments, which can occur due to high oxygen consumption of infiltrated water, and oxic conditions are re-established from the deeper sediment layers upwards to the lake bottom. Unsaturated sediments beneath the lake lead to a reduction of the hydraulic potential and to a decrease of the infiltration rates.

In contrast to RBF intensive water level changes with high flood rates and re-opening of the interstice does not occur in LBF, but in the shallow littoral zone of a lake, wind induced waves lead to a resuspension of fine sediment components and a re-opening of the interstice; a wave with an amplitude of 20 cm still affects the sediment in a water depth of 2 m with a resting orbital movement of 0.6 cm. The resuspended sediment particles will be distributed in the whole lake by wind induced currents and will finally settle in deeper lake areas.

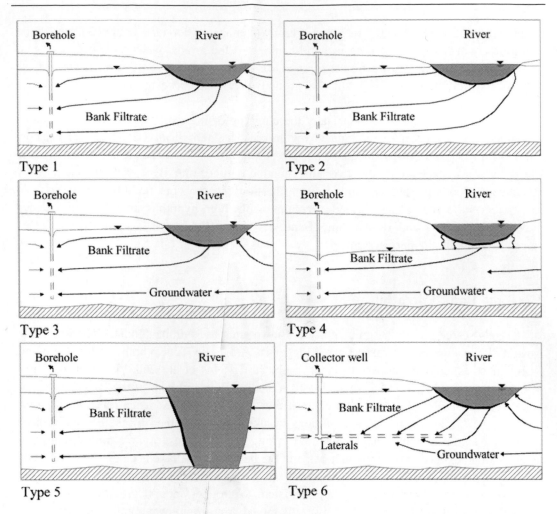

Figure 2. Schematic representation of different types of flow conditions at river bank filtration sites. Type 1: River serve as drainage system with infiltration conditions, but at the filtration site, exfiltration occurs. Type. 2: River with exfiltration conditions. Type 3: River with exfiltration and infiltration and a strong groundwater flow beneath the river. Type 4: Formation of unsaturated conditions beneath the river occurs if groundwater abstraction rates are not adapted to the hydraulic conductivity of the river bed or if the hydraulic conductivity of the river bed material becomes clogged due to surface water pollution inputs. Type 5: The river bed cuts into the confining layer and the filtration site is completely isolated from the groundwater flow. Type 6: Bank filtration with lateral wells. From Hiscock & Grischek (2002).

At Lake Tegel infiltration site, a two years investigation was carried out at different local positions (in front of *Phragmites*, erosion and water lily stands) and indicated a small scale differentiation of the hydraulic permeability (Figure 4), the hydraulic potential k_f varied from $3 \times 10^{-5} - 8 \times 10^{-8}$ m sec^{-1}; the mean infiltration rate were 9 L m^{-2} h^{-1} (0.7 – 27.0 L m^{-2} h^{-1}), leading to a infiltration velocity of 0.5 m day^{-1} with a variance of 0.05 – 1.8 m day^{-1}. In general, summer infiltration rates were about 10^1 times higher than in winter period. The well abstraction rate during normal production conditions did not influence significantly the infiltration rate, a consequence of the subsoil water mixing processes as well as of the very important clogging factor. Exceeded water abstraction rates led to unsaturated soil beneath the lake and to a decreased hydraulic conductivity as mentioned above (Hoffmann & Gunkel, 2009a).

At Lake Tegel the absolute infiltration area for one well is 5,500 m^2, which means about 0.3 m^2 inhabitant^{-1}. Detailed sediment investigations at Lake Müggelsee, Berlin, Germany, confirms, that only a small littoral zone < 5 m depth with k_f > 10^{-5} can support bank filtration (Massmann et al. 2008a).

A hydrogeological model of the bank filtration site in Lake Tegel is given by Massmann et al. (2008b) pointing out a vertical differentiation of the groundwater: The shallow groundwater observation wells capture bank filtration water with a transit time of at least 4 - 5 month, deeper groundwater observation wells deliver water with an age of some years to a few decades. Groundwater underflow of the lake occurs, even being more then 1 km in width. Due to mixing of these distinct groundwater flows, the portion of bank filtration water from the proximate lake shore amounts only 48 %. Field studies on the fate and transport of pharmaceutical residues in Lake Tegel bank filtration site confirm these water transit times (Heberer et al., 2004).

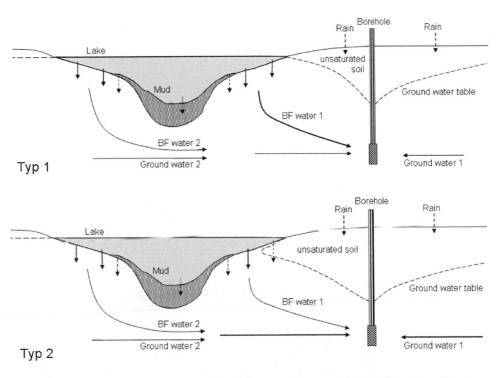

Figure 3. Schematic representation of two different types of flow conditions at lake shore filtration sites. Type 1: The lake littoral zone serve as a drainage system and bank filtration water is mixed with groundwater and infiltration water of the opposite lake side. Type 2. Formation of unsaturated conditions beneath the lake littoral zone occurs if groundwater abstraction rates are not adapted to the hydraulic conductivity of the littoral zone or if the hydraulic conductivity of the lake shore becomes clogged, an aeration of the lake sediment from beneath occurs.

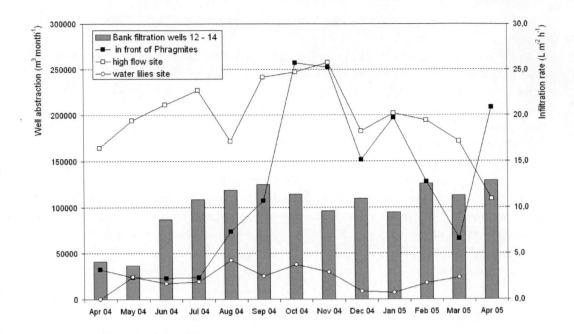

Figure 4. Annual variation of Lake Tegel, Berlin, Germany, bank infiltration rates at different locations (in front of *Phragmites* stands, at a palisade passage with high water flow, and in a water lilies stand, close together < 20 m), and the variance of well abstraction rate (well no. 12 – 14, 100 m beside the lake shore, in front of the experimental area).

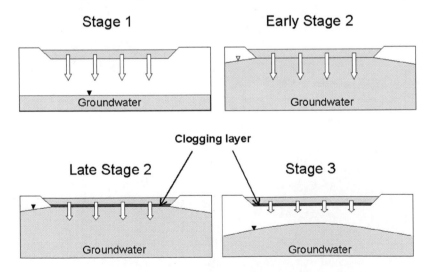

Figure 5. Schematic infiltration situation of an artificial groundwater recharge pond near Lake Tegel, Berlin, Germany. Stage 1: Surface water infiltrates at the beginning of the recharge cycle, the infiltration rate is maximal. Early stage 2: Saturated conditions have been established. Late stage 2: A clogging layer has formed at the bottom of the ground water recharge pond and infiltration rate is slightly reduced, still saturated conditions occur. Stage 3: Hydraulic resistance of clogging layer is too high, and an unsaturated zone develops beneath the pond and infiltration rate decreases rapidly, the groundwater table declines. From Greskowiak et al. (2005).

3.3. Ground Water Recharge Ponds and Slow Sand Filters

Worldwide ground water recharge ponds are used to stabilize and increase the ground water level as well as to guarantee sufficient water purification during infiltration process (DVGW, 2006). But the nutrient content of the raw water (mainly N and P) lead to a high primary production in infiltration ponds and more eutrophic conditions will be established in the ponds. Due to low water depth of 1 to 2 m and light penetration to the bottom, dominantly benthic algae develop and form a dense layer at the sand surface, the so called schmutzdecke. This leads to clogging of the water-sand-interface, 1) mechanical by the dense growth of algae, and 2) by the precipitation of calcium carbonate due to increase of the pH during photosynthesis (= biological Ca-precipitation).

The infiltration conditions of artificial recharge ponds are characterized by the time period after start to work, and three phases have to be distinguish: An open sand filter without bioactivity, the development of an interstitial biocoenosis with rapid clogging and reduced infiltration capacity, and a complete clogging, that makes necessary a removal of the upper sand layer. Greskowiak et al. (2005) verified this cycle for the Tegel recharge pond, Berlin, Germany (Figure 5). At the beginning of the recharge cycle the infiltration rate is maximal and saturated conditions have been established beneath the pond. During GWR working, a clogging layer has been formed at the bottom of the GWR pond and infiltration rate is reduced: With an increasing hydraulic resistance of the clogging layer, an unsaturated zone develops beneath the pond and infiltration rate decreases rapidly, a removal of a few centimetres of the surficial sand layer becomes necessary. This surficial sand is washed and re-deposited in the pond several times per year.

In GWR ponds infiltration velocities range from $0.1 - 0.2$ m h^{-1}, and as maximum 0.5 m h^{-1} can be reached. These high infiltration rates are realized by the water pressure of 1 - 2 meters of the infiltration pond.

A similar, small scale construction can be used as SSF with the so called direct filtration, being used for decentralized water treatment; two different technologies are applied, SSF with up flow or down flow (Graham & Collins, 1996; Gimbel et al., 2006; Bernardo et al., 2006).

4. CLOGGING OF THE INFILTRATION ZONE

Clogging is a well known process in slow sand filters, where the hydraulic permeability decreases during working period and lead to an extreme reduction of the infiltration capacity; visually this process leads to the formation of the schmutzdecke, a surface layer of micro-organisms, algae, especially filamentous algae and particulate organic matter (POM; Figure 6). Clogging can be caused by some mechanical, chemical and biological processes, and is the limiting factor for infiltration capacity in many bank filtration sites.

Well-known mechanical factors for clogging are the input of fine sand particles (silt, clay), fine particulate organic matter (FPOM), and the oversaturation and development of gas bubbles (oxygen by primary production or methane by methanogenese) in the sand pore system. Too, occurrence of unsaturated sand filter conditions leads to a severe decrease of the hydraulic permeability.

Chemical processes for clogging are mainly the precipitation of calcium carbonates, iron oxyhydroxy compounds, sulphur polymers and iron sulphides. Significant biological processes for clogging are the development of biomass within the interstice (algae, bacteria) and the excretion of extra-cellular polymeric substances (EPS), in most cases polysaccharides and polypeptides (Flemming etal., 1999). The clogging processes are strongly determined by photoautotrophic and heterotrophic production, but too by conversion and decomposition of FPOM in the interstice. Different pathways for the entry of FPOM into the sediments have to be distinguished, namely passive settling on the sediment surface layer, the bioconversion of POM to FPOM, passive transport by convergent infiltration currents and an episodically burial after sediment movements (Rinck-Pfeiffer et al., 2000; Langergraber et al., 2003). This intrusion of sestonic[5] matter, suspended in river or lake water, is often considered to be the most important factor for clogging (Hiscock & Grischek, 2002; Langergraber et al., 2003). According to Okubo & Matsumoto (1983), the concentration of suspended solids should not exceed 2 mg L^{-1} to avoid rapid mechanical clogging. But too, re-opening of the interstice is observed, in rivers by high flood and occurrence of bed load, and in lakes by the resuspension of fine particles by wind waves in the littoral zone.

Clogging of the interstice under natural, but induced infiltration conditions at Lake Tegel, Berlin, Germany, is caused to a large extent by POM, which is composed of living biomass such as epipsammic diatoms, forming the biofilm, fine rhizomes of macrophytes and detritus[6]. This biological clogging reaches down to a sediment depth of least 10 cm. The interstice of the sandy littoral zone is filled up to about 50 % with POM such as detritus, living bacteria and algae cells (Figure 7).

Figure 6. Schutzdecke of a groundwater recharge pond near Lake Tegel, Berlin, Germany, after 4 months service with removal of the surface sand layer.

---

[5] Seston = suspended inorganic and organic (dead and alive) matter.
[6] Dead organic matter.

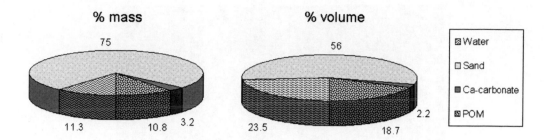

Figure 7. Mean mass and volume fractions of the water-sand-boundary layer (0 - 8cm depth) at the bank filtration site at Lake Tegel, Berlin, Germany (mean of 1 year, n = 16).

Within the interstice a biofilm is developed, consisting of an adapted biocoenosis of bacteria, algae and small invertebrates, living in the pore system of the sand (Beulker & Gunkel, 1996). An excretion of EPS occurs, build up of carbohydrates and/or proteins, which form complex three-dimensional structures within the interstice (Figure 8).

Figure 8. Water-sand-boundary layer at the bank filtration site at Lake Tegel, Berlin, Germany, accumulation of planktonic algae cells, mostly diatoms, and detritus flocs on the superficial sand layer, 0 - 1 cm depth, Lake Tegel bank filtration site, Berlin, Germany (27.04.2004).

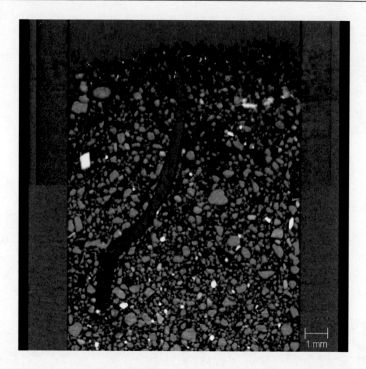

Figure 9. CT of a sediment core at the bank filtration site at Lake Tegel, Berlin, Germany, blue = overlying water and water in the interstice, white = Ca-carbonate precipitations, grey = sand grains, a tunnel built by an invertebrate and filled with water is recognizable. The CT technology was developed and applied by the BAM. Berlin, Germany.

In both, rivers and lakes, the burrowing activity (bioturbation) of the meiofauna is of very high significance to maintain a hydraulic permeability (Figure 9). Due to the mobility of the meiofauna and some migration upwards and downwards interstice is re-opened. Furthermore parts of the POM, being in the pore system, are consumed by these organisms. The CT[7] graph of Lake Tegel, sediment demonstrates the nearly complete filling of the interstice by POM and shows a typical tunnel formed by sediment living organisms.

At Lake Tegel infiltration site, complete clogging did not occur during the two years investigation period, but the in situ hydraulic permeability decreases about 10^{-2} m sec^{-1} compared with the inert sand interstice, thus a steady state of clogging processes by the development of the biocoenosis and the re-opening by bioturbation occurs, that means, a local high accumulation or development of POM is more attractive for sediment feeders, and they will migrate into this area, a phenomenon well known as patchiness in river and lake littoral ecology (Yamamuro & Lamberti, 2007). However, experimental data about this dynamic interaction, the build-up of EPS and the destruction of POM and EPS by interstitial fauna are scarce. Most of the available information concerns artificial GWR systems, where the inflow conditions and the turn over processes in the sediment boundary layer are significantly different from those of natural littoral zones.

---

[7] Computer tomography.

5. SEDIMENT BOUNDARY LAYER

Over the past few years the benthic boundary layer has been studied intensively, as chemical gradients are most obvious and microbial activities are maximal in this layer supporting biofilm growth (Beulker & Gunkel, 1996; Hiscock & Gruschek 2002). The biofilm is a complex community of algae, bacteria, fungi and invertebrates, living in the interstice system: bacteria and algae produce extra-cellular polymeric substances (EPS), forming a dense three-dimensional structure (Lawrence et al., 1998, 2002; Paterson, 2001). The development of a high structured biofilm, consisting of several groups of organisms occurs only under oxic conditions; the limiting oxygen concentration is not yet knows. Do to the occurrence of algae, a daily cycle with oxygen enrichment during day and oxygen depletion during night must be considered, thus critical oxygen concentrations are reached only at night. Thus, the development of a dense layer of superficial filamentosus algae like the schmutzdecke in GWR ponds will in general impact the development of a high diversity biofilm because of light limitation in the interstice and, as a consequence, decrease of the interstitial algae as well as by an increased probability of anoxic conditions during night (oxygen depletion due to respiration of the schmutzdecke's algae). Concerning oxygen balance, two different states have to be clearly distinguished, anoxic or anaerobic infiltration water and the formation of a micro-zoning in the pore system with small anoxic and anaerobic zones. Microscopic analyses point out, that too under oxic conditions a micro-zoning occurs with some anaerobic areas maybe in dead end pores, verified by the presence of iron sulphide (pyrite).

Biofilms and the associated algae are of special importance for water purification processes due to a net oxygen production as well as an adsorption of ions like toxic metals and of dissolved organic matter (DOM). The bacterial community consume and metabolize DOC, and the excretion of exoenzymes leads to an elevated mineralization efficiency of DOM and some inorganic polymers (e. g. polyphosphate, colloids; Decho, 2000; Wingender & Flemming, 2001). Other well-known properties of biofilms are the high water binding capacity and the stabilization of the surrounding sediment grains (Yallop et al., 2000). The biofilm has a three dimensional structure and is the basis for a micro-zoning of the filter area (Flemming et al., 1999), dead end pores occur and serve in retention of small particles (Auset & Keller, 2006), but too enable the formation of oxygen mirco-zoning in the interstice; some degradation processes, e. g. of drugs, are linked to a small scale change of oxidative and reductive conditions in the filter.

The most important factor for the spatial build up and the function of the biofilm is the excretion of EPS by bacteria and algae as 'housing'. Structural determinations of EPS forming biofilms have shown substantial progresses during the last decades, and neutral and polyanionic polysaccharides as well as peptids (glycolproteins, lipoproteins) form EPS (Characklis et al., 1990; Wingender et al., 1999; Flemming & Wingender, 2001; Sutherland, 2001). The EPS is a secretion of algae (e. g. diatoms excrete mucus for movement or cell sheaths) or part of the cell structure (e. g. lipopolysaccharides of gram negative bacteria). EPS from hydro-gel to viscose elastic structures like filaments, nets and plaques (Figure 10). A conversion of EPS, that means a re-construction due to enzymatic degradation by hydrolase and condensation of the formed oligosaccharides can occur, but in general the stability of the

EPS is very high, and EPS structure consist over some weeks to months, which mean the EPS outlive the builders for a long-time.

The biofilm biocoenosis is built up by bacteria, fungi and algae, whereas the vertical distribution of the algae is limited by the transparency, which means the light penetration into the interstice. Bacteria and fungi settle too in deeper, aphotic zones; bacteria build up an adapted community, on one hand related to the DOC input and its degradability and on the other hand by the micro-zoning in the three dimensional interstice system. It must be assumed that the bacterial community is to a high degree a local one, but up to now only few investigations are available (Kolehmainen et al., 2007). The abundance of bacteria at Lake Tegel, Berlin, Germany, was very high, and they reached up to 2×10^9 cells gram^{-1} sediment, determined with DAPI[8] fluorescence technique; the vertical distribution showed highest cell numbers at the surface layer of 0 – 5 cm, while in depth of > 20 cm, still 0.2×10^9 cells gram^{-1} sediment were found. Further specification of these bacteria can be done using FISH[9] or PCR[10] (Spring et al., 2000, Emtiazi et al., 2004).

The bacterial community in the interstice seems to be a complex system of linked species, adapted to DOM characteristics, fixed in the three dimensional EPS structure and being at least partly located in dead end pores, this give a analogous community to the bacterial flocs in surface water (Zimmermann-Timm, 2002).

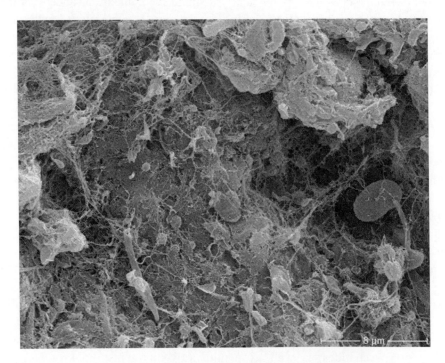

Figure 10. Biofilm with some algae and bacteria cells and a dense fibrillose net structure of extracellular polymeric substances (EPS) in the interstice of the water-sand-boundary layer at the bank filtration site at Lake Tegel, Berlin, Germany (depth of 3 – 4 cm, 17.05.2004).

---

[8] 4'6-diamidino-2-phenylindole-2HCl, a fluorescent dye for DNA.
[9] Fluorescence in situ hybridisation.
[10] Polymerase chain reaction.

Beside bacteria cells activity, the bacterial exo-enzyme activity (e.g. aminopeptidase, glucosidase, phosphatise) is of high significance for degradation processes of DOM, too (Miettinen et al., 1996, Hendel et al., 2001).

The investigations carried out in the fine sand infiltration site of Lake Tegel, Berlin, Germany, offered a high portion of POM with a maximum at the surface layer with 15 mg g^{-1} sediment, stretching down to 50 cm (Figure 11; Gunkel et al. 2008). Epipsammic algae[11] occur with a high biomass in the upper interstitial zone of about 0 – 6 cm depth, and it has to be pointed out that planktonic algae species from the lake water are only transported into the interstice to a small extent (Beulker & Gunkel, 1996; Gunkel et al., 2008). Thus this water-sediment-boundary layer serves as a mechanical filter for these algae cells – a process being of very high importance for the behaviour and fate of all types of POM in bank filtration. This leads to

1. the surficial accumulation of algae cells,
2. an easy resuspension of surficial deposited algae,
3. an insignificant penetration of algae cells into the interstice,
4. a high attraction of the surficial sediment layer for herbivorous interstitial fauna with its burial activity,

oxygen concentrations of deeper sediment layers are not reduced by POM mineralisation but only by DOM concentration and mineralisation.

Concerning toxic cyanobacteria, too, an accumulation at the surficial sediment layer must be expected (see Chapter 6.6).

The occurrence of interstice algae lead to a natural bioproduction in the small photic surface layer of a few centimetres, and both, POC as well as DOC of the infiltrating water is influenced by the in situ production of POC and DOC (Hoffmann & Gunkel 2009b). The vertical distribution of chlorophyll (Chl a) confirmed that interstitial algae biomass forms a significant part of the total POM in the upper interstitial zone of 0 – 6 cm (Figure 12), and primary production is assumed to be the most important source of organic carbon in the interstice. Up to now only biomass and no turnover data of interstitial algae are available, too a lack of information exists concerning pico-algae[12] (Dittrich et al., 2004).

At Lake Tegel the Chl a concentrations in the upper 5 centimetres of sediment were very high (21 to 28 μg cm^{-3}) and decreased with depth, only traces of Chl a were detected below 5 cm depth. A Chl a concentration of about 25 μg cm^{-3} must be evaluated as extremely high compared with the lake water, which even under eutrophic conditions contains only about 20 μg L^{-1} Chl a. Thus, the total algal biomass in the upper sandy layer of the interstice was about 1000 times higher than in the corresponding water body of Lake Tegel (Gunkel et al., 2008).

---

[11] Algae attached to sand grains.
[12] Pico-algae are cells between 0.2 and 2 μm that can be either photoautotrophic of heterotrophic.

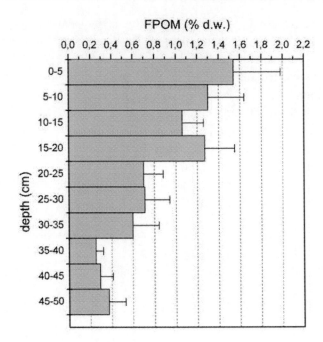

Figure 11. Depth distribution of FPOM (< 1000 μm) in the interstice of the sandy lake sediments at the bank filtration site at Lake Tegel, Berlin, Germany (means + SD, n = 6, 2004). From Gunkel et al. (2008).

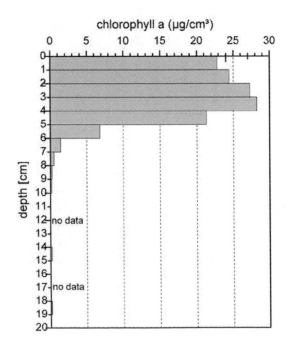

Figure 12. Depth distribution of chlorophyll-a concentrations in the interstice of the sandy lake sediments at the bank filtration site at Lake Tegel, Berlin, Germany (means + SD, n = 1, March to August 2004). From Gunkel et al. (2008).

The origin of these algae were epipsammic[13] species, mostly diatoms such as *Fragilaria* spp., *Achnanthes* spp., *Cocconeis* spp., *Amphora pediculus*, *Cymbella* spp., *Gomphonema* spp., *Rhoicosphenia abbreviata* and *Cymatopleura* spp.; some other algae classes such as Chlorophyceae, Cryptophyceae and Cyanobacteria occurred (Gunkel et al., 2008). Of high interest is the low significance of planktonic[14] algae of the lake water, only three times during the investigation period of one year a significant portion of planktonic algae was observed within the interstice with < 24 % of the total algae abundance; in general the relative abundance of the planktonic algae within the interstice was < 3 % (Gunkel et al., 2008).

In the interstice a food web is build up due to the presence of the meiofauna, small invertebrates living in the interstice, being bacteria, algae or detritus feeder, some of them are carnivorous, too (Wotton et al., 1996; Beulker et al., 1996; Gibert et al., 1998). Fishes like carp serve as predator to the meiofauna. Thus a cycling of organic carbon is build up within the infiltration stretch, and DOC is transferred to POC, but the total amount of organic carbon is reduced due to respiration of the organisms, which means a transformation of organic carbon to CO_2. A first analysis of turnover processes point out the high significance of the interstitial food web (Hoffmann & Gunkel, 2009b).

The abundance of the meiofauna in an undisturbed bank filtration site is very high: In Lake Tegel 19,600 ind. dm^{-2} sediment area were found as maximum, the annual mean was 8,300 ind. dm^{-2}; this meiofauna abundance corresponds to that of some other Berlin lake littoral zones with a maximum of >20.000 ind. and a mean of 4,400 ind. dm^{-2} during a three years study (Beulker & Gunkel, 1996). The vertical distribution of the meiofauna showed two different zones, one with very high abundances in 0 – 10 cm depth and the other with a significant decrease in abundance and diversity in depths > 15 cm (< 200 ind. dm^{-3} in 15 – 20 cm). Thus the abundance of the meiofauna amounts up to 5,000 – 15,000 ind. L^{-1} sediment – an animal density never reached in surface water.

The meiofauna consists of many different species belonging to different classes and orders of fauna such as flatworms (Tubellaria), rotifera (Rotatoria), roundworms (Nematoda), snails (Gastropoda), mussels (Bivalvia), worms (Annelida), water bears (Tardigrada), seed shrimps (Ostracoda), water fleas (Cladocera), copepods (Copepoda) and insect larvae (Chironomidae, Nematocera), a detailed description of the meiofauna of Lake Tegel is given in Beulker & Gunkel (1996).

6. REMOVAL OF CONTAMINANTS

Bank filtration is used as a pre-treatment as well as a simple and unique treatment of surface water for drinking water supply, and nowadays these two different applications are of increasing significance as a first step to reduce some pathogens like faecal bacteria and protozoa, especially the new protozoa *Cryptosporidium parvum*[15] and *Giardia intestinalis*[16].

---

[13] Algae species attached to sand grains or other particles.
[14] Algae species suspended in surficial water.
[15] *Cryptosporidium parvum* is a one-celled microscopic and obligate intracellular parasite, and a life cycle exists with asexual and sexual reproductive cycle in the intestine of an individual host (humans and domestic mammals), robust oocysts are excreted in the faeces and infect host. Waterborne Cryptosporidiosis was detected in 1984 and up to now many outbreaks of waterborne Cryptosporidiosis are known. Humans impacts

In developing countries bank filtration can serve as unique treatment for a local, not centralized water supply. Thus, bank filtration water purification processes are of high significance, and the retention of contaminants, the filter capacity and the ecological stability of the self purification process by the interstice community must be evaluated, and in a suitable bank filtration management optimized.

Water purification processes during bank filtration are influenced mainly by

- the infiltration rate and the extension of the active sandy filter zone, that together determine the effective contact time of contaminants with the micro organisms,
- geological and hydrogeological conditions (e.g. pH, redox potential, sand grain surface characteristics),
- characteristics of the interstice, especially the filling of the interstice with inorganic and organic matter as well as the formation of dead end pores,
- the interstitial flora and fauna as a highly adapted community of bacteria, fungi, algae and small invertebrates.

Bank filtration is of high complexity, and different physical, chemical and biological water purification processes must be distinguished,

- mechanical filtration of the water by the superficial sand layer, sand granulation determines the filter pores and particle retention,
- physical-chemical adsorption of FPOM such as viruses and bacteria to the interstice surface, the so called straining,
- ingestion of POM and FPOM by detritus feeder and build up of biomass, that means the carbon conversion into the intersticial food web,
- assimilation of DOC by heterotrophic bacteria and by some mixotrophic protozoa and algae,
- microbial degradation of DOC by bacteria and their exo-enzymes,
- adsorption of dissolved inorganic and organic components (ions, chelated ions, DOC) of the water to surfaces of the interstice,
- assimilation and turn over of dissolved inorganic compounds (mainly NH_4^+, NO_3^-, SO_4^{2-}) by micro-organisms,
- precipitation and co-precipitation of inorganic and organic contaminants, mainly as calcium carbonate and Fe-oxihydroxo complexes,
- degradation of some anthropogenic organic substances (drugs) during the groundwater passage by redox chemical processes.

Use of surface water for drinking water production is common in many regions, especially due to the limited resources of aquifers, but in general raw water quality is inferior

---

are gastro-intestinal crytosporidiosis, a severe watery diarrhea, and pulmonary and tracheal crytosporidiosis with coughing and fever.

[16] *Giardia intestinalis* is a one-celled microscopic parasite and cause diarrhea illness, the parasite lives in the intestine of humans or animals and is passed in the faeces, it can survive outside the body for long periods of time. During the past 2 decades, *Giardia* infection has become recognized as one of the most common causes of waterborne disease in humans in the United States, *Giardia* is found worldwide.

to that of ground water. Contaminations occur by direct emissions (sewage, waste water) as well as by diffuse emissions (agricultural land use, animal farming, percolation of contaminated soils, air pollution). Additionally, contamination is caused by some natural processes like emission of humic acids (mineralisation of plant biomass and of soils rich in organic) and by eutrophication processes after introduction of nutrients (N, P and organic matter) in lakes and low land rivers. Thus many different contaminants are of significance for drinking water quality and have to be analysed and evaluated for the behaviour in bank filtration processes, especially

- physical effects like surface water heating,
- inorganic contaminants, especially metal ions, NH_4^+, NO_3^- and SO_4^{2-},
- DOC as a summary parameter for natural and anthropogenic compounds (e. g. humic acids, excretions products of bacteria and algae, degradation products of POM and sewage), this includes the disinfection by-product precursors, and drugs as a functional group of special interest,
- POM such as leaves and other plant rests,
- pathogens like viruses, bacteria (coliform bacteria) and new protozoa like *Giardia intestinalis* and *Cryptosporidium parvum*,
- toxic cyanobacteria and extracellular cyanotoxins.

6.1. Temperature Equalisation

In most bank filtration sites surface water has a yearly oscillation of water temperature, but due to the climate change, water temperatures increases up to now about 2 °C (as annual mean) in Central European countries in the last three decades, but in rivers and lakes, especially in the epilimnion[17] of lakes water temperatures increase significantly, and such high temperatures as 30 °C were observed[18] in the epilimnion of Lake Müggelsee, Berlin, Germany (Adrian et al., 2006; Gunkel, 2009; Groß-Wittke et al., 2009). During bank filtration a mixing of surface water and ground water occurs, leading to lower temperatures of raw water compared to surface water in summer, and vice versa in winter. This temperature equalization amount several degrees in German regions (Schmidt et al., 2003).

6.2. Retention of Inorganic Compounds

Inorganic trace elements such as iron, manganese and several heavy metals are eliminated during sand filter passage mainly by adsorption processes to negatively loaded clay surfaces, precipitation with amorphous iron oxi-hydroxides and alumina hydroxide, as well as to POM, especially the EPS, in the interstice; additionally co-precipitation with calcium carbonates and iron oxi-hydroxides occurs. Under anoxic conditions chemical reduction of Fe^{3+} and Mn^{4+} occurs, and the water soluble ions Fe^{2+} and Mn^{2+} are transported by the water flow into the groundwater; furthermore co-precipitated heavy metal ions that are

---

[17] Warm superficial water layer.
[18] Adrian, Leibniz-Institute of Freshwater Ecology and Inland Fisheries, Berlin, pers. comm.

also mobilized, and metal ion migration into groundwater will take place. Under anaerobic conditions a precipitation of metal ions occurs as sulphides. Most heavy metal sulphides are poorly soluble in water, and the metal sulphides are more inactive, which means the re-oxidation to the soluble ions occurs slowly. During bank filtration reduction of heavy metal concentration was significant in River Rhein with > 80 % for Cr, As, Zn, with 50 – 80 % for Pb, Ni, Cd, Cu) and with < 50 % for Hg, Sn, Se (Schmidt et al. 2003).

Too, some other ions, characteristic for surface water, underlie redox-chemical and bio-chemical reactions during bank filtration: Nitrate is reduced by microbial activity to ammonia (nitrate reduction), done by many bacteria species, or nitrate is reduced to nitrogen (N_2) as denitrification process done by some bacteria genera, like *Nitrococcus denitrificans* (Grischek et al. 1998). Denitrification is a nitrate respiration occurring in an anoxic micro milieu, and dissolved organic substances are used as a carbon source. Thus, this process leads to a decrease of NO_3^- and DOC, and a rough calculation gives a decrease of 1 mg L^{-1} DOC by bacterial reduction of 1 mg L^{-1} NO_3^-.

If the redox potential of the sediment decreases, sulphate reduction (sulphate respiration) by bacteria of the genera *Desulfovibrio* and *Desulfotomaculum* occurs, strictly anaerobic bacteria, and the very toxic sulphide ion is produced. Under these anaerobic conditions CO_2 as well some acetic acids are reduced to methane by *Methanobacterium*, *Methanobacillus*, *Methanococcus* and others bacteria, too, and the redox potential in the sand filter becomes of high significance for water purification (Groß-Wittke et al., 2009). Both processes, the sulphate reduction and the methanogenese, will lead to a severe impact respectively to a complete break down of the interstitial biocoenosis as well as of the groundwater fauna. Increased input of DOC as well as water heating with increased bioactivity leads to more oxygen depletion and the risk of anaerobic processes during bank filtration increases, being observed already in a GWR system (Massmann et al,. 2008a).

6.3. Degradation of DOC

The input and turn over of DOM respectively DOC during bank filtration is of high significance because raw water should not be contaminated with a high concentration level of DOC, because in general DOC serves as desinfection byproduct precursors. DOC is part of natural organic matter (DOM) and removal is done by abiotic processes like adsorption and flocculation as well as by biotic processes like carbon assimilation by heterotrophic and mixotrophic organisms (bacteria, some algae and invertebrates). Investigations carried out in aquatic systems point out that a very high turn over rate of DOC occurs, the so called microbial loop, and it must be assumed that this microbial loop is the most significant process in interstice, too (Gibert et al., 1994). DOC concentration decreases during bank filtration, and investigations point out, that the most effective decrease occurs during the first meters or even decimetres of infiltration stretch; this was observed in sand columns experiments as well as in observations wells during bank filtration, but only parts of the DOC is removed, the easily degradable DOC. During the following groundwater passage some additional DOC is removed, but only by a low rate (Brugger et al., 2001). The polysaccharide LC-OCD[19] fraction of the DOC is very fast biodegraded during infiltration, while other LC-OCD

---

[19] Liquid Chromatography - Organic Carbon Detection.

fractions like humic acids and low molecular weight acids are only partly removed (Grünheid et al., 2005; Hoffmann & Gunkel, 2009b).

In Lake Tegel, Berlin, Germany, the DOC is removed from about 8 mg L^{-1} to 4 mg L^{-1} during bank filtration (Grünheid et al., 2005), about 20 % of the DOC is removed after a sediment passage of 26 cm (Hoffmann & Gunkel, 2009b), whereas in the Enns River, Austria, the DOC decreases from 1.2 mg L^{-1} to 0.6 mg L^{-1} (Brugger et al., 2001).

The occurrence of drugs is of high significance in waters influenced by sewage input, emissions from animal farming or introduction of clear water from sewage treatment plants. Investigations carried out at Lake Tegel pointed out that the main factor for the removal of drugs is the biodegradability under oxic or anoxic conditions, and some drugs occurring in surface waters like primidone (anti-epileptic drug) are not removed during bank filtration and ground water passage, while others like prophenazone (anti-inflammatory drug) are removed already within the first infiltration section (Heberer et al., 2004). Most drugs are water soluble and persistent to allow a stable concentration in the organisms (animals, humans), thus the removal in sewage treatment plants is limited, and drugs in general possess a high environmental risk potential.

6.4. Vertical Transport of Particulate Organic Matter

In bank filtration three vertical transport processes for POM (which includes bacteria, algae cells, and protozoa) have to be carefully separated from each other,

1. POM particles can be transported by water currents towards deeper sediment layers,
2. fragmentation of crude POM to FPOM, and the favoured vertical transport of the smaller particles, and
3. transport of the FPOM by organisms after ingestion of organic material and excretion of faeces after migration of the organisms to deeper sediment zones is, at present, still an unknown factor.

In situ sediment core experiments in Lake Tegel, Berlin, Germany, indicated significant retention of inert resin particles, used as tracers, regardless of whether positively or negatively charged particles were used, or whether the time of exposure was 3 or 14 days; the resin particles had a diameter of 2.44 μm and were therefore of a similar size to that of very small algae cells. Compared with the effective mean pore diameter (59 – 154 μm) and pore water velocity of 0.5 m d^{-1}, a similar particle transport was expected, however the resin particle transport was about 55 times slower than the pore water velocity, and a considerable proportion of particles (about 75 %) is retained permanent in the upper sediment layer of 2 cm (Hoffmann & Gunkel, 2009a). The results indicate that the vertical transport of particles is restricted to the upper sediment layer, even microfine resin particles undergo much slower migration through the pore system than expected from the calculated pore water velocity.

The observed high restriction of small particles to the sediment surface stands in contradiction to the pore size of the sediment and the infiltration velocity of the pore water. This retention effect was too proved by natural FPOM of different size classes (leave fragments, algae cells), labelled with fluorescence dye, the vertical transport of FPOM with the water flow is without significance (Gunkel et al., 2008).

At sediment depths down to 20 cm, POM fragmentation due to detritivorous activity by meiofauna and macrozoobenthos was observed the most probable mechanism leading to the vertical distribution of POM (Figure 11). The interstice built an effective filter system for particle retention, and clogging mainly induced by biological processes such as bacteria and algae production with formation of three-dimensional EPS structures as well as adsorptive processes on EPS, the straining cause this phenomenon (see Chapter 6.5).

These results confirm that the water-sediment-boundary layer with a thickness of a few centimetres functions as an interface with high biological activity, a system with high filtration capacity and a complex three-dimensional structure, composed of algae, fungi and bacteria, embedded in an EPS matrix. Recent investigations carried out on pathogens (see below) support the model of the biological filter function and clearly point out the high significance of the active, undisturbed water-sediment boundary layer.

6.5. Vertical Transport of Pathogens (Virus, Bacteria and Protozoan)

The vertical transport of pathogens during bank filtration is of main importance, due to the high reduction rates needed for some of the pathogens like viruses (*Hepatitus A*, *Polio-virus*), bacteria (e.g. *Salmonella typhimurium*) and protozoa (*Giardia intestinalis* and *Cryptosporidium parvum*), where reduction rates of 10^3 up to 10^7 are needed. The maximum concentration of *Giardia intestinalis* and *Cryptosporidium parvum* in drinking water should be 10^{-5} ind. L^{-1} to keep the infection risk below 10^{-4} per person and year. (Dutch Drinking Water Decree; VROM 2001), the US EPA gives a guiding level of zero for drinking water (EPA 2003), the WHO (2004) of 6.3×10^{-4} ind. L^{-1}. For water treatment technique US EPA requires a removal of 10^2 (*Cryptosporidium*) respectively 10^3 (*Giardia*). Surface water are contaminated by *Giardia* and *Cryptosporidium* in a range of $1 - 100$ ind. L^{-1}, that means a removal of 10^4 to 10^7 is required to reach the maximum infection risk of $< 1 : 10,000$ (Schijven et al., 2002).

A common approach to reduce the infection risk in bank filtration is a long ground water transport of > 50 days to eliminate or inactivate bacteria and other micro-organisms, even if the main retention of pathogens occurs at the biologically active water-sediment-boundary layer, because disturbances such as erosion (after floods) or desiccation (at low flow rate) can lead to a breakthrough into deeper sediment layers and a contamination of groundwater by micro-organisms, even at high concentrations (Matthess & Pekdeger, 1985). Normally inactivation of viruses and survival of bacteria is regulated by temperature and pH, but typical groundwater conditions are not in the range of toxic levels for bacteria. Furthermore, the regular survival of bacteria like *E. coli* in water is up to 3 months (Schijven et al. 2002). Thus, a breakthrough of micro-organisms to groundwater can occur, even if a 50 day transit time is given; it rather depends on disturbances of the water-sediment-boundary layer. In all systems disturbances are expected, in RBF by high floods and desiccation, in LBF by desiccation of the littoral zone and erosion by waves, recreation and water sports, and in GWR ponds by removal of the schmutzdecke. A reduction of the breakthrough risk can be reached only by a good and sufficient management of the bank filtration sites, which means water protection zones e. g. like in Germany (Table 1).

Table 1. Water protection zone of the German Federal Water Act (WHG 2002)

1) Water protections zones	2) Protection concept
3) Protection zone I	4) the extraction point and it's immediate environment must be free from any contamination and any land use is prohibited
5) Protection zone II	6) an inner protection zone which covers the area for 50 day flow time to the well to protect groundwater for microbial contamination and any disturbance of the soil layer is prohibited and the application of contaminants is regulated
7) Protection zone III	8) the complete water micro basins of the well, any contamination is regulated.

The removal of all types of pathogens (viruses, bacteria, protozoa) during bank filtration was demonstrated in many experiments with sand columns as well as in field studies with observations wells (Schijeven et al., 2002), and the mean efficiency of removal amounts 10^2 to 10^6. For bacteria and bacteriophages a removal of $1.4 \times 10^1 - 10^3$ is reported (Hijnen et al., 2004, Dizer et al., 2004, López-Pila et al., 2006, Wand et al. 2007), while the protozoa and their oocysts respectively cysts are eliminated to more than 10^5 (Hijnen et al., 2004, 2007; Medema et al., 2003). According to the WHO (2004) the baseline removal of micro-organisms during bank filtration is 10^4 within 4 m of the infiltration stretch.

The effective layer for removal is the water-sediment-interface, which means a filter layer of a few centimetres up to a few decimetres (Medema et al., 2003, Wand et al., 2007); The high efficiency of the water-sediment-boundary layer related to the removal of pathogens becomes evident after a removal of the schmutzdecke, leading to a decrease of the elimination capacity of about 10^2 (Hijnen et al., 2004), and a breakthrough of coliphages, respectively (Dizer et al., 2004).

One mechanism for removal of microorganisms is a mechanical filtration, but as mentioned above, the pores of the sand filter are large compared to the organisms (the infective stage of *Cryptosporidium*, the oocyst, has only a diameter of 3 µm), and other capture mechanisms like straining becomes more significant for smaller organic particles. Straining is a physicochemical filtration process, that is based in a first step on a successful collision between a particle and the sand grain surface, and in a second step on an immobilization of the particles. The cell membrane respectively virus surface as well as the sand grain surface are negatively charged (= negative zeta potential) in most natural aquatic systems, and repulsive electrostatic interactions are expected, but collision and immobilisation lead to bonding. It has to be assumed that the three dimensional structure of EPS within the sand grain pores enhance straining by its surface structure (area, zeta potential). The ratio of cell diameter (= strained particle) to sand grain diameter (= collector) can be as small as 0.05 or even 0.005 (Tufenkji et al., 2004, 2006; Auset & Keller, 2006).

The process of straining gives a good understanding of the high efficiency of the water-sediment-interfaces for removal of all types of pathogens, and also gives an idea about the differences in removal of different microbe species (e. g. of coliphages138 and 241; López-Pila et al. 2006), which can be caused by membrane surface structures. A new modelling of microbial transport in porous media points out the high importance of straining for microbe motility in bank filtration (Tufenkji 2007).

6.6. Vertical Transport of Cyanobacteria and Cyanotoxins

Cyanobacteria are worldwide spread, but the increasing eutrophication as well as the increasing temperature and solar radiation due to climate change promote the mass development of cyanobacteria (Bouvy et al., 2000; Wiedner et al., 2007a, b). The knowledge of the high toxic potential of cyanobacteria and their toxins has been increased significantly the last during decade (Jochimsen et al., 1998; Chorus & Bartram 1999), and in many regions it has become a central problem in raw water as well as drinking water quality (Ueno et al., 1996; McGregor & Fabbro, 2000; Molica et al., 2005; CRC, 2006, Fastner et al., 2007). The development of scientific based cyanotoxin concentration limits in drinking water is very complex due to the high number of toxic compounds of different structure as well as to the high number of congeners of a toxin (more than 80 microcystins are already detected). Thus the WHO guiding value for microcystin in drinking water of 1 μg L^{-1} microcystin-LR is a provisory one (WHO 2004). In view of cyanotoxins the pre-treatment of surface water is nowadays focus of interest, promoting studies of removal of cyanobacteria and cyanotoxins by bank filtration, because other water treatment technologies did not give good results in cyanotoxin elimination (Hrudey et al., 1999; Rapala et al., 2002); in general the oxidation of DOC by chlorine or ozone cannot result in such a low level as 1 μg L^{-1} microcystin-LR, which would comply a DOC removal of 95 % if the microcystin concentrations in raw water amounts 20 μg L^{-1}.

Cyanobacteria and their cell bound cyanotoxins have to be distinguished from cyanotoxins that are released in the surface water as extra-cellular cyanotoxin. Two different mechanisms lead to high soluble cyanotoxin concentration in water, the excretion of cyanotoxins from living cells as a natural metabolism and the lyses of dead cyanobacteria cells.

The removal of cyanobacteria cells by bank filtration is successful, and the principal mechanism is as mentioned by the removal for other pathogens, the straining. The removal of algae cells in the upper layer of sand filters was proved in sand columns and filed experiments by Grützmacher et al. (2002) and Hijnen et al. (2007). The survival of the algae cells in the water-sediment-interface is limited, and after cell death a release of cell bound cyanotoxins occurs. Probably some of the cyanobacteria cells are consumed by the meiofauna, but on one hand little is known about the fate of cyanobacteria and cyanotoxins during digestion, and on the other hand cyanobacteria seem to be an unattractive food source for organisms, already proved for filtering zooplankton species.

The fate of the cyanotoxins during bank filtration depends on the microbial degradation of the compounds. It has been demonstrated, that microcystin-RR and -LR are degradable, and moreover that *Sphingomonas paueimobilis,* a strictly aerobic, chemo-organotrophic, gram-negative bacteria with high temperature tolerance is responsible for degradation (Park et al. 2001); degradation of microcystin-LR occurred fast within four days. This is confirmed by microcystin degradation studies in sand filters under oxic conditions, whereas under anoxic conditions the results are still contradictory (Holst et al., 2003; Ho et al., 2006, Grützmacher et al., 2007).

The degradation of other cyanotoxins is not yet studied in detail, saxitoxin seems to be degradable (Sens & Dalsasso 2007), and recent investigations on the degradability of cylindrospermopsin indicate a high persistence.

7. CONCLUSION

RBF and LBF are two techniques which can contribute to a clean and safe water supply. Benefits are the temperature equalization (in temperate regions), the removal of fine particles, the protection against shock loads due to floods or erosion after extreme precipitation and the degradation of DOM. The artificial GWR and SSF are artificial bank filtration systems, which too enable water purification, but these systems have to be managed in a good way, as some problems occur like the development of the schmutzdecke, which limits the capacity and effectiveness of water purification. Underground passage is a natural process in the water cycle and improves the water quality – this is a first step to secure good drinking water quality. Currently contaminants like cyanobacteria with the excretion of the cyanotoxins, the protozoa *Giardia* and *Cryptosporidium* are vitally important and makes it necessary to apply new water treatment steps.

In all bank filtration sites, an advanced protection system has be developed to reduce destructive impacts on the water-sediment-boundary layer by water sports and shipping, mainly by surface and underwater waves, as well as on surface and raw water quality by eutrophication and contamination or depletion of oxygen during infiltration.

REFERENCES

Adrian, R., Wilhelm, S. & Gerten, D. (2006). Life-history traits of lake plankton species may govern their phenological response to climate warming. *Global Change Biology 12*, 652-661.

Auset, M. & Keller A. A. (2006). Pore-scale visualization of colloid straining and filtration in saturated porous media using micromodels. *Wat. Resources Res. 42*, W12S02. Doi: 10.1029/2005WR004639.

Bernardo, L. D., Sens, M. L., Kuroda, E. K., Dalsasso, R. L., de Melo Filho, L. C., Mondardo, R. I. & Dantas, Â. D. B. (2006). Filtração directa. In: V. L: Pádua (Ed.) Contribução ao Estudo da Remoção de Cianobactérias e Microcontaminanes orgânicos por meio de Técnicas de Tratamento de Água para Consumo Humano. Projecto PROSAB 4, Vol. 1 Água, ABES, RJ, 275-334.

Bertin, C. & Bourg, A. C. (1994). Radon-222 and chloride a natural tracers of the infiltration of river water into an alluvial aquifer in which there is significant river/groundwater mixing. *Environ. Sci. Technol. 28*, 794-798.

Beulker, C. & Gunkel, G. (1996). Studies on the occurrence and ecology of meiofauna and benthic algae in the littoral sediment of lakes. *Limnologica 26*, 311-326.

Beyer, W. (1964). Zur Bestimmung der Wasserdurchlässigkeit von Kiesen und Sanden aus der Kornverteilungskurve. *Wasserwirtschaft Wassertechnik 14*, 165-168.

Bouvy, M., Falcao, D., Marinho, M., Pagano, M. & Moura, A. (2000). Occurrence of Cylindospermopsis (Cyanobacteria) in 39 Brazilian tropical reservoirs during the 1998 drought. *Aquatic Microbial Ecology 23*, 13-27.

Bretschko, G. & Klemens, W. E. (1986). Quantitative methods and aspects in the study of the interstitial fauna of running waters. *Stygologia 2*, 297-316.

Brugger, A., Reitner, B., Kolar, I., Quéric, N. & Herndl, G. J. (2001). Seasonal and spatial distribution of dissolved and particulate organic carbon and bacteria in the bank of an impounding reservoir on the Enns River, Austria. *Freshwater Biology 46*, 997-1016.

Characklis, W. G. & Marshall, K. C. (Eds.) (1990). Biofilms. Wiley & Sons, New York. 796 pp.

Chorus, I. & Bartram, J. (1999). Toxic Cyanobacteria in Water: A guide to public health significance, monitoring and management. WHO. Chapman and Hall, London. 400 pp.

CRC (2006). Cyanobacteria – Management and Implications for Water Quality. Cooperative Research Centre for Water Quality and Treatment, Salisbury. Fact Sheet, 32 pp.

Dash, R. R., Mehrotra, I., Kumar, P. & Grischek, T. (2008). Lake bank filtration at Nainital, India: water quality evaluation. *Hydrogeology J. 16*, 1089-1099.

Decho, A. W. (2000). Microbial biofilms in intertidal systems: an overview. *Cont. Shelf Res. 20*, 1257-1273.

Dittrich, M., Kurz, P. & Wehrli, B. (2004). The role of autotrophic picocyanobacteria in calcite precipitation in an oligotrophic lake. *Geomicrobiology Journal 21*, 45-53.

Dizer, H., Grützmacher, G., Bartel, H., Wiese, H. B., Szewzyk, R. & López-Pila, J. M. (2004). Contribution of the colmation layer to the elimination of coliphages by slow sand filtration. *Water Science Technology 50*, 211-214.

DVGW (Ed.) (2006). Exportorientierte F&E auf dem Gebiet der Wasserver- und -entsorgung Teil I: Trinkwasser. Eigenverlag DVGW, Karlsruhe. ISBN 3-00-015478-7.

Emtiazi, F., Schwartz, T., Marten, S.M., Krolla-Siedenstein, P. & Obst, U. (2004). Investigation of natural biofilms formed during the production of drinking water from surface water embankment filtration. *Water Res. 38*, 1197-1206.

EPA (2003). Natioanl primary drinking water standards. Available from: URL: http://www.epa.gov/OGWDW/contaminants/index.html.

Fastner, J., Rücker, J., Stüken, A., Preußel, K., Nixdorf, B., Chorus, I. Köhler, A. & Wiedner, C. (2007). Occurrence of the cyanobacterial toxin cylindrospermopsin in Northeast Germany. *Environmental Toxicology*, DOI 10.1002, 26-32.

Flemming, H.-C. & Wingender, J. (2001). Relevance of microbial extracellular polymeric substances (EPSs) – Part I: Structural and ecological aspects. *Water Science and Technology 43*, 1-8.

Flemming, H.-C., Wingender, J., Moritz, R., Borchard, W. & Mayer, C. (1999). Physico-chemical properties of biofilms: A short review. In: C. W. Keevil, A. Godfree, D. Holt and C. Dow. (Eds.). Biofilms in the aquatic environment. Royal Society of Chemistry, Spec. Publ. 242, 1-12.

Gibert, J., Danielopol, D. & Stanford, J. A. (1994). Groundwater Ecology. Academic Press, 571 pp.

Gibert, J., Marmonier, P. & Plénet, S. (1998). Efficiency of bank filtration: biotic processes. *Verh. Internat. Verein. Limnol. 26*, 1027-1031.

Gimbel, R., Graham, N. J. D., Collins, M. R. (2006) Recent Progress in Slow Sand and Alternative Biofiltration Processes. IWA Publishing, 600 pp.

Graham, N. J. D. & Collins, M. R. (1996). Advances in Slow Sand Filters and Alternative Biology Filtration, John Wiley & Sons, Chichester. 461 pp.

Graham, R. & Collins, N. (Eds.) (1996). Advances in Slow Sand and Alternative Biological Filtration. John Wiley & Sons Ltd., Chichester, 462 pp.

Greskowiak, J., Prommer, H., Massmann, G., Johnston, C. D., Nützmann, G. & Pekdeger, A. (2005). The impact of variably saturated conditions on hydrogeochemical changes during artificial recharge of groundwater. *Applied Geochemistry 20*, 1409-1426.

Grischek, T., Hiscock, K.M., Metschies, T., Dennis, P. & Nestler, W. (1998). Factors affecting denitrification during infiltration of river water into a sand and gravel aquifer in Saxony, Germany. *Water Res. 32*, 450-460.

Groß-Wittke, A., Hoffmann, A. & Gunkel, G. (2009). Influence of temperature on redox conditions and physicochemical parameters in pore water during bank filtration in the sandy littoral zone of Lake Tegel. *Water Res.* (in prep.).

Grünheid, S., Amy, G. & Jekel, M. (2005). Removal of bulk dissolved organic carbon (DOC) and trace organic compounds by bank filtration and artificial recharge. *Water Res. 39*, 3219-3228.

Grützmacher, G., Bartel, H. & Chorus, I. (2007). Cyanobakterientoxine bei der Uferfiltration. *Bundesgesundheitsblatt – Gesundheitsforschung – Gesundheitsschutz 50*. Doi: 10.1007/s001303-007-0161-6.

Grützmacher, G., Böttcher, G., Chorus, I. & Bartel, H. (2002). Removal of microcystin by slow sand filtration. *Environ. Toxicol. 17*, 386-394.

Gunkel, G. (2009). Vulnerability of aquatic systems to climate chance – Findings from European waters. In: L. Duarte and P. Pinto (Eds.). Sustainable development: energy, environment and natural disasters. Fundação Luis de Molina, Lisboa. (in press).

Gunkel, G., Beulker, C., Hoffmann, A. & Kosmol, J. (2008). Fine particulate organic matter (FPOM) transport and processing in littoral interstice – use of fluorescent markers. *Limnologica*. Doi:10.1016/j.limno.2008.11.001.

Hakenkamp, C. C., Morin, A. & Strayer, D. L. (2002). The functional importance of freshwater mciofauna. In: S. D. Rundle, A. L. Robertson and J. M. Schmid-Araya, (Eds.). Freshwater Meiofauna: Biology and Ecology. Backhuys, Leiden, The Netherlands, 321-335.

Heberer, T., Mechlinski, A., Fanck, B., Knappe, A., Massmann, G., Pekdeger, A. & Fritz, B. (2004). Field studies on the fate and transport of pharmaceutical residues in bank filtration. *Ground Water Monitoring & Remediation 24*, 70-77.

Hendel, B., Marxen, J., Fiebig, D. & Preuß, G. (2001). Extracellular enzyme activities during slow sand filtration in a water recharge plant. *Water Res. 35*, 2484-2488.

Higgins, R. P. & Thiel, H. (1988). Introduction to the Study of Meiofauna. Smithsonian Institution Press. ISBN 0-87474-488-1, 488 pp.

Hijnen, W. A. M., Dullemont, Y. J., Schijven, J. F. & Hanzens-Brouwer, A. J. (2007). Removal and fate of Cryptosporidium parvum, Clostridium perfringens and small-sized centric diatoms (Stephanodiscus hantzschii) in slow sand filters. *Water Res. 41*, 2151-2162.

Hijnen, W. A. M., Schijven, J. F., Bonné, P., Visser, A. & Medema, G. J. (2004). Elimination of viruses, bacteria and protozoa oocycts by slow sand filtration. *Water Science Technology 47*, 241-247.

Hiscock, K. M. & Grischek, T. (2002). Attenuation of groundwater pollution by bank filtration. *J. Hydrology 266*, 139-144.

Ho, L., Meyn, T. Keegan, A., Hoefel, D., Brookes, J. Saint, C. P. & Newcombe, G. (2006). Bacterial degradation of microcystin toxins within a biologically active sand filter. *Water Res. 40*, 768-774.

Hoffman, A. & Gunkel, G. (2009a). Bank filtration in the sandy littoral zone of Lake Tegel (Berlin): Biological-chemical structure of the interstices and clogging processes. *Limnologica* (subm.)

Hoffmann, A. & Gunkel, G. (2009b). Bank filtration in the sandy littoral zone of Lake Tegel (Berlin): Carbon production, turnover and related biological-chemical processes. *Limnologica* (in prep.)

Holst, T., Jørgensen, N. O. G., Jørgensen, C. & Johansen, A. (2003). Degradation of microcystin in sediments at oxic and anoxic, denitrifying conditions. *Water Res. 37*, 4748-4760.

Hrudey, S., Burch, M., Drikas, M. & Gregory, R. (1999). Remedial measures. In: I. Chorus and J. Bartram (Eds.). *Toxic Cyanobacteria in Water: A guide to public health significance, monitoring and management.* WHO. Chapman and Hall, London. Chap. 9.

Jochimsen, E. M., Carmichael, W. W., Cardo, D. M., Cookson, S. T., Holmes, C. E. M., Antunes, M. B., de Melo Filho, D. A., Lyra, T. M., Barreto, V. S. T., Azevedo, S. M. F. O. & Jrvis, W. R. (1998). Liver failure and death after exposure to microcystins at a hemodialysis center in Brazil. *New England Journal of Medicine 338*, 873-878.

Kolehmainen, R. E., Langwaldt, J. H. & Puhakka, J. A. (2007). Natural organic matter (NOM) removal and structural changes in the bacterial community during artificial groundwater recharge with humic lake water. *Water Res. 41*, 2715-2725.

Kühn, W. & Müller, U. (2000). Riverbank filtration. An overview. *J. AWWA, 92*, 60-69.

Langergraber, G., Haberl, R., Laber, J. &. Pressl, A. (2003). Evaluation of substrate clogging processes in vertical flow constructed wetlands. *Wat. Sci. Technol. 48*, 25-34.

Lawrence, J. R., Neu, T. R. & Swerhone, G. D. W. (1998). Application of multiple parameter imaging for the quantification of algal, bacterial and exopolymer components of microbial biofilms. *J. Microbiol. Methods 32*, 253-261.

Lawrence, J. R., Scharf, B., Packroff, G. & Neu, T. R. (2002). Microscale evaluation of the effects of grazing by invertebrates with contrasting feeding modes on river biofilm architecture and composition. *Microb. Ecol. 43*, 199-207.

López-Pila, J. M., Szewzyk, R. & Dizer, H. (2006). Die Elimination von viralen Krankheitserregern bei der Uferfiltration. Available from: URL: http://www.kompetenz-wasser.de/fileadmin/user_upload/pdf/veranstaltungen/Grundwasserkonferenz_2006/wasser_berlin_lopez.pdf.

Massmann, G., Heberer, T., Grützmacher, G., Dünnbier, U., Knappe, A., Meyer, H., Mechlinski, A. & Pekdeger, A. (2007). Trinkwassergewinnung in urbanen Räumen – Erkenntnisse zur Uferfiltration. *Grundwasser – Zeitschr. Fachsektion Hydrogeologie* 12, 232-245.

Massmann, G., Nogeitzig, A, Taute, T. & Pekdeger, A. (2008a). Seasonal and spatial distribution of redox zones during lake bank filtration in Berlin, Germany. *Environ. Geol. 54*, 53-65.

Massmann, G., Sültenfuß, J., Dünnbier, U., Knappe, A., Taute, T., Pekdeger, A. (2008b). Investigation of groundwater residence times during bank filtration in Berlin – a multi-tracer approach. *Hydrol. Proc. 22*, 788-801.

Matthess, G. & Pekdeger, A. (1985). Survival and transport of pathogenic bacteria and viruses in ground water. In: C. H. Ward and P. McCarty (Eds.) Ground Water Quality. John Wiley and Sons, New York. pp. 472-482.

McGregor, G. B. & Fabbro, L. D. (2000). Dominance of Cylindrospermopsis raciborski (Nostacales, Cyanoprokaryota) in Queensland tropical and subtropical reservoirs: Implications for monitoring and management. *Lakes & Reservoirs Research & Management 5*, 195-205.

Medema, G. J., Hoogenboezem, W., Veer, A. J. van der, Ketelaars, H. A. M., Hijnen, W. A. M. & Nobel, P. J. (2003). Quantitative risk assessment of cryptosporidium in surface water treatment. *Water Science Technology 47*, 241-247.

Miettinen, I. T., Vartiainen, T. & Martikainen, P. J. (1996). Bacterial enzyme activities in ground water during bank filtration of lake water. *Water Res. 30*, 2495-2501.

Molica, R. J. R., Oliveira, E. J. A., Carvalho, P. V. V. C., Costa, A. N. S. F., Cunha, M. C. C., Melo, G. L. & Azevedo, S. M. F. O. (2005) Occurrence of saxitoxins and an anatoxin-a(s)-like anticholinesterase in a Brazilian drinking water supply. *Harmful Algae 4*, 743-753.

Okubo, T. & Matsumoto, J. (1983). Biological clogging of sand and changes of organic constituents during artificial recharge. *Water Res. 17*, 813-821.

Pádua, V. L. (2006). Contribução ao Estudo da Remoção de Cianobactérias e Microcontaminanes orgânicos por meio de Técnicas de Tratamento de Água para Consumo Humano. *Projecto PROSAB* 4, Vol. 1 Água, ABES, RJ, 504 pp.

Park, H.-D., Sasaki, Y., Maruyama, T. Yanagisawa, E., Hiraishi, A. & Kato, K. (2001). Degradation of cyanobacterial hepatoxin microcystin by a new bacterium isolated from a hypertrophic lake. *Environ Toxicol 16*, 337-343.

Paterson, D. M. (2001). The fine structure and properties of the sediment surface.- In: B. P. Boudreau and B. B. Jorgensen (Eds.). *The benthic boundary layer. Transport processes and biogeochemistry.* Oxford Univ. Press. 127-143.

Rapala, J., Lahti, K., Räsänen, L. A., Esala, A.-L., Niemelä, S. I., Sivonen, K. (2002). Endotoxins associated with cyanobacteria and their removal during drinking water treatment. *Water Res. 36*, 2627-2635.

Ray, C. (2008). Worldwide potential of riverbank filtration. *Clean Techn. Environ. Policy 10*, 223-225.

Ray, C., Melin, G. & Linsky, R. B. (2003). *Riverbank Filtration. Improving Source-Water Quality. Water Science Technology Library,* Vol. 43, Kluwer Acad. Publ., Dordrecht. 364 pp.

Rinck-Pfeiffer, S., Ragusa, S., Sztajnbik, P. & Vandevelde, T. (2000). Interrelationships between biological, chemical, and physical processes as an analog to clogging in aquifer storage and recovery (ASR) wells. *Water Res.* 34, 2110-2118.

Schijeven, J., Berger, P. & Miettinen, I. (2002). Removal of pathogens, surrogates, indicators, and toxins using riverbank filtration. In: C. Ray, G. Melin and R. B Linsky (Eds.). *Riverbank Filtration. Improving Source-Water Quality.* Water Science Technology Library, Vol. 43, Kluwer Acad. Publ., 73-116.

Schmidt, K. C., Lange, F. T., Brauch, H.-J. & Kühn, W. (2003). Experiences with riverbank filtration and infiltration in Germany. *Proc. Int. Symp. Artificial Recharge of Groundwater,* Daejon, Korea. pp. 115-141.

Seiler, K.-P. & Gat, J. R. (2007). Groundwater Recharge from Run-Off, Infiltration and Percolation. Springer, Dordrecht. 241 pp.

Sens, M. L., Dalsasso, R. L. (2007). Bank Filtration of Reservoir as an efficient technique for water supply treatment. In: G. Gunkel and M. C. Sobral (Eds.). *Reservoir and River Basin Management: Exchange of Experiences from Brazil, Portugal and Germany.* Universitätsverlag TU Berlin. pp. 253-264.

Spring, S., Schulze, R., Overmann, J. & Schleifer, K.-H. (2000). Identification and characterization of ecologically significant prokaryotes in the sediment of freshwater lakes: molecular and cultivation studies. *FEMS Microbiology Reviews 24,* 573-590.

Sutherland, I. W. (2001). Exopolysaccharides in biofilms, flocs and related structures. *Wat. Sci. Tech. 43,* 77-86.

Tufenkji, N. (2007). Modeling microbial transport in porous media: Traditional approaches and recent developments. *Adv. in Water Resources 30,* 1455-1469.

Tufenkji, N., Dixon, D. R., Considine, R. & Drummond, C. J. (2006). Multi-scale Cryptosporidium/sand interactions in water treatment. *Water Res. 40,* 3315-3331.

Tufenkji, N., Miller, G. F., Ryan, J. N. Harvey, R. W. & Elimelech, M. N. (2004). Transport of Cryptosporidium oocysts in porus media: Role of Straining and physicochemical filtration. *Environ. Sci. Technol. 38,* 5932-5938.

Ueno, Y., Nagata, S., Tsutsumi, T., Hasegawa, A., Watanabe, M. F., Park, H.-D., Chen, G.-C. & Yu, S.-Z. (1996). Detection of microcystin, a blue-green algal hepatotixin, in drinking water sampled in Haimen and Fusui, endemic areas of primary liver cancer in China, by highly sensitive immunoassay. *Carcinogenesis 17,* 1317-1321.

VROM (2001) Waterleidingbesluit (*Drinking Water Decree*). Ministry of Housing, Physical Planning and the Environment. The Hague, The Netherlands.

Wand, H., Vacca, G., Kuschk, P., Krüger, M. & Kästner, M. (2007). Removal of bacteria by filtration in planted and non-planted sand columns. *Water Res. 41,* 159-167.

WHG (2002) Gesetz zur Ordnung des Wasserhaushalts (Wasserhaushaltsgesetz – WHG). BGBl I, 3245.

WHO (2004) Guidelines for Drinking water Quality. World Health Organization, Geneva, 3. ed.

Wiedner, C. Rücker, J. & Weigert, B. (Eds) (2007a). Cylindrospermopsis raciborskii and Cylindrospermopsin in Lakes of the Berlin Area – Occurrence, Causes, Consequences. Schr. Kompetenzzentrum Wasser Berlin, Vol. 6, 89 S.

Wiedner, C. Rücker, J., Brüggemann, R. & Nixdorf, B. (2007b). Climate change affects timing and size of populations of an invasive cyanobacterium in temperate regions. *Oecologia,* Doi: 10.1007/s00442-007-0683-5, 12 pp.

Wingender, J., Neu, T. R., Flemming, H.-C. (1999). What are bacterial extracellular polymeric substances? In: J. Wingender, T. R. Neu and H.-C. Flemming (Eds.). Microbial Extracellular Polymeric Substances: Characterization, Structure & Function. Springer-Verlag, Berlin. pp. 1-19.

Wotton, R. S., Chaloner, D. T. & Armitage, P. D. (1996). The colonisation, role in filtration and potential nuisance value of midges in slow sand filter beds. In: N.

Graham and R. Collins (Eds.). Advances in Slow Sand and alternative Biological Filtration. Wiley & Sons, Chichester. pp. 149-157.

Yallop, M. L., Paterson, D. M. & Wellsbury, P. (2000). Interrelationships between rates of microbial production, exopolymer production, microbial biomass, and sediment stability in biofilms of intertidal sediments. *Microb. Ecol. 39*, 116-127.

Yamamuro A. M. & Lamberti, G. A. (2007). Influence of organic matter on invertebrate colonization of sand substrata in a northern Michigan stream. *J. North American Benthol. Soc.* 26, 2, 244–252.

Zimmermann-Timm, H. (2002). Characteristics, Dynamics and Importance of Aggregates in Rivers - An Invited Review. *Internat. Rev. Hydrobiol. 87*, 197-240.

In: Water Purification   ISBN 978-1-60741-599-2
Editors: N. Gertsen and L. Sønderby   © 2009 Nova Science Publishers, Inc.

Chapter 4

CHITOSAN FOR PESTICIDE CONTROL ON ENVIRONMENTAL PROTECTION AND WATER PURIFICATION

Cristóbal Lárez Velásquez and *Enrique Millán Barrios*

Departamento de Química, Facultad de Ciencias
Universidad de Los Andes, Mérida 5101, Venezuela

ABSTRACT

The present review is focused on the diverse approaches employing chitin and/or chitosan as either active or passive components which have been employed to pesticide control in environmental protection and water purification. In some cases, ideas are presented in order to contribute in the solution of particular problems associated with this topic. The first section is dedicated to introduce the most important basic elements in the work as for example classification of the pesticides (according to their toxicological effects, chemical similarities and type of plague to control), description of the chitin and chitosan and some general applications of these biomaterials on removal of pollutants. Second section briefly analyzes some strategies employed in order to minimize the effects of the indiscriminate application of pesticide, through previous actions, as controlled release and protected dosing system. Third part is dedicated to discuss the diverse action modes which chitin or chitosan can be used for pesticide removal, including their uses in coagulation/flocculation process, filtration membranes and as adsorbents. Some applications of these materials as supports of pesticide-degradating agents are presented in the section four including enzymes micro-organisms and catalysts. Finally, a section dedicated to trends in this topic is presented. In general, the review shows that these materials posses high potentialities for use in pollutants removal. Applications as adsorbent appear to be the most promising at short term, especially to either anionic or acid pesticides, considering that perhaps chitosan is the unique natural polycation. Similarly, some proposal to produce new composite materials with

---
* clarez@ula.ve; ejmb@ula.ve

applications on pesticide removal, using processes involving chitin and chitosan are presented.

1. INTRODUCTION

Daily significant volumes of pesticides are discharged from countless sources creating a serious environmental problem, especially when these substances are poured into rivers and lakes and other water reservoirs. Pesticides accumulation in water sources used for human consumption is also highly worrying for human health, because, in general, conventional water treatment methods (i.e., coagulation-flocculation, sedimentation, and conventional filtration) do not seem to facilitate either removal or transformation of pesticides in drinking water (U.S. Environmental Protection Agency, 2000).

From the environmental and human health point of view, discovery of new methods and materials that allow improving level of control, removal or transformation of these pollutants has cardinal importance. As in other areas, a lot of effort has been paid to increasingly use environmental friendly materials. Chitosan, a biopolymer coming from diverse natural sources but mainly obtained by chemical transformations of chitin (one of the renewable materials more abundant in nature) has been extensively studied during the last twenty years by its high potential in water treatment.

This review focuses on the principal systems based in chitin and/or chitosan (and their derivatives) that have been reported for the controlled and/or protected release, removal or biodegradation of pesticides. Due to the present relevance of this topic, which no doubt will continue in the near future, some ideas on the production of either materials or methods for removal of pesticide based on these polymeric materials are briefly presented.

**Table 1. Recommended WHO pesticide classification according
to their toxicological effects**

Class		LD_{50} for the rat (mg/Kg weigth)			
		Oral		Dermal	
		Solids	Liquids	Solids	Liquids
Ia	Extremely hazardous	5 or less	20 or less	10 or less	40 or less
Ib	Highly hazardous	5 – 50	20 – 200	10 – 100	40 – 400
II	Moderate hazardous	50 – 500	200 – 2000	100 – 1000	400 – 4000
III	Slightly hazardous	Over 500	Over 2000	Over 1000	Over 4000

The LD_{50} value is a statistical estimate of the number of mg of toxicant per kg of bodyweight required to kill 50% of a large population of test animals.

Table 2. Pesticide classification according to their chemical similarities

Code	Type of compound	Code	Type of compound
AS	Arsenic compounds	OP	Organophosphorus compounds
BP	Bipyridylium derivatives	OT	Organotin compounds
C	Carbamates	PAA	Phenoxyacetic acid derivatives
CO	Coumarin derivatives	PZ	Pyrazoles
CU	Copper compounds	PY	Pyrethroids
HG	Mercury compounds	T	Triazine derivatives
NP	Nitrophenol derivatives	TC	Thiocarbamates
OC	Organochlorine compounds		

Table 3. Pesticide classification according to type of controlled plague

Acaricides	Algicides	Antifeedants
Avicides	Bactericides	Bird repellents
Herbicides	Fungicides	Herbicide safeners
Insecticides	Insect attractants	Insect repellents
Molluscicides	Mammal repellents	Nematicides
Rodenticides	Virucides	

1.1. Classification of Pesticides

Pesticides can be organized in groups according to diverse considerations, ranging from their toxicological effects to chemical properties, including type of plagues to be controlled. Thus, considering their toxic effects in the human beings, the pesticides have been classified by the World Health Organization (WHO, 2004) in three main groups (Table 1). Table 2 shows pesticide classification according to their similar chemical properties; use of the same antidote is among the properties more important to consider in this classification. Table 3 shows pesticide classification considering the type of controlled plague.

1.2. Chitin and Chitosan

Chitosan is a well studied linear polysaccharide which can be considered, from the point of view of their repetitive units, a copolymer: poly-(β-1,4-glucosamine-co-N-acetyl- β-1,4-glucosamine). It occurs naturally in several fungi, especially *Mucor* species. However, commercial chitosan is usually prepared by a chemical chitin N-deacetylation reaction. It has been demonstrated that acetylated (GlcNAc; β (1$\rightarrow$4) 2- acetamido – 2 – deoxy – b – D 4 - glucopyranose) and deacetylated units (Glc; β (1$\rightarrow$4) 2-amino-2-deoxy-b-D-glucopyranose) are randomly distributed along the chains, in proportions defined by a parameter known as deacetylation degree (DD) (i.e., DD is the fraction of deacetyladed units in the polymer chain) as it is illustrates in Figure 1.

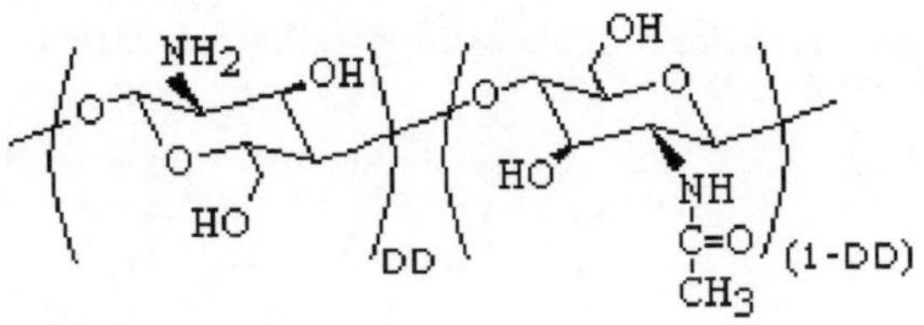

Figure 1. Repetitive structural chemical units of chitosan. Monomer units are randomly distributed in the chain.

Chitosan physicochemical properties are markedly depending on intrinsic characteristics of the polymer (molar mass, DD, distribution of repetitive units) and the experimental conditions in which the material is studied (temperature, pH, ionic strength, solvent, associated counterions, etc.). Thus, chitosan is insoluble in neutral and basic aqueous medium but it is very soluble in acidic aqueous solutions. Similarly, materials with DD < 0.50 are insoluble even in aqueous acid solutions but solubility increases when DD get higher values. Protonation of amine moieties on the deacetylated units is responsible for solubilization of chitosan in acidic aqueous medium because quaternary amine salt formation (Figure 2) destroys the intra and intermolecular hydrogen bonds in which these participate.

Due to its natural origin, chitosan possesses intrinsic advantages such as low cost, biodegradability, biocompatibility, non-toxicity, good sorption properties and film forming capacity. In particular, protonated amino groups enable properties such as antibacterial, protein affinity and water solubility. On the other hand, non-protonated amine groups and hydroxyl groups favor heavy metal chelation and facilitate some polymer modification reactions.

1.3. Recovery of Pesticides Using Chitin and Chitosan

Applications of chitin and chitosan (and their derivatives) in water treatments have extended rapidly due to the competitive advantages of these materials (Kawuamura, 1991; Majeti & Ravi, 2000), i.e., their natural origin, biodegradability and low cost. Similarly, facility to chemical modification makes them extremely attractive materials for development of friendly environmentally systems for water purification. In that sense, these biopolymers have been used in different activities related with water remediation such as: (a) removal of diverse pollutants including heavy metals (Varma *et al.*, 2004) (considering its high selectivity to transition metallic ions of the group III - but not for metallic ions of the groups I and II - at low concentrations (Muzzarelli, 1973)), dyes (Cestari *et al.*, 2004), oils (Ahmad *et al.*, 2004); (b) coagulant agent (Bratskaya *et al.*, 2002); (c) flocculant agent (Divakaran & Pillai, 2001); (d) adsorbent (Crini, 2005); (e) ultra-filtration membranes (Verbych *et al.*, 2005); etc.

Chitin and chitosan possesses important features which enable them to be used in agriculture as biocide agent (Devlieghere *et al.*, 2004), growth stimulant (Barka *et al.*, 2004), elicitor (Prapagdee *et al.*, 2007), etc. Also, they have been employed as matrix in controlled pesticide release (Hirano, 1978; McCormik *et al.*, 1982; Texeira *et al.*, 1990; Larez, 2008). Nevertheless, information about their utilization in systems for pesticide removal is scarce, in spite of this type of pollutants has been included in the list of hazardous compounds for the environment and human health (Häggblom & Valo, 1995).

Table 4 shows some group-representative pesticides on the WHO classification, where chitosan has been employed as a component of the formulation for release (controlled and/or protected), removal and/or biodegradation.

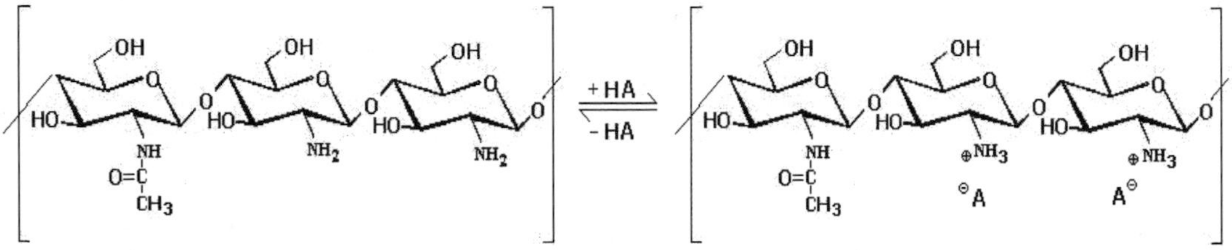

Figure 2. Protonation reaction of chitosan in acidic aqueous medium.

Table 4. Some representative-group pesticides on recommended WHO classification related to chitosan studies

Group IA	Group IB	Group II	Group III
Brodifacoum	**Dichlorvos**	**Carbaryl**	**Isoproturon**
Rodenticide	Insecticide	Insecticide	Herbicide
Parathion	**Warfarin**	**Diquat**	**Malathion**
Insecticide	Rodenticide	Herbicide	Insecticide
Ethoprophos	**Nicotine**	**Paraquat**	**MCPA**
Insecticide (soil)		Herbicide	Herbicide
Hexachlorobenzene		**Lindane**	**Dicamba**
Fungicide (seed)		Insecticide	Herbicide
Mercuric chloride		**7Imidacloprid**	**Oxadixyl**
Fungicide (soil)		Insecticide	Fungicide

2. PESTICIDES CONTROLLED DOSING EMPLOYING CHITIN AND CHITOSAN

2.1. Controlled Release of Pesticides

Substitution of traditional agrochemical formulations by controlled release systems helps, among other benefits, to avoid the employment of excessive quantities of active substances, constituting this action by itself an environmental protection strategy. The principal aims pursued with the use of these systems are:

- Protection of active agents
- To allow the automatic release of the proper active agent at the selected place and at appropriate rate.
- Ensure concentration level at optimal limits and time providing higher specificity and persistence.

- Sustained release in time is one of the goals more pursued in the agriculture because this allows assuming the control of diverse problems as:
- The effects of the released substances are spread in time, which produces substantial economic savings due to it is possible to exercise a better control of the employed quantities.
- The release happens when the plant needs it, generally in minor doses that those obtained when the active agent is added alone.
- Reduction of the number of applications, diminishing the contact of the workers with agrochemicals and the hours dedicated to this labor, as well as stress in the plants.
- Diminish the risk of human beings and animals by toxic contamination since localized release control is assured.
- Use of right agrochemical doses, which obviously carries minors economic costs.
- More friendly environmental systems, especially when degradation of the biomaterial used as support not affect the quality of the soil.

One of the firsts reports proposing the use of chitosan and chitosan derivatives (as membranes) for controlled release of agrochemicals was carried out by Hirano (1978). Likewise, chitin was proven as matrix for release of an agrochemical in the beginnings of the eighties, when the chemical union of the pesticide Metribuzin, and its subsequent release, to this biopolymer was reported (McCormik *et al.*, 1982). Later, Texeira *et al.*, (1990) reported the controlled release of Atrazine (a common herbicide used in corn fields) using films and pearls made with chitosan (and derivatives) hydrogels. These studies showed that herbicide covered with chitosan presents a rapid initial release, followed for a rate constant stage; similarly, chitosan hydrogels achieved to extend release period up to 7 months, compared with only 4 days obtained with the herbicide without covering.

More recently, the chemical coupling of the insecticide Carbaryl (1-naphthyl methylcarbamate) onto chitosan has been reported either by direct conjugation on the glucosamine ring or through a spacer between this and the insecticide moiety (Chirachanchai *et al.*, 2001). Table 5 shows some chitin/chitosan-based systems related to controlled release of agrochemical.

2.2. Protected Dosing of Pesticides

Some very ingenious systems that allow a relatively safe application – or protected dosing – of a chemical agent to a specific plague have been designed, as for example the "anti-escape traps" used to apply localized pesticides without affecting other susceptible species by lethal effects of the active agent. The rodenticide Brodifacoum has been employed in this type of application (Creeger & Fakiro, 2007). Chitosan has been used in these systems, taking advantage of the adhesive properties of their solutions, as one of the carrier polymers for the pesticide. Thus, it adheres to the rodent and is then eaten during the animal's grooming.

Table 5. Some chitin/chitosan-based systems related to controlled release of agrochemical

Matrix	Pesticida	References
Chemically substituted chitin	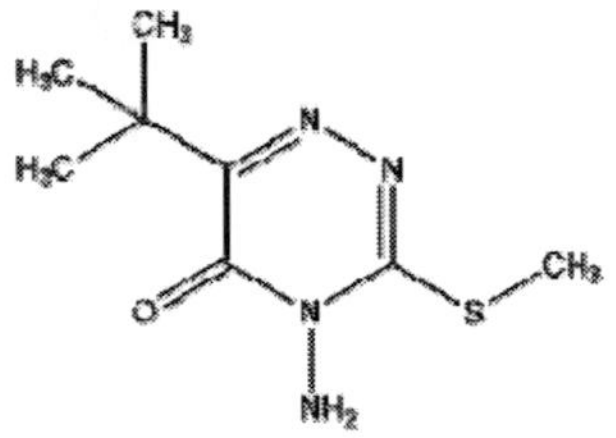	McCormick *et al.*, 1982

Metribuzin

| Films and pearls of chitosan | 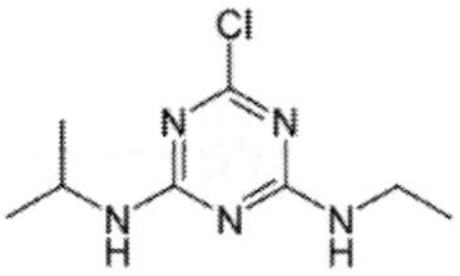 | Texeira *et al.*, 1990 |

Atrazine

| Microcapsules of chitosan prepared by interfacial reaction | 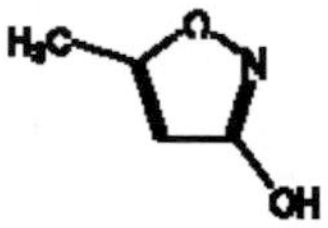 | Yeom *et al., 2000* |

3-methyl-5-hydroxyisoxazole

| Pesticide chemically modified chitosans | 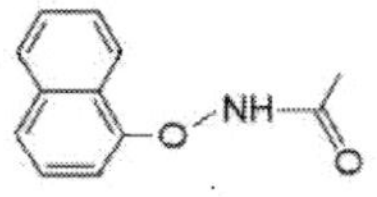 | Chirachanchai *et al.*, 2001 |

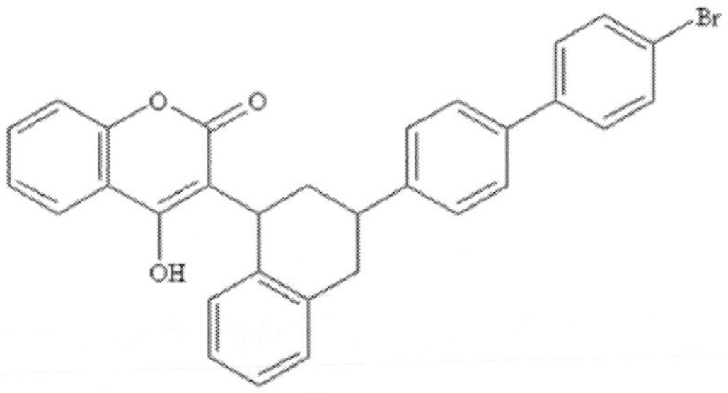

Carbaryl

Brodifacoum

On the other hand, very recent studies have shown that these materials can also be microencapsulated by the formation of polymer surface layers around pesticide microparticles, in order to protect other susceptible species from their action. In that sense, the insecticide Imidacloprid has been protected with alternating layers of chitosan and alginate, using the technique known as layer by layer auto-assembling (Guan *et al.*, 2008). Additionally, photocatalysts that promote the degradation of the micro-encapsulated pesticide were studied.

Imidacloprid

3. REMOVAL OF PESTICIDES USING CHITIN AND CHITOSAN

The search to low-cost systems for contaminants removal in water purification processes is a very active and extensive scientific research area because of the complexity involved in the diversity of pollutants that can water possess, considering that most of the times coming from dissimilar sources. Additionally, when considering the situation of water contaminated with pesticides, the picture becomes more worrying, especially in relation to the human health because, as it has been mentioned at the beginning, methods generally used in conventional water treatments seem to have little or no effect on the successful removal of them (U.S. Environmental Protection Agency, 2001).

On the other hand, although in certain cases some additional purification processes, including disinfection, irradiation and/or softening of the water, could promote the conversion of pesticides in less toxic products, there will always be the possibility that this procedures could also induce the formation of more toxic sub-products, whose effects are unknown because of the scarce information on them.

3.1. Coagulation/flocculation

Chitosan has been studied as a coagulant or flocculant for a wide variety of aqueous suspensions (Table 6) (Pan *et al.*, 1999; Huang & Chen, 1996) In many of these studies the mechanism responsible for separation have not been clearly established and the terms coagulation and flocculation can be used indistinctly (Roussy *et al.*, 2005), with some results pointing towards mechanisms of charge neutralization (Ashmore & Hearn, 2000) whereas others indicate that chitosan can operate by means of the bridging mechanism (Chen *et al.*, 2003; Guibal *et al.*, 2006). These discrepancies have been explained, in general terms, considering the effective density of charge of the polyelectrolyte employed (Strand *et al.*, 2003).

The ability of chitosan for coagulation and flocculation has been related to: a high content of –OH groups, which makes the polymer hydrophilic and contributes to chelating effects; cationic charge derived from protonation of the amine groups at acidic pH; an electron pair for each amine groups (more available at pH close to or greater than the pKa =6.5). In addition, chitosan with moderate to high molecular weight can provides bridging mechanisms for coagulation/flocculation process.

Table 6. Some systems in which chitosan has been studied as coagulant/flocculant agent

Suspension material	References
Yeast	Weir *et al.*, 1993
Proteins	Savant & Torres, 2000
Humic susbtances	Bratskaya *et al.*, 2002
'model colloid' polymer latices	Ashmore *et al.*, 2001
Mammalian cells	Riske *et al.*, 2007
Bacteria	Hughes *et al.*, 1990; Strand *et al.*, 2003
Oil/water emulsions	Bratskaya *et al.*, 2006
Latex particles	Ashmore & Hearn, 2000
Silt	Divakaran & Pillai, 2002

The use of chitosan as clarifying agent for water purification, regarding to the removal of turbidity, has been well documented for several years. Pioneering studies (Penistone & Johnson, 1970) rapidly associated a major efficiency in the clarification of suspensions of montmorillonite with a higher content of deacetylated groups on chitosan. Afterwards, studies showed that this biomaterial was also effective to reduce turbidity in suspensions of organic material proceeding from vegetable sources (Bough, 1975a) as well as of proteins-containing suspensions (Bough, 1975b).

Chitosan has been studied as flocculant either alone (Divakaran and Pillai, 2001), or in combination with other cationic flocculants (Pinotti *et al.*, 2001) or modified to increase its content of cationic groups (Jian-Ping *et al.*, 2007) or to increase the ionic character of these groups (Lárez *et al.*, 2003). Table 6 shows some of the systems for which coagulation/flocculation has been studied using chitosan as clarifying agent.

In spite of chitosan had been reported as an effective coagulant in surface water treatment, there is little information on its employment as coagulant agent for pesticides. Nevertheless, it would be expected that solutions of chitosan in acidic aqueous medium could work as coagulant for water-insoluble pesticide applied as a very fine dust (but that can later form colloidal suspensions in water) as for example the Methyl-parathion (solubility in water 50 ppm).

Methyl parathion

Some of these systems are really interesting considering that they might generate new materials with high potential to be used as adsorbents in processes of water purification, according to recent trends in the employment of immobilized biomass for these purposes (Aksu, 2005), including pesticide removal as discussed later on. Figure 3 shows a basic scheme for generation of new adsorbents, employing chitosan in coagulation/flocculation processes of aqueous biomass suspensions.

3.2. Filtration

Membrane separation technology has gained increasing interest due to new developments of highly selective materials which possess also good chemical and mechanical stabilities. Some of the more interesting applications of membranes related with separation of azeotropic and gaseous mixtures, and proteins, have been recently review by Xu *et al*. (2008).

The use of membranes for filtration processes generally result effective on organic pollutants removal when are applied after a coagulation (Leiknes *et al*., 2004) or adsorption (Ericsson & Trägardh, 1996) previous step. Although both processes (adsorption or coagulation) by themselves are little effectives to remove this type of pollutants, they are generally necessary to avoid the rapid plugging of the membrane by removing larger particles. Some similar systems that substantially improve the level of pesticide rejection have been reported (Van der Bruggen *et al*., 1998; Kosutic *et al*., 2005).

Chitosan has been studied in filtration systems for purifying of contaminated water with persistent organic pollutants (POP´s). In 1986, Thomé & van Daele reported the use of a chitosan filter as a complement to traditional activated carbon filter, which completely eliminated the components of the mixture known as Aroclor 1260, a polychlorinated biphenyl (PCB) mixture containing approximately 38% $C_{12}H_4Cl_6$, 41% $C_{12}H_3Cl_7$, 8% $C_{12}H_2Cl_8$, and 12% $C_{12}H_5Cl_5$ with an average chlorine content of 60%.

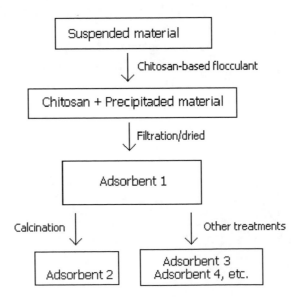

Figure 3. A basic scheme for generation of new adsorbents employing chitosan for precipitation of suspended biomass.

More recently, Assis & de Britto (2008) have reported the modification of glass porous membranes with complex polyelectrolytes formed by alternating deposition of chitosan and carboxymethylcellullose for removal of the herbicide Atrazine. These authors have proposed a tentative model for the binding of the herbicide to the polyelectrolyte complex implying the formation of hydrogen bridge links and/or complex charge transfer to non-specific sites, which are dynamically available as a function of conformational changes of the deposited polymers.

Atrazine

A potential application of chitosan based membranes for selective removal of chemical substances, including pesticides, is related to the use of molecular recognition sites in the DNA. These sites can generally bind some specific substances in a preferential form and may even discern between optical isomers. It is known that chitosan forms polyelectrolyte complexes with the DNA (Mao *et al.*, 2001) and, in general, the properties of such polycomplexes can be controlled by means of the selection of appropriated conditions for their formation (Higuchi *et al.*, 1997). A very elegant application of membranes formed with this type of complexes, which could serve as model for application in the removal of pesticides, is the chiral separation of phenylalanine with the chitosan/DNA polycomplex (DNA proceeding from testicles of salmon) (Matsuoka *et al.*, 2006).

Another possibility for pesticide removal using chitosan based membranes is related to current trend towards the formation of the so called molecularly imprinted membranes. During the manufacture of this type of membranes the molecule to retain is employed as a mold inside the polymer; later on it is taken out by repeated rinse with solvent and then an imprinted polymeric matrix is obtained, which preserves memory on the shape, size and interactions of the "printed" molecule (Andersson *et al.*, 1996; Silvestri *et al.*, 2004).

A similar system to the previously described can be formed by placing the molecule to retain between layers of polyelectrolytes of opposite charges, deposited layer by layer, as already it has been done in systems of liberation of drugs (Manna & Patil, 2008). Posterior removal of the mold molecule leaves available the sites for molecular recognition

3.3. Adsorption

Methods based on the adsorption of POP's have probably been the most effective route for removal them, having been tried a great variety of material adsorbents. Among the numerous studied adsorbents activated coal has been catalogued as the most effective due to its great superficial area, though also it has more elevated costs (Aksu, 2005). Table 7 presents a list of other materials that have been tested as adsorbents for pesticides.

Table 7. Several materials studied as adsorbents for pesticides removal

Adsorbent	Adsorbed pesticide	Reference
Zeolites	Atrazine, lindane and diazinone	**Lemić et al. (2006).**
Clays	Penconazole, linuron, alachlor, atrazine and metalaxyl	Sanchez-Martin et al. (2006)
Peat	Paraquat, Diquat, and Amitrole	Mac Carthy & Djebbar (1986)
Ocean peat moss	Azinphosmethyl	Aroguz (2006)
Fly ash bagasse	Lindane and malathion	Gupta et al. (2002)
Cork	α-Cypermethrin	Domingues et al. (2007)
Rice husk	Methyl parathion	Akhtar et al. (2007)
Diatomaceous earth	Atrazine and organo-phosphorus pesticides	Agdi et al., (2000)
Iron oxides	Atrazine, isoproturon, mecoprop, 2,4-D, and bentazone	Clausen & Fabricius (2001)
Pine bark	Lindane and heptachlor	Ratola et al. (2003)
Almond shell residues	Pentachlorophenol	Estevinho et al. (2006)
Rubber granules	2,4-D and Atrazine	Alam et al. (2007)
Pyrolized crab shells	Azinphosmethyl	Gulen et al. (2005)

Studies on Adsorption of Pesticides Using Chitin and Chitosan

Chitosan has received considerable attention as a possible adsorbent regarding metal removal due to its recognized chelating activity and, specially, because it can be obtained to low cost from a diversity of natural sources (Babel & Kurniawan, 2003). Similarly, it has also been extensively studied as adsorbent for dyes (Wu et al., 2001). Nevertheless, in spite of the increasing interest in these applications, there are a few references directly related to the adsorption of pesticides and mechanisms related to these processess.

Probably the first report on the capacity of these materials for pesticide adsorption was the work of Richards and Cutkomp (1946). These authors, associating the major sensibility for DDT poisoning to the possession of a chitinous cuticle and considering the capacity of chitin to adsorb DDT from colloidal suspensions, proposed that chitinous cuticles could concentrate DDT selectively to produce a higher dose inside the animals. Two years later Lord (1948) demonstrated that DDT – and some structurally related compounds (showed below) – can be adsorbed on this biomaterial in quantities approximately equal and with similar rates.

DDT (1:l:l-Trichloro-2:2-bis (4'-chlorophenyl)ethane

1:l:l-Trichloro-2:2-bis (4'-fluorophenyl)ethane

1:1:1-Trichloro-2:2-bis (4'-bromophenyl)ethane

1:1:1-Trichloro-2:2-bis (4'-iodophenyl)ethane

1:1:1-Trichloro-2:2-diphenyl)ethane

1:1:1-Trichloro-2:2-bis (4'-ethoxyphenyl)ethane

At 1972, Masri & Friedman compared the capacity of some materials, natural and synthetic (like chitosan, polymanines, cellulose derivatives, starch derived polyamines, and copolymers of styrene/amine-styren), to bind the fungicide mercuric chloride ($HgCl_2$), obtaining promising results with all of them. Their studies pointed towards the free amine groups in the chitosan and the polyamines as the binding sites for Hg^{+2} ions and, more importantly, towards the possible role of the natural polyamincs on distribution of the mercury in the environment.

Later, Davar & Wightman (1981) showed that as it should be expected because the cationic nature of the protonated amino groups, chitosan can strongly adsorb some chlorinated herbicides derived from the fenoxi-acetic acid, as for example MCPA; dicamba; 2,4-D; 2,4,5-T; whose structures are shown down below. The driven force for the fast adsorption of these substances on chitosan is related to the strong electrostatic attraction between the positively charged amine groups on the biopolymer ($-NH_3^+$) and the negatively charged carboxylic groups ($-COO^-$) from the acids. The occurrence of these interactions have been shown for several systems including: (a) linear chitosan/long hydrophobic chain carboxylic acid, with precipitation of the simplex formed (Wei & Hudson, 1993); (b) crosslinked chitosan/long hydrophobic chain, with the occurrence of the phenomenon denominated gel collapse (Lárez et al., 1995; Barreiro-Iglesias et al., 2005); (c) linear chitosan/linear negatively charged polyelectrolyte, with formation of the so called polyelectrolyte complexes (Lárez et al., 2002; Peniche and Arguelles, 2001); etc.

An important consideration for application of chitosan as adsorbent on the removal of this type of pesticides should have present that binding of the adsorbate can become very strong, specially in systems where may occur hydrophobic interactions through hydrophobic zones in the polymer (i.e. introduced by means of chemical modification of the original material (Jiang et al., 2006) or the hydrophobic tail of the adsorbates. These interactions favor the formation of hydrophobic domains, after of the initial ionic interaction (Lárez et al., 1995), which can to avoid the reutilization of the material.

The adsorption-desorption processes of insecticide Lindane to chitin have been studied as a model organic phase for simulating pesticide adsorption-desorption in marine systems (Gonzalez et al., 1992). These processes were studied as a function of chitin concentration, temperature, pH and salinity. Results have shown the existence of different classes of sites, with different accessibility, which at higher lindane concentration could be well described with a two sites Langmuir isotherm. At low lindane concentrations a single adsorption isotherm can be used properly. Increasing temperature and decreasing salinity resulted in both low lindane adsorption and in a more reversible process. On the other hand, an increase of pH resulted in lower adsorption of insecticide.

Lindane

Chemical modification of chitosan with substances that confer it a major hydrophobicity looks like a strategy to consider in the adsorption of hydrophobic pesticides. Chemical modification of chitosan to obtain more hydrophobic matrix has already been studied in other areas, as for example in systems for drug release, formation of complexes with surfactants, etc. In this sense, Martin *et al.* (2003) have studied the controlled release of the hydrophobic drug denbufylline with a glycol-chitosan (soluble in water) modified with palmitoyl chloride. Glycol-modified chitosan showed higher affinity to the drug than unmodified glycol-chitosan, and it can additionally to form physical hydrogels due to hydrophobic interactions generated through the hydrocarbonated tails of the palmitoyl residues. Similar studies for the chemical modification of chitosan with oleoyl chloride have showed that the quantity of modifier introduced in these systems can be well controlled (Lárez *et al.*, 2007). Another very important advantage of these systems is that this type of modifier agents is obtained of natural sources.

MCPA   Dicamba   2,4-D   2,4,5-T

Oleoylated chitosan

On the other hand, Tanada *et al.* (1993) realized studies on the *in vitro* removal of the extremely toxic herbicide known as Paraquat (Gramoxone) using granulate chitosan. The aim of the study was obviously to use it as primary treatment for acute poisonings with this substance, which is generally mortal after being consumed by human beings. Results showed that the quantity of Paraquat adsorbed on chitosan, as well as its rate of adsorption, are higher in a saline normal medium that in either gastric artificial solution or in pure water. The superficial area of the chitosan results to be a determinant factor in all the studied solutions.

Similarly, when the hydroxyl groups in the C-6 of the chitosan is replaced by carboxymethyl or sulfonic groups, obtained materials showed a better capacity to paraquat adsorption in pure water but an inhibition of the adsorption was observed when the studies were carried out in aqueous NaCl solutions (Nakamura *et al.*, 1993). Interestingly, opposite results were obtained when these modified chitosans were crosslinked, being observed an important increase in the removal of the paraquat when NaCl concentration in the external solution is smaller than 1.8%.

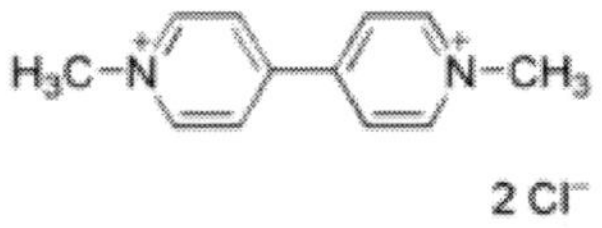

2 Cl⁻

Paraquat dichloride

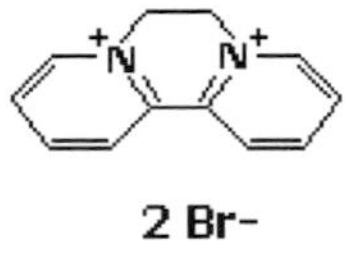

2 Br–

Diquat dibromide

Nakamura *et al.* (1993) also showed that polarity of these substances is very important to their adsorption on chitosan, being observed a bigger adsorption for diquat that for paraquat, due to lower polarity of paraquat. In this work was also shown the inhibitory effect of NaCl on the adsorption of the herbicides on modified chitosan with carboxylic and sulfonic pendant groups. Chitin has also been employed as adsorbent in aqueous medium for these herbicides (Bakasse *et al.* (2005). An additional study relative to pesticide adsorption on chitin is related to removal of the fungicide 4,4-*iso*-propylidene diphenol (Bisphenol A, BPA) and its derivate diphenylolpropane 4,4-dioxyaceticacid (BPAc) (Sismanoglu , 2007)

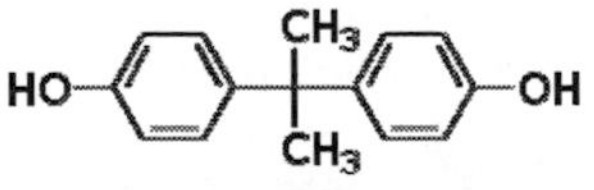

Biphenol A

Some studies have also been reported on the removal of the insecticide Methyl Parathion (Lou *et al.*, 1998; Yoshizuka *et al.*, 2000) employing glutaraldehyde and epichlorhydrine crosslinked chitosan based micro-particles (CMP´s), as well as the silver complexes of them (CMP-S´s). Results showed that epichlorhydrine- crosslinked CMP´s were superior for Ag⁺ ions adsorption that those crosslinked with glutaraldehyde but the glutaraldehyde-crosslinked CMP-S´s were more effectives for adsorption of methyl parathion. Latter MPC-S´s showed better capacity to be re-used.

In recent comparative studies related with the capacity of some drugs to bind to natural polymers it has been demonstrated that the rodenticide Warfarin (Coumadin) bind strongly to chitosan (Hiroki *et al.*, 2005). These results are interesting because later on Shao-Sung *et al.*

(2007) reported that ingestion of chitosan (as dietary complement) by people subjected to medicals treatments with this substance (used as anticoagulant) seems to promote its effect.

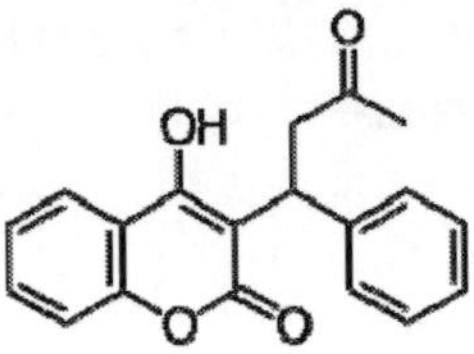

Warfarin

Adsorption of algicide Isoproturon has been studied as an intermediate step in a multiple water treatment study that included initial coagulation with poly(aluminum chloride) (Sarkar *et al.*, 2007), in an attempt to diminish the levels of pesticides, which, as mentioned previously, are little affected by conventional processes. Although treatment with chitosan produces better results than when using bentonite as adsorbent, the results obtained with activated carbon are significantly better.

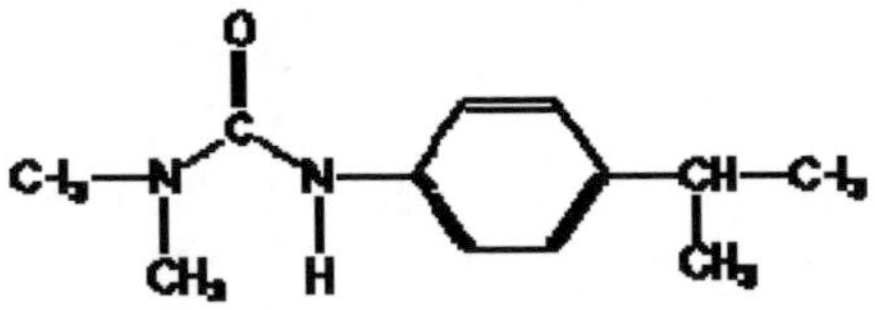

Isoproturon

More recently, in a very interesting study dos Santos *et al.* (2008) have shown that adsorption of the herbicide Trifluralin perfectly fits the Lagmuir isotherm model. Authors, based on the experimental values obtained for adsorption enthalpy (ΔH_{ads}^o = -10 kJ.mol^{-1}), have suggested that process involves the electrostatic attraction of the negative dipole in the nitro group of Trifluralin and protonated amino group of chitosan, although they do not completely rule out interaction of these groups with the tertiary aromatic amine of Trifluralin. Adsorption is less affected by the addition of NaCl to the aqueous medium when the herbicide concentration is low but it is inhibited to greater extent at higher concentrations, which may also indicate that there could be some effect associated to cooperative interaction. The results were elegantly complemented with the development of an electrochemical method for analysis of Trifluralin using a glassy carbon electrode modified with chitosan.

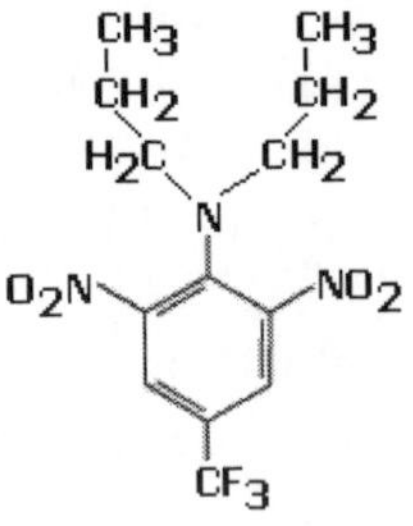

Trifluralin

Identification of electroactive species using the ability of chitosan to generate an electrode/solution interface with a greater partition coefficient for the analyte (Cruz *et al.*, 2000), has become an important tool in quantification of biomedical, such as dopamine (Wang *et al.*, 2006)and environmental related substances, such as pesticides. Some works of this type could provide valuable initial information on the removal of pesticides by using chitosan as an adsorbent although in some cases the ability of chitosan as adsorbent is not considered the important factor (Dua *et al.*, 2008).

Biosorption Using Absorbents Containing-Chitin/Chitosan Biomass

"Biosorption" is defined as the uptake of pollutants from aqueous solutions by non-growing or non-living microbial mass. The term has been related to diverse metabolism-independent processes such as physical and chemical adsorption, electrostatic interaction, ion exchange, complexation, chelation, and microprecipitation, which occur essentially in the cell wall rather than oxidation through anaerobic or aerobic biodegradation (Aksu, 2005). Although the potential of biomass related to pesticide adsorption has been known for several years (Voerman & Taemmen, 1969), we have considered adequate including in this review a small section related to the adsorption studies on these materials because its actual interest. Biomass can be usually obtained, at low cost, as wastes from a number of biotechnological processes, especially those implying inactivate biomaterials (heat killed, dried, acid and/or otherwise chemically treated). Table 8 shows some biomass-based system studied for pollutant removal. Although most of the studies on biomass-based adsorbents have been directed toward the removal of metallic ions (Mcafee *et al.*, 2001; Yan & Viraraghavan, 2008) there are also some reports on the adsorptive capacity of these materials for organic pollutants, which has been attributed, on part, to the presence of chitin and chitosan (Banks & Parkinson, 1992; Zhou & Banks, 1993; Gallagher *et al.*, 1997).

An interesting work in this direction, related with pesticide removal, is reported by Saiano & Ciofalo (2007). They have obtained an effective absorbent for removal of the fungicide Oxadixyl in aqueous solutions, which was prepared by alkaline hydrolysis of the *Phomopsis helianthi* mycelium. Chemical treatment originates a phase consisting of insoluble fractions of chitosan and glucans which may adsorb up to 6 mg of pesticide/gram of adsorbent. An excellent fitting to Langmuir isotherm model was obtained for the adsorption process.

Table 8. Some biomass-based systems studied for pollutant removal

Biomass source	Pollutant	References
Baker's yeast	Lindane, Dieldrin	Voerman & Taemmen, 1969
White rot fungi	Pentachlorophenol	Logan *et al.*, 1994
Rhizopus oryzae	Bromophenol blue	Gallagher *et al.*, 1997
Bacillus subtilis	2,4,6-Trichlorohenol	Daughney & Fein, 1998
Cladosporium sp	DDT, DDD and DDE	Juhasz & Naidu, 2000.
Sargassum muticum	Phenol; 2-chloro-phenol (2-CP), and 4-chlorophenol (4-CP)	Rubín *et al.*, 2006

Oxadixyl

To conclude this section it is important to indicate that the uses of chitin/chitosan-containing biomass as pollutant adsorbent surely will grow in the next years due to the lower costs of these materials and, especially, because information on recycling of chitin/chitosan based adsorbents is even scarce. Thus, studies on process related to this topic, such as desorption, ionic exchange, etc., for these materials will also be necessary in the immediate future.

Adsorption Of by-products Derived of Pesticide Transformations

In order to finalize this section related to adsorption of pesticides on chitin/chitosan is important to mention some works relatives to the adsorption of by-products coming of pesticide-degradation reactions, which have also been studied as a controlling mechanism of these pollutants. As it has been previously mentioned, information about pesticide degradation by-products, and their effects on environment and human health, is even insufficient. However, there are a lot of works related to the adsorption of phenol-related compounds (some of which are by-products coming from degradation reactions of certain pesticides) that could be considered as initial model systems to learn on control of these substances. Some few ones will be mentioned and briefly discussed.

Some interesting results have been reported by Ngah & Fatinathan (2006) in their studies on the capacity of chitosan flakes and chitosan pearls (glutaraldehyde-crosslinked chitosan) to adsorb p-nitrophenol (PNP), a by-product from the enzymatic degradation of the well know pesticides methyl parathion and parathion. According these authors, both systems were well described by the Freundlich isotherm model and results showed that crosslinked hydrogels have a greater capacity for adsorption (2.48 mg.g^{-1}) that the chitosan flakes (0.63 mg.g^{-1}) while adsorbate binds more rapidly to chitosan flakes (3.82×10^{-1} g.mg^{-1}.min^{-1}) that chitosan pearls (3.34×10^{-1} g.mg^{-1}.min^{-1}). Results were explained considering that chitosan pearls show higher BET surface area (0.69 m^2/g) than flakes chitosan (0.42 m^2/g), which facilitates the adsorption of p-nitrophenol due to its more loose pore structure.

Another study related to PNP adsorption on chitosan-based adsorbents has been reported by Uzun & Guzel (2004). They prepared a monocarboxymethyl-ated chitosan derivative (MCM-chitosan) by a reaction similar to:

$QNH_2 + Cl-CH_2-COOH \rightarrow QNH-CH_2-COOH + HCl$

and then compared its adsorption capacity with unmodified chitosan. Results demonstrated that chitosan is more effective to remove PNP due to the MCM-chitosan lacks on promoting PNP ionization reaction:

$NO_2-Bz-OH \leftrightarrow NO_2-Bz-O^- + H^+$

According these authors, MCM-chitosan behaves as typical zwitterions,

$QNH-CH_2-COOH \leftrightarrow QNH_2^+-CH_2-COO^-$

and the amino moieties exist mainly in the aminium form, which has a less proton-attracting character.

4. CHITIN AND CHITOSAN AS SUPPORT TO PESTICIDE DEGRADATION AGENTS

Degradation of organic pollutants in water has been carried out by chemical, biological and/or biochemcal processes. Chemical methods generally involve the use of strong oxidant compounds, ozonation, fenton reaction or heterogeneous photocatalysis, while biological degradation has been achieved by employing various enzymes or microorganisms (biofilters or activated sludge plants). Some approaches have involved (a) capturing of the pollutant present in the wastewater with an organic matrix (ranging from sugar cane bagasse (Crisafully et al., 2008) to biopolymers) followed by disposal and/or transformation of the used matrix; (b) continuous adsorption and later hydrolysis of the pollutant until decay.

The following sections intend to present a view of the use of chitin and/or chitosan as supports of biological active materials (i.e., enzymes, microorganisms) and catalysts for pesticide degradation with the intention that these can be considered as models of similar systems.

4.1. Enzymes Supported on Chitin and Chitosan

Kim et al. (2007) have reported the use of various enzymes to treat waste waters. Enzymes like peroxidase (in the presence of hydrogen peroxide) and laccases (in the presence of oxygen) catalyze the oxidation of a wide variety of pollutants compounds like phenol, biphenols, anilines, benzidines, and other aromatic compound (Durante et al., 2004). In particular, laccases have been used for the oxidation of phenolic dyes, phenols and chlorophenols, lignin-related diphenylmethanes, and organophosphorus compounds.

A number of oxidative enzymes from bacteria, fungi and plants have also been reported to play an important role in numerous waste treatment applications (Durán & Esposito, 2000). Peroxidases and phenoloxidases (horseradish peroxidase (HRP), lignin peroxidase and manganese peroxidase) can act on specific pollutants, transforming them into by-products easier to treat and disposal. Degradations of 2,4-dichlorophenol, 4-chlorophenol and 2-chlorophenol catalyzed by laccase (Zhanga *et al.*,2008) is a rgood example of an enzyme pesticide removing system. Despite of the advantage of these systems, enzyme recycling remains as a problem to solve. A partial solution commonly considered involves the enzyme immobilization on biopolymer matrix which usually improves its useful life and thereby a reduction in treatment cost is achieved.

Chitin and chitosan have been used as support (entrapment or immobilization) for enzyme, cells and microorganisms (Krajewska, 2004), in order to eliminate pesticides or toxicants in water. A wide variety of enzymes have been reported to be immobilized on/in chitin- and chitosan-based gels which can be incorporated by physical and chemical ways. For example, phenol-related contaminants in water can be chemiadsorbed on chitosan and then oxidize to quinone by means of an enzymatic process employing mushroom tyrosinase immobilized on this biopolymer (Dursun & Kalayci., 2005; Sun *et al.*, 1992).

Peroxidases can also be used to remove polychlorinated phenols (PCP's) from polluted wastewater (Quintanilla-Guerrero *et al.*, 2008) trough a process based on the redox reaction involving hydrogen peroxide. These enzymes are able to catalyze the oxidative polymerization of PCP's to form insoluble polymers (Ward *et al.*, 2001). This enzymatic treatment offers some advantages as a high degree of specificity, operation under mild conditions, high reaction rate, and, very important, low concentrations of soluble phenols are reached (Karam & Nicell, 1997). Nevertheless, during the removal process, a decrease in peroxidase activity has been observed, a common problem which in some cases has been solved by an adequate immobilization system with chitosan (Girelli *et al.*, 2006). Additionally, peroxidases from other sources such as soybean, turnip roots and bitter gourd, have been proposed as alternative to HRP, opening news an exciting opportunities in this field.

4.2. Microorganisms Supported on Chitin and Chitosan

Immobilization of cells and microorganisms on polymeric membranes and other materials seems to be an appropriated strategy to pollutant degradation. Some examples found in the recent literature including microalgae immobilization on diverse polymeric materials as polyurethane, polyacrylamide, polysaccharides, chitosan, etc., (Moreno-Garrido, 2008) This could be achieved by bead entrapment, carrier binding, adsorption techniques, encapsulation, cell coating, and film attachment (Chena *et al.*, 2007). Immobilization techniques may increase stability of biological organism to toxic environments, induce the retention of a higher concentration of microorganisms within the reactor media and help the separation of suspended biomass from waste effluents (dos Santos *et al.*, 2009). Interestingly, chitin and chitosan have been proposed as a convenient material to keep microbial cells alive (Odaci *et al.*, 2008). Some systems that have been already studied are briefly presented.

R. corynebacteriorides QBTo cells were immobilized in chitin and chitosan flakes by cultured together in both biopolymers to be employed to treat crude oil-contaminated seawater. The results showed that bioremediation was just significant when the strain immobilized on chitin and chitosan flakes were employed. Authors explained results considering the protective effect of the biopolymers which promoted biofilm formation and allowed the strain to survive (Gentili *et al.*, 2006). Similarly, algae *Scenedesmus sp. and*

Scenedesmus obliquus cells immobilized on chitosan beads have shown promising results on viability, growth and nutrients uptake and they were was efficient in removing phosphate and nitrate (Fierro *et al.*, 2008). However, authors have indicated that further studies are needed to prevent nitrite build up in water containing chitosan beads before its use for water quality management.

Another related work has reported that *Acidithiobacillus ferrooxidans*, a acidophilic bacterium capable of oxidizing ferrous sulfate, was immobilized on chitosan and crosslinked chitosan beads as a system for bioproduction of ferric iron, which could then be employed to desulfurize gases or in the treatment of acid mine drainages (Giaveno *et al.*, 2008).

Chena *et al* (2007) reported degradation of phenols on suspended and chitosan immobilized *Pseudomonas putida* cells forms. They observed that on the immobilized culture cell pH plays and important role for phenol degradation efficiency which may be attributed to surface properties of chitosan bed.

To conclude this section it is important to mention that studies related to the fabrication of biosensors for pesticide detection could be excellent sources of information to design biodegradation systems for these pollutants. A good example to take advantage of this type of information could be the recent paper published by Odaci *et al* (2008), where *Pseudomonas fluorescens* and *P. putida* cells immobilized on chitosan were placed at carbon and carbon nanotubes electrodes, obtaining good analytical responses when galactose, mannose and xylose were analysed (demonstrating that cells keep its activity even under the operation conditions of the biosensor). Thus, the employment of a system using *Pseudomonas fluorescens* cells immobilized with chitosan for some pesticide should be the following expected step considering that immobilized *P. putida* has already been reported to phenol removal using different chitosans (Chena *et al.*, 2007; Hsieh *et al.*, 2008).

4.3. Catalysts Supported on Chitin and Chitosan

Chitosan is characterized by a strong affinity for transition metals. The polymer can be used as a support for the preparation of heterogeneous catalysts in the form of colloids, flakes, gel beads, fibers (including hollow fibers), or immobilized on inorganic supports (alumina, silica, or other metal oxides). It has been valued as a suitable support because of its high sorption capacities for the catalytic metals, stability of metal ions (such as Pt and Pd), and physical (and chemical) versatility (Guibal *et al.*, 2005). It is a material with small specific surface area and low porosity. Additionally, the conformation of the polymer (together with its flexibility) is an important advantage for this kind of application. Despite of the work done for catalysts deposition on chitosan surfaces, at laboratory scale, water treatment information with this system is scarce.

On the other hand, in order to solve some disadvantages associated with the difficulty to separate chitosan particles when it is employed in powdered form, attempt to synthesize magnetic particles have been carried out. In fact, magnetic-modified particles of chitin, chitosan and alginate are materials suitable for PCP´s removal (Qu *et al.*, 2008). Also chitosan-bound Fe_3O_4 magnetic adsorbent has been used for acid dyes - crocein orange G (AO12) and acid green 25 (AG25)- adsorption in water (Qu *et al.*, 2008; Chang & Chen, 2005).

Some similar systems, containing an adequate catalyst, have been prepared for pollutant degradation applications. Thus, nanoscale Pd–Fe/chitosan and Pd–Fe/silica particles has been prepared and successfully employed, reaching the complete dechlorination of 1,2,4-trichlorobenzene. Also, TiO_2–chitosan/glass system has been proposed as a promising photodegradation–adsorption system to dye and monoazo wastewater treatment (Zainal *et al.*, 2008). Finally, a cross-linked chitosan-supported palladium catalyst (Guibal *et al.*, 2005) is claimed to be successfully used to degrade nitrophenol in aqueous dilute solutions using sodium formate as the hydrogen donor.

Other practical applications of these systems include electrochemical sensors based in similar systems have been reported to detect organophosphates insecticides such as parathion at nanogram scale (Qu *et al*, 2008).

5. FUTURE

The trend towards the increasing use of friendly environmentally substance seems to be an irreversible necessity in order to attempt preserving our planet. The consequences of the use of indiscriminate quantities of deleterious substances by mankind appear every time with major intensity (as it is being associated to the recent increment of natural disasters), forcing more drastic official restraints in many countries. Thus, chitin and chitosan seemed to have assured in the future an important position among the materials to be considered for environmental and human health applications owing to the numerous advantages that they posses, as it has previously been discussed.

Relative to chitin, its future use in this type of applications will largely depend on the discovery of more practical solvent systems than the known nowadays (i.e., 5% LiCl/N,N-dimehtyformamide), which only permit to dissolve small quantities of biopolymer and leave this contaminated with salts after a difficult solvent evaporation process. One of the major limitations of this abundant and cheap material has just been the lack of solvent systems that allow its processing, in major scale, for many other applications, including the formulation of systems related to the removal of pesticides such as adsorbents and filtration membranes. Thus, discovery of new solvent systems would facilitate preparation of chitin-based nanosystems containing immobilized either enzymes or microorganisms or catalysts, etc., which could be highly efficient in processes like biodegradation or monitoring of pollutants.

On the other hand, the processing of chitosan has less solvent limitations because its water solubility in acid medium. This fact provides to chitosan with a bigger number of applications that those of chitin. Thus, it has been used as a soluble coagulant/flocculant in water purification treatments. Limitations related to precipitation, observed when it is employed in basic medium, can be solved by an adequate chemical modification (i.e., N,N,N-trimethylated chitosan is soluble in a wide pH range). The use of a more hydrophobic or higher molecular weight chitosan can be a solution for systems requiring flocculants that work by bridging mechanism.

Flocculation of inactive biomass with chitosan can also generate some methods to produce a numerous variety of new adsorbents, as it has been schematically proposed in Figure 3. It is quite interesting that chitosan, besides acting as flocculant in these systems, can protect the co-precipitated biomass due to their biocide properties.

Regards to membrane filtration, chitosan appears as one of the best options to be considered due to the unlimited preparation possibilities that it can offer. Thus, chitosan could be used to construct membranes by diverse approaches, including the use of different chemical modifiers, physical and chemical crosslinking, molecularly imprinted membranes, polyelectrolyte complexes formation, etc., all of which can be applied to pesticide removal.

We also believe that studies in sorption/desorption process of pesticides on chitosan should be increased in the next years because nowadays it is necessary more information on these systems. This information may offer new insights on the reuse of these materials, diminishing costs. In the same direction, more research surely will be made on use of biomass-based adsorbents, including the search of new methods originating materials with higher chitin and/or chitosan proportions.

Perhaps the more exciting field on the use of chitosan for pesticide removal is that related to its use for immobilization. Certainly, it is a very active area with more and more investigations trying to build systems that imitate the nature. All of these developments are largely supported on the fast growth of biotechnology and genetic engineering, which has allowed create new cell-biosensors with more specific biorecognition elements. Thus, it is possible to obtain a biosensor that self-identify the type of pesticide present and, through electrochemical methods, liberates the appropriated microorganism or specific segment enzyme to the pesticide target.

Another possibility related to these intelligent systems should consider the use of model-based computer-aided (bioinformatic) designs for controlled pesticide release , which may become a tool to orientate the synthesis of intelligent and biomimetic systems (Muro-Suñe *et al.*, 2005). Thus, synthesis of new chitosan derivatives involving chelating and grafting groups will reinforce the fabrication of intelligent hydrogels where could coexist, covalently immobilized in different segments of the polymer, redox catalytic centers (HRP, lactase) and complexes molecular structures, like cyclodextrines (with adequate cavities to support metal ions and microorganism) (El-Tahlawy *et al.*, 2006).

Finally, bi and tri-metallic catalysts nanoparticles could be dispersed in the biopolymer matrix or biopolymers membranes in order to ensure pesticide catalytic conversion to non-toxic by-products (Ghauch & Tuqan, 2008). Thus, similar new nanoparticle systems, via encapsulation or electrostatic/ magnetic interaction with the biopolymer, should be developed in the next years.

REFERENCES

Agdi, K., Bouaid, A., Martin, A., Fernández, P., Azmani, A., Camara, C. (2000). Removal of atrazine and four organo-phosphorus pesticides from environmental waters by diatomaceous earth-remediation method. *Journal of Environmental Monitoring, 2*, 420-423.

Ahmad, A., Sumathi, S., Hammed, B. (2004). Chitosan: A natural biopolymer for the adsorption of residue oil from oily wastewater. *Adsorption Science & Technology, 22*, 75-88.

Aksu, Z. (2005). Application of biosorption for the removal of organic pollutants: a review. *Process Biochemistry, 40*, 997–1026.

Akhtar, M., Moosa, Hasany, S., Bhanger, M., Iqbal, S. (2007). Low cost sorbents for the removal of methyl parathion pesticide from aqueous solutions. *Chemosphere, 66*, 1829-1838.

Alam, J., Dikshit, A., Bandyopadhyay M. (2007). Kinetic Study of Sorption of 2,4-D and Atrazine on Rubber Granules. *Journal of Dispersion Science and Technology, 28*, 511-517.

Andersson, L., Nichols, I., Mosbach, K. (1996). Molecular imprinting: the current status and future development of polymer-based recognition. *Advance Molecular Cell Biology, 15B*, 651-670.

Aroguz, A. (2006) Kinetics and thermodynamics of adsorption of azinphosmethyl from aqueous solution onto pyrolyzed (at 600°C) ocean peat moss (*Sphagnum* sp.). *Journal of Hazardous Materials, 135*, 100-105.

Ashmore, S. & Hearn, J. (2000). Flocculation of model latex particles by chitosans of varying degrees of acetylation. *Langmuir 16*, 4906-4911.

Ashmore, S., Hearn, J., Karpowicz, F. (2001). Flocculation of latex particles of varying surface charge densities by chitosans. *Langmuir, 17*, 1069-1073.

Assis, O. & de Britto, D. (2008). Formed-in-place Polyelectrolyte Complex Membranas for Atrazine Recovery from Aqueous Media. *Journal of Polymers and the Environment*, [DOI 10.1007/s10924-008-0101-z]

Babel, S. & Kurniawan, T. (2003). Low-cost adsorbents for heavy metals uptake from contaminated water: a review. *Journal of Hazardous Materials, B97*, 219-243.

Bakasse, M., Choukry, N., El Gaini, L., Hatim, Z., Tabyaoui, B., Brahmi, R. (2005). *Adsorption of Paraquat, Doquat and Methylene Blue onto Chitin in aqueous solution.* The Africans Materials Research Society, Third International Conference. Marrakeck; Morroco,

Banks, C.J. & Parkinson, M.E. (1992). The mechanism and application of fungal biosorption to colour removal from raw water. *Journal of Chemical Technology & Biotechnology, 54*, 192-196.

Barka, E., Eullaffroym P., Climentm C., Vernetm G. (2004). Chitosan improves development and protects *Vitis vinifera* L. against *Botrytis cinerea. Plant Cell Reports, 22*, 608-614.

Barreiro-Iglesias B., Alvarez-Lorenzo, C., Concheiro, A. (2005). Chitosan/sodium dodecylsulfate interactions. Calorimetric titration and consequences on the behaviour of solutions and hydrogel beads. *Journal of Thermal Analisis and Calorimetry, 82*, 499-505.

Bough, W. (1975a). Reduction of suspended solids in vegetable canning waste effluents by coagulation with chitosan. *Journal of Food Science, 40*, 297-301.

Bough, W. (1975b). Coagulation with chitosan: an aid to recovery of byproducts from egg breaking wastes. *Poultry Science, 54*, 1904-1912.

Bratskaya, S., Avramenko, V., Sukhoverkhov, S., Schwarz, S. (2002). Flocculation of humic substances and their derivatives with chitosan. *Colloid Journal, 64*, 681-685.

Bratskaya, S., Avramenko, V., Schwarz, S., Philippova, I., (2006) Enhanced flocculation of oil-in-water emulsions by hydrophobically modified chitosan derivatives. *Colloids and Surfaces A, 275*, 168-176.

Cestari, A., Vieira, E., dos Santos, A., Mota, J., de Almeida, V. (2004). Adsorption of anionic dyes on chitosan beads. 1. The influence of the chemical structures of dyes and temperature on the adsorption kinetics. *Journal of Colloid and Interface Science, 280*, 380–386.

Chang, Y.C. & Chen, D.H. (2005). Preparation and adsorption properties of monodisperse chitosan-bound Fe_3O_4 magnetic nanoparticles for removal of Cu(II) ions. *Journal of Colloid and Interface Saciencie, 283*, 446-451.

Chen, L., Chen, D., Wu, C. (2003). A New Approach for the Flocculation Mechanism of Chitosan. *J. Polymers and the Environment, 11*, 87-92.

Chena, Y-M., Lin, T-F., Huangb, C., Lin, J-C., Hsieh, F-M. (2007). Degradation of phenol and TCE using suspended and chitosan-bead immobilized *Pseudomonas putida*. *Journal of Hazardous Materials 148*, 660–670.

Chirachanchai, S., Lertworasirikul, A., Tachaboonyakiat, W. (2001). Carbaryl insecticide conjugation onto chitosan via iodochitosan and chitosan carbonyl imidazolide precursors. *Carbohydrate Polymers, 46*, 19-27.

Clausen, L. & Fabricius, I. (2001). Atrazine, isoproturon, mecoprop, 2,4-D, and bentazone adsorption onto iron oxides. *J. Environmental Quality, 30*, 858–869.

Creeger, S. & Fakiro, U. (2007). US Patent 20070251139.

Crini, G. (2005). Recent developments in polysaccharide-based materials used as adsorbents in wastewater treatment. *Progress in Polymer Science, 30*, 38–70.

Crisafully, R., Milhome, M. , Cavalcante, R., Silveira, E., Keukeleire, D., Nascimento, R. (2008). Removal of some polycyclic aromatic hydrocarbons from petrochemical wastewater using low-cost adsorbents of natural origin. *Bioresource Technology 99*, 4515–4519.

Cruz, J., Kawasaki, M., Gorski, W. (2000). Electrode Coatings Based on Chitosan Scaffolds. *Analytical Chemistry, 72*, 680-686.

Daughney, C.J. & Fein, J.B. (1998). Sorption of 2,4,6-Trichlorophenol by Bacillus subtilis. *Environmental Science & Technology, 32*, 749-752.

Davar, P. & Wightman, J.P. (1981). In: P. H. Tewari (Eds.), *Adsorption from Aqueous Solutions* (pp. 163–177). New York: Plenum Publ. Corp.

Devlieghere, Г., Vermeulen, A., Debevere, J. (2004). Chitosan: antimicrobial activity, interactions with food components and applicability as a coating on fruit and vegetables. *Food Microbiology, 21*, 703–714.

Divakaran, R. & Pillai, V. (2001). Flocculation of kaolinite suspensions in water by chitosan. *Water Research, 35*, 3904–3908.

Divakaran, R. & Pillai, V. (2002). Flocculation of river silt using chitosan. *Water Research, 36*, 2414–2418.

Domingues, V., Priolo, G., Alves, A., Cabral, M., Delerue-Matos, C. (2007). Adsorption behavior of α -cypermethrin on cork and activated carbon. *J. Environmental Science and Health, B42*, 649-654.

dos Santos, A., Valentim, A., Goulart, M., Caxico, F. (2008). Adsorption Studies of Trifluralin on Chitosan and its Voltammetric Determination on a Modified Chitosan Glassy Carbon Electrode. *Journal of Brazilian Chemistry Society, 19*, 704-710.

dos Santos, V. L., de Souza Monteiroa, A., Telles Bragaa, D., Matos Santoro, M. (2009). Phenol degradation by *Aureobasidium pullulans* FE13 isolated from industrial effluents. *Journal of Hazardous Materials 161*, 1413–1420.

Dua, D., Ye, X., Zhang, J., Zeng, Y., Tu, H., Zhang, A., Liu, D. (2008). Stripping voltammetric analysis of organophosphate pesticides based on solid-phase extraction at zirconia nanoparticles modified electrode. *Electrochemistry Communications, 10,* 686–690.

Durán, N. & Esposito, E. (2000). Potential applications of oxidative enzymes and phenoloxidase-like compounds in wastewater and soil treatment: a review. *Applied Catalysis B: Environmental, 28,* 83–99.

Durante, D., Casadio, R., Martelli, L., Tasco, G., Portaccio, M., De Luca, P., Bencivenga, U., Rossi, S., Di Martino, S., Grano, V., Diano, N., Mita, D.G. (2004). Isothermal and non-isothermal bioreactors in the detoxification of wastewaters polluted by aromatic compounds by means of immobilized laccase from *Rhus vernicifera, Journal of Molecular Catalysis B: Enzymatic 27,* 191–206.

Dursun, A.Y. & Kalayci, C. S. (2005). Equilibrium, kinetic and thermodynamic studies on the adsorption of phenol onto chitin. *Journal of Hazardous Materials, B123,* 151–157.

El-Tahlawy, K., Gaffar, M.A., El-Rafie, S. (2006). Novel method for preparation of b-cyclodextrin/grafted chitosan and its application. *Carbohydrate Polymers 63,* 385–392.

Ericsson, B. & Trägårdh, G. (1996). Treatment of surface water rich in humus — Membrane filtration vs. conventional treatment. *Desalination, 108,* 117–128.

Estevinho, B.N., Ratola, N., Alves, A., Santos. L. (2006). Pentachlorophenol removal from aqueous matrices by sorption with almond shell residues. *Journal of Hazardous Materials, B137,* 1175-1181.

Fierro, S., Sánchez-Saavedra, M., Copalcúa, C. (2008) Nitrate and phosphate removal by chitosan immobilized Scenedesmus. *Bioresource Technology, 99,* 1274–1279.

Gallagher, K.A., Healy, M.G., Allen, S.J. (1997). Biosorption of synthetic dye and metal ions from aqueous effluents using fungal biomass. In: Wise DL. (Ed.). Global Environmental Biotechnology (pp. 27–50). UK: Elsevier.

Gentili, A.R., Cubitto, M.A., Ferrero, M., Rodriguéz, M.S. (2006). Bioremediation of crude oil polluted seawater by a hydrocarbondegrading bacterial strain immobilized on chitin and chitosan flakes. *International Biodeterioration & Biodegradation 57,* 222–228.

Ghauch, A. & Tuqan, A. (2008). Catalytic degradation of chlorothalonil in water using bimetallic iron-based systems. *Chemosphere, 73,* 751–759.

Giaveno, A., Lavalle, L., Guibal, E., Donati, E. (2008). Biological ferrous sulfate oxidation by A. ferrooxidans|immobilized on chitosan beads. *Journal of Microbiological Methods, 72,* 227–234.

Girelli, A.M., Mattei, E., Messina, A. (2006). Phenols removal by immobilized tyrosinase reactor in on-line high performance liquid chromatography. *Analytica Chimica Acta, 580,* 271–277.

Gonzalez, M., Perez, J., Santana, J. (1992). Lindane Adsorption-Desorption on Chitin in Sea Water. *International Journal of Environmental Analytical Chemistry, 46,* 175-186.

Guan, H., Chi, D., Yu, J., Li, X. (2008). A novel photodegradable insecticide: Preparation, characterization and properties evaluation of nano-Imidacloprid. *Pesticide Biochemistry & Physiology, 92,* 83-91.

Guibal, E. (2005). Heterogeneous catalysis on chitosan-based materials: a review. *Progress in Polymer Science, 30,* 71-109.

Guibal, E., Van Vooren, M., Dempsey, B., Roussy, J. (2006). A Review of the Use of Chitosan for the Removal of Particulate and Dissolved Contaminants. *Separation Science and Technology, 41,* 2487-2514.

Gulen, J., Aroguz, A.Z., Dalgın, D. (2005). Adsorption kinetics of azinphosmethyl from aqueous solution onto pyrolyzed Horseshoe sea crab shell from the Atlantic Ocean. *Bioresource Technology, 96*, 1169-1174.

Gupta, V.K., Jain, C.K., Ali, I., Chandra, S., Agarwal, S. (2002). Removal of lindane and malathion from wastewater using bagasse fly ash—a sugar industry waste. *Water Research, 36*, 2483-2490.

Häggblom, M.M. & Valo, J.R. (1995). Bioremediation of chlorophenol wastes. In: Young LY, Cerniglia CE, (Eds.) Microbial transformation and degradation of toxic organic chemicals (pp. 389–434). New York: Wiley-Liss Inc;.

Higuchi, A., Hashimoto, T., Yonehara, M., Kubota, N., Watanabe, K., Uemiya, S., Kojima, T., Hara, M. (1997). Effect of surfactant agents and lipids on optical resolution of amino acid by ultrafiltration membranes containing bovine serum albumin. *Journal of Membrane Science, 130*, 31–39.

Hirano, S. (1978). A facile method for the preparation of novel membranes from *N*-acyl and *N*-arylidine chitosan gels. *Agricultural and Biological Chemistry, 42,* 1938.

Hiroki, S., Tooru, M., Mitsutoshi, A., Setsu, Y., Atsushi, M., Kenji, K., Yoshiro, O., Keishi, Y., Makoto, A., Masaka, O. (2005). Investigation of Binding of Drugs with Natural Polymer Supplements. *Japanese Journal of Pharmaceutical Health Care and Sciences, 31*, 744-748.

Hsieh, F.M., Huang, C., Lin, T.F, Chen, Y.M., Lin, J.C. (2008). Study of sodium tripolyphosphate-crosslinked chitosan beads entrapped with *Pseudomonas putida* for phenol degradation. *Process Biochemistry, 43*, 83–92.

Huang, C. & Chen, Y. (1996). Coagulation of colloidal particles in water by chitosan. *Journal of Chemical Technology & Biotechnology, 66*, 227-232.

Hughes, J., Ramsden, D., Symes, K. (1990). The flocculation of bacteria using cationic synthetic flocculants and chitosan. *Biotechnology Techiques, 4*, 55–60.

Jiang, G.B., Quan, D., Liao, K., Wang H. (2006). Novel Polymer Micelles Prepared from Chitosan Grafted Hydrophobic Palmitoyl Groups for Drug Delivery. *Mol. Pharm., 3,* 152–160

Jian-Ping, W., Yong-Zhen, C., Xue-Wu, Ge., Han-Qing, Y. (2007). Optimization of coagulation–flocculation process for a paper-recycling wastewater treatment using response surface methodology. *Colloids and Surfaces A: Physicochemical and Engineering Aspects, 302*, 204-210.

Juhasz, A.L. & Naidu, R. (2000). Extraction and recovery of organochlorine pesticides from fungal mycelia. *Journal of Microbiological Methods, 39,* 149-58.

Karam, J. & Nicell, J.A. (1997). Potential Applications of Enzymes in Waste Treatment. *Journal of Chemical Technology & Biotechnology, 69,* 141-153.

Kawamura, S. (1991). Effectiveness of Natural Polyelectrolytes in Water Treatment. *Journal of the American Water Works Association, 83*, 88-91.

Kim, E.Y., Chae, H.J., Chu, K.H. (2007). Enzymatic oxidation of aqueous pentachlorophenol. *Journal of Environmental Sciences 19*, 1032–1036.

Kosutic, K., Furac, L., Sipos, L., Kunst, B. (2005). Removal of arsenic and pesticides from drinking water by nanofiltration membranes. *Separation and Purification Techniques, 42*, 137–144.

Krajewska, B. (2004). Application of chitin- and chitosan-based materials for enzyme immobilizations: a review, *Enzyme and Microbial Technology 35*, 126–139.

Lárez, C., Crescenzi, V., Dentini, M., Ciferri, A. (1995). Assemblies of amphiphilic compounds over rigid polymers: 2. Interaction of sodium dodecyl-benzenesulfonate with chitosan/scleraldehyde gels. *Supramolecular Science, 2*, 3-4.

Lárez, C., Canelón, F., Millán, E., Katime, I. (2002). Interpolymeric complexes of poly(itaconic acid) and chitosan. *Polymer Bulletin, 48*, 361-366.

Lárez, C., Lozada, L., Millán, E., Katime, I., Sasía, P. (2003). La densidad de carga de polielectrolitos y su capacidad de neutralización en sistemas coloidales. *Revista Latinoamericana de Metalurgia y Materiales, 23*, 16-20.

Lárez, C., Rivas, A., Velásquez, W., Bahsas, A. (2007). Amidación del quitosano con cloruro de oleoilo. *Revista Iberoamericana de Polímeros, 8*, 229-240.

Lárez, C. (2008). Algunas potencialidades de la quitina y el quitosano para usos relacionados con la agricultura en Latinoamérica. *UDO Agrícola, 8*, 1-22.

Leiknes, T., Odegaard, H., Myklebust, H. (2004). Removal of natural organic matter (NOM) in drinking water treatment by coagulation–microfiltration using metal membranes. *Journal of Membrane Science, 242*, 47–55.

Lemić, J., Kovacević, D., Tomasević-Canović, M., Kovacević, D., Stanić T., Pfend, R. (2006). Removal of atrazine, lindane and diazinone from water by organo-zeolites. *Water Research, 40*, 1079-85.

Logan, B.E., Alleman, B.C., Amy, G.L., Gilbertson, R.L., (1994). Adsorption and removal of pentachlorophenol by white rot fungi in batch culture. *Water Research, 28*, 1533–8.

Lord, K.A. (1948). The Contact Toxicity of a Number of D.D.T. Analogues and of Four Isomers of Benzene Hexachloride to Macrosiphoniella San Born I and Oryzaephilus Surinamensis. *Annals of Applied Biology, 35*, 505–526.

Lou, Z., Yoshizuka, K., Inoue, K. (1998). Adsorption behavior of methyl parathion on silver-complex chitosan microspheres. *Nippon Ion Kokan Gakkai, Nippon Yobai Chushutsu Gakkai Rengo Nenkai Koen Yoshishu, 79*, 14-17.

Mac Carthy, P. & Djebbar, K. (1986). Removal of Paraquat, Diquat, and Amitrole from Aqueous Solution by Chemically Modified Peat. *Journal of Environmental Quality, 15*, 103-107.

Majeti, N.V. & Ravi, K. (2000). A review of chitin and chitosan applications. *Reactive & Functional Polymers, 46*, 1-27.

Manna, U. & Patil, S. (2008). Encapsulation of Uncharged Water-Insoluble Organic Substance in Polymeric Membrane Capsules via Layer-by-Layer Approach. *Journal of Physical Chemistry B, 112*, 13258-13262.

Mao, H., Roy, K., Troung-Le, V., Janes, K., Lin, K., Wang, Y., August, J., Leong, K. (2001). Chitosan-DNA nanoparticles as gene carriers: synthesis, characterization and transfection efficiency. *Journal of Controlled Release, 70*, 399-421.

Martin, L., Wilson, C., Koosha, F., Uchegbu, I. (2003). Sustained buccal delivery of the hydrophobic drug denbufylline using physically cross-linked palmitoyl glycol chitosan hydrogels. *European Journal of Pharmaceutics and Biopharmaceutics, 55*, 35–45.

Masri, M. & Friedman, M. (1972). Mercury Uptake by Polyamine-Carbohydrates. *Environmental Science & Technology, 6*, 745-746.

Masri, M., Reuter, F., Friedman, M. (1974). Binding of metal cations by natural substances. *Journal of Applied Polymer Science, 18*, 675-681.

Matsuoka, Y., Kanda, N., Lee, Y.M., Higuchi, A. (2006). Chiral separation of phenylalanine in ultrafiltration through DNA-immobilized chitosan membranes. *Journal of Membrane Science, 280,* 116-123.

Mcafee, B.J., Gould, W.D., Nadeau, J.C., da Costa, A.C.A. (2001). Biosorption of metal ions using chitosan, chitin and biomass of *Rhizopus Oryzae. Sep Sci Technol, 36,* 3207–3222.

McCormick, C.L. Anderson, K.W., Hutchison, B.II. (1982). Controlled Activity Polymers with Pendently Bound Herbicides. *Polymer Reviews, 22,* 57-87.

Moreno-Garrido, I. (2008). Microalgae immobilization: Current techniques and uses. *Bioresource Technology, 99,* 3949–3964.

Muro-Suñe, N.., Gani, R., Bell, G., Shirley, I. (2005). Model-based computer-aided design for controlled release of pesticides. *Computers and Chemical Engineering 30,* 28–41.

Muzzarelli, R.A.A. (1973). Alginic acid, chitin and chitosan. In: *Natural chelating polymers.* (1st edition, pp. 177–227). Englewood Cliffs, NJ: Pergamon Press.

Nakamura, T., Kyotani, S., Kawasaki, N., Tanada, S., Nishioka, Y. (1993). In vitro adsorption of paraquat onto substituted chitosan beads. *Nippon Eiseigaku Zasshi., 48,* 973-979.

Ngah, W. & Fatinathan, S. (2006). Chitosan flakes and chitosan–GLA beads for adsorption of *p*-nitrophenol in aqueous solution. *Colloids and Surfaces A: Physicochemical and Engineering Aspects, 277,* 214–222.

Odaci, D., Timur, S., Telefoncu, A. (2008). Bacterial sensors based on chitosan matrices. *Sensors and Actuators, B134,* 89–94.

Pan, J., Huang, C., Chen, S., Chung, Y. (1999). Evaluation of a modified chitosan biopolymer for coagulation of colloidal particles. *Colloids and Surfaces A: Physicochemical and Engineering Aspects, 147,* 359–364.

Peniche, C. & Arguelles-Monal, W. (2001). Chitosan based polyelectrolyte complexes. *Macromolecular Symposia, 168,* 103–116.

Penistone, Q. & Jonson, E. (1970). Treating an aqueous medium with chitosan and derivative of chitin to remove a impurity. US Patent 3,533,940.

Pinotti, A., Bevilacqua, A., Zaritzky, N. (2001). Comparison of the Performance of Chitosan and a Cationic Polyacrylamide as Flocculants of Emulsion Systems. *Journal of Surfactants and Detergents, 4,* 57-63.

Prapagdee, B., Kotchadat, K., Kumsopa, A., Visarathanonth, N. (2007). The role of chitosan in protection of soybean from sudden death syndrome caused by *Fusarium solani* f. sp. *Glycines. Bioresource Technology, 98,* 1353-1358.

Qu, Y., Min, H., Wei, Y., Xiao, F., Shi, G., Li, X., Jin, L. (2008). Au–TiO2/Chit modified sensor for electrochemical detection of trace organophosphates insecticides. *Talanta 76,* 758–762.

Quintanilla-Guerrero, F., Duarte-Vázquez, M.A., García-Almendarez, B.E., Tinoco, R., Vazquez-Duhalt, R., Regalado, C. (2008). Polyethylene glycol improves phenol removal by immobilized turnip peroxidase. *Bioresource Technology, 99,* 8605–8611.

Ratola, N., Botelho, C., Alves, A. (2003). The use of pine bark as a natural adsorbent for persistent organic pollutants - study of lindane and heptachlor adsorption. *Journal of Chemical Technology & Biotechnology, 78,* 347-351.

Richards, A. & Cutkomp, L. (1946). Correlation between the possession of a chitinous cuticle and sensitivity to DDT. *Biological Bulletin of Woods Hole, 90,* 97-108.

Riske, F., Schroeder, J., Belliveau, J., Kang, X., Kutzko, J., Menon, M. (2007). The use of chitosan as a flocculant in mammalian cell culture dramatically improves clarification throughput without adversely impacting monoclonal antibody recovery. *Journal of Biotechnology, 128,* 813-823.

Roussy, J., Van Vooren, M., Dempsey, B.A., Guibal, E. (2005). Influence of chitosan characteristics on the coagulation and the flocculation of bentonite suspensions. *Water Research, 39,* 3247-3258.

Rubín, E., Rodríguez, P., Herrero, R., Sastre, M.E. (2006). Biosorption of phenolic compounds by the brown alga *Sargassum muticum. J Chem Technol & Biotechnol, 81,* 1093-1099.

Saiano, F. & Ciofalo, M. (2007). Removal of Pesticide Oxadixyl from an Aqueous Solution. *Bioremediation Journal, 11,* 57–60.

Sanchez-Martin, M.J., Rodriguez-Cruz, M.S., Andrades, M.S., Sanchez-Camazano, M. (2006). Efficiency of different Clay Minerals modified with cationic Surfactant in the adsorption of Pesticides: Influence of Clay type and Pesticide Hydrophobicity. *Applied Clay Science, 31,* 216-228.

Sarkar, B., Venkateswralub, N., Nageswara Raob, R., Bhattacharjeec, C., Kalea, V. (2007). Treatment of pesticide contaminated surface water for production of potable water by a coagulation–adsorption–nanofiltration approach. *Desalination, 212,* 129–140.

Savant, V. & Torres, J. (2000). Chitosan-based coagulating agents for treatment of cheddar cheese whey. *Biotechnology Progress, 16,* 1091–1097.

Shao-Sung, H., Shih-Hsien, S., Chern-En, C. (2007). Chitosan Potentiation of Warfarin Effect. *The Annals of Pharmacotherapy, 41,* 1912-1914.

Silvestri, D., Cristallini, C., Ciardelli, G., Giusti, P., Barbani, N. (2004). Molecularly imprinted bioartificial membranes for the selective recognition of biological molecules. *Journal of Biomaterials Science, Polymer Edition, 15,* 255-278.

Sismanoglu, T. (2007). Removal of some fungicides from aqueous solution by the biopolymer chitin, *Colloids and surfaces. A, Physicochemical and engineering aspects. 297,* 38-45.

Strand, S., Varum, K., Østgaard, K. (2003). Interactions between chitosans and bacterial suspensions: adsorption and flocculation. *Colloids and Surfaces B: Biointerfaces, 27,* 71-/81.

Sun, W-Q., Payne, G.F., Moas, M.S., Chu, J. H., Wallace, K.K. (1992). Tyrosinase Reaction/Chitosan Adsorption for removing phenols from wastewater. *Biotechnology Progress, 8,* 179-186

Tanada, S., Kyotani, S., Nakamura, T., Nishioka, Y. (1993). In vitro paraquat removal with granular chitosan. *Journal of Environmental Science and Health. Part A, Environmental science and engineering, 28,* 671-682.

Teixeira, M.A., Paterson, W.J., Dunn, E.J., Li, Q., Hunter, B.K., Goosen, M.F. (1990). Assessment of Chitosan gels for the controlled release of agrochemicals. *Industrial & Engineering Chemistry Research, 29,* 1205–1209.

Thomé, J. & Van Daele, Y. (1986). Adsorption of polychlorinated biphenyls (PCB) on chitosan and application to decontamination of polluted stream waters. In: Muzzarelli, R., Jeuniaux, C. & Gooday, G.W. (Eds.), Proceedings of the Third International Conference on Chitin and Chitosan (pp. 551–554).Italy: Senigallia.

U.S. Environmental Protection Agency (2000). Summary of Pesticide Removal/Transformation Efficiencies from various Drinking Water Treatment Processes. Committee to Advise on Reassessment And Transition (CARAT). October 3.

Washington DC, USA. Available on line in: http://www.epa.gov/oppfead1/carat/2000/oct/dw4.pdf

U.S. Environmental Protection Agency. (2001). The Incorporation of Water Treatment Effects on Pesticide Removal and Transformations in Food Quality Protection, Act (FQPA) Drinking Water Assessments. Office of Pesticide Programs. October 25. Washington, USA. Available on line in: http://www.epa.gov/pesticides/trac/science/water_treatment.pdf

Uzun, I. & Güzel, F. (2004). Kinetics and thermodynamics of the adsorption of some dyestuffs and *p*-nitrophenol by chitosan and MCM-chitosan from aqueous solution. *Journal of Colloid and Interface Science, 274*, 398–412.

Van der Bruggen, B., Schaep, J., Maes, W., Wilms, D., Vandecasteele, C. (1998). Nanofiltration as a treatment method for the removal of pesticides from ground waters. *Desalination, 117*, 139–147.

Varma, A.J., Deshpandea, S.V., Kennedy, J.F. (2004). Metal complexation by chitosan and its derivatives: a review. *Carbohydrate Polymers, 55*, 77–93.

Verbych, S., Bryk, M., Alpatova, A., Chornokur, G. (2005). Ground water treatment by enhanced ultrafiltration. *Desalination, 179*, 237-244.

Voerman, S. & Tammes, P.M.L. (1969). Adsorption and desorption of lindane and dieldrin by yeast. *Bulletin of Environmental Contamination and Toxicology; 45*, 271–7.

Wang, C.Y., Wang, Z.X., Zhu, A.P., Hu, X.Y. (2006). Voltammetric Determination of Dopamine in Human Serum with Amphiphilic Chitosan Modified Glassy Carbon Electrode. *Sensors, 6*, 1523-1536.

Ward, J.H., Bashir, R., Peppas, N.A. (2001). Micropatterning of biomedical polymer surfaces by novel UV polymerization techniques. *Journal of Biomedical Materials Research, A56*, 351-360.

Wei, Y.C & Hudson, S.M. (1993). Binding of Sodium Dodecyl Sulfate to a Polyelectrolyte Based on Chitosan. *Macromolecules, 26*, 4151-4154.

Weir, S., Ramsden, D., Hughes, J., Le Thomas, F. (1993). The flocculation of yeast with chitosan in complex fermentation media: the effect of biomass concentration and mode of flocculant addition. *Biotechnology Techniques 7*, 199–204.

World Health Organization (2004). The WHO recommended classification of pesticides by hazard and guidelines to classification: 2004. ISBN 92 4 154663 8 (NLM classification: WA 240)

Wu, F., Tseng, R., Juang, R. (2001). Enhanced abilities of highly swollen chitosan beads for color removal and tyrosinase immobilization. *Journal of Hazard. Materials, B81*, 167–177.

Xu, D., Hein, S., Wang, K. (2008). Chitosan membrane in separation applications. *Materials Science and Technology, 24*, 1078-1087.

Yan, G. & Viraraghavan, T. (2008). Mechanism of Biosorption of Heavy Metals by *Mucor rouxii. Engineering in Life Sciences, 8*, 363-371.

Yeom, C.K., Kim, Y.H., Lee, J.M. (2000). Microencapsulation of water-soluble herbicide by interfacial reaction. II. Release properties of microcapsules, *Journal of Applied Polymer Science, 84*, 1025–1034.

Yoshizuka, K., Lou, Z., Inoue, K. (2000). Silver-complexed chitosan microparticles for pesticide removal. *Reactive and Functional Polymers, 44*, 47-54.

Zainal, Z., Hui, L.K., Hussein, M.Z., Abdullah,A.H., Moh'd Khair, I., Hamadneh, E. (2008). Characterization of TiO2–Chitosan/Glass photocatalyst for the removal of a monoazo dye via photodegradation–adsorption process. *J. Hazard. Mater.*, doi:10.1016/j.jhazmat. 2008.07.154

Zhanga,J., Liua, X., Xua, Z., Chena, H., Yang, Y. (2008). Degradation of chlorophenols catalyzed by laccase. *International Biodeterioration & Biodegradation 61,* 351–356.

Zhou, J.L. & Banks, C.J. (1993). Mechanism of humic acid colour removal from natural waters by fungal biomass biosorption. *Chemosphere, 27,* 607–620.

In: Water Purification
Editors: N. Gertsen and L. Sønderby

ISBN 978-1-60741-599-2
© 2009 Nova Science Publishers, Inc.

Chapter 5

SILICA-SUPPORTED CdS-SENSITIZED TiO$_2$ PARTICLES IN PHOTO-DRIVEN WATER PURIFICATION: ASSESSMENT OF EFFICIENCY, STABILITY AND RECOVERY FUTURE PERSPECTIVES

Ahed H. Zyoud and Hikmat S. Hilal[*]

Department of Chemistry, Najah N. University,
P. O. Box 7, Nablus, West Bank, Palestine

ABSTRACT

Surfaces of Rutile TiO$_2$ particles have been modified with CdS particles. The TiO$_2$/CdS system has been used as catalyst in water purification by photo-degradation of organic contaminants such as methyl orange (a commonly encountered contaminant dye). Both UV and visible ranges have been investigated. CdS sensitization of TiO$_2$ to visible region has been observed, as the TiO$_2$/CdS system showed higher catalytic efficiency than the naked TiO$_2$ system in the visible region. The TiO$_2$/CdS system was unstable under neutral, acidic conditions and basic conditions. Leaching out, of CdS into hazardous aqueous Cd^{2+} ions, while working at pH 7 or lower occurred. This imposes limitations on future usage of CdS-sensitized TiO$_2$ photo-catalytic systems in water purification processes.

In an effort to solve out the leaching difficulties, and to make catalyst recovery easier, the TiO$_2$/CdS system has been supported onto insoluble silica particles giving Silica/TiO$_2$/CdS systems for the first time. The silica/TiO$_2$/CdS system showed lower efficiency than TiO$_2$ and TiO$_2$/CdS systems in UV regions. In the visible region, the Silica/TiO$_2$/CdS was less efficient than the TiO$_2$/CdS but more efficient than naked TiO$_2$. The silica support has an added application value of making catalyst recovery much easier, after reaction completion. Unfortunately the difficulty of the Cd^{2+} ion leaching out has been solved out partly only in basic media. Pre-annealing of the catalyst systems did

---

[*] Corresponding author. Fax: +970-9-2387982; E-mail: hikmathilal@yahoo.com

not give significant effect on stability. Despite the numerous literature reports, on using CdS as sensitizer in degradation studies, its tendency to leach out puts a limitation on its future usage. Should this tendency not be solved out completely, replacement with other more safe dyes should be considered. Effects of catalyst concentration, catalyst recovery, contaminant concentration, temperature and pH, on catalyst efficiency, have also been studied.

Keywords: Photo-degradation, TiO_2, Silica Support, Cadmium Sulfide, Methyl Orange

1. INTRODUCTION

TiO_2, a stable, abundant, low cost and non-hazardous material, has been widely investigated as catalyst for photo-degradation of organic contaminants in water. Complete mineralization of methyl orange has been manifested by different forms of TiO_2 powders with UV light [1]. This is due to its ability to convert water soluble oxygen into OH radicals, when excited by light [2-3]. Moreover, the highly positive potential of its valence band edge, makes TiO_2 a very powerful oxidizing agent which practically photo-degrades any water contaminant. Despite these advantages, TiO_2 has a large band gap, 3.2 eV, which demands radiations with 390 nm and shorter. Since oncoming solar light has wavelengths mostly longer than this value, the use of TiO_2 in large scale water purification applications by solar light will be limited. For this purpose, research has been active to sensitize TiO_2 to visible regions. Different types of dyes, with band gap values ~1.5-2.5 eV, and conduction band edges less positive than that of TiO_2 have been suggested. Scheme I summarizes the concept of TiO_2 sensitization.

Scheme I shows that TiO_2 demands UV light to excite the electrons from the valence band to the conduction band of TiO_2 itself. In dye sensitized TiO_2, the dye molecules themselves are excited in the visible region, where the electrons are excited from the valence band to the conduction band of the dye itself. Thus TiO_2 activity enhancement, by dye molecules in the visible region, is attributed to sensitization.

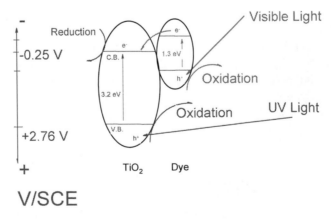

Scheme 1. Comparison between sensitized and naked TiO_2 excitation processes.

One example of suggested sensitizers is CdS which is typically deposited as small particles onto TiO_2 particle surface. TiO_2 sensitization with CdS is still being heavily investigated in photo-degradation studies [4-5]. Many reports described using the TiO_2/CdS system in photo-degradation of water contaminants, but few referred to the hazards associated with CdS stability and Cd^{2+} ion leaching out to solution [5, 6] and/or techniques to eliminate such difficulties. The other technical difficulty of recovering TiO_2/CdS system from solutions after reaction completion has not also been seriously targeted by literature. Separating small TiO_2 particles from water after use is a difficult process, due to their hydrophilic nature [7] and needs to be achieved by economic techniques.

For these reasons, TiO_2/CdS system has been revisited here to assess the potential value of the CdS sensitized TiO_2. Taking into account the toxic effects of Cd^{2+} ions, it is intended to investigate the possibility of CdS to degrade under photo-degradation conditions. [8].

Supporting TiO_2 onto insoluble solid surfaces has been widely suggested for photo-degradation study. In addition to many other solid supports, silica has been widely described for such purposes [9- 17].

Supported TiO_2 systems are widely studied [5]. Other than photo-electrochemical solar cells, in which dye-sensitized TiO_2 systems were supported onto conducting thin films, [18], only very few reports described supporting the combined dye-sensitized TiO_2 systems in degradation experiments [19]. We hereby wish to see if using solid supports, such as silica, may help in TiO_2/CdS system recovery from reaction mixtures after reaction completion. Moreover, we wish to see if silica particles may reduce the tendency of CdS to leach out. Silica was chosen here because it is widely abundant in natural waters, such as lakes, and it may thus interact with dye-sensitized TiO_2 particles suggested for use as contaminant photo-degradation catalysts. This study will shed more light on the future feasibility of using CdS as a potential sensitizer for supported and unsupported TiO_2 in water purification with solar light.

2. EXPERIMENTAL

2.1. Chemicals

Commercial Rutile TiO_2 powder (less than 5 micrometer in diameter) was purchased from Aldrich. Rutle TiO_2 was intentionally used here due to its known lower catalytic activity compared to Anatase TiO_2 [20-21]. Thus it is desirable to enhance the activity of the less active Rutile system.

Thiourea, $CdCl_2$ and organic solvents were all purchased from either Aldrich-Sigma Co. or Frutarom Co. in pure form. Methyl Orange, with the structural formula shown below, has been purchased from Aldrich.

Methyl Orange

Preparation of TiO$_2$/CdS System

CdS was deposited onto commercial TiO$_2$ particles by Chemical Bath Deposition (CBD) techniques as described earlier [22-27]. TiO$_2$ particles (10.0 g) were suspended in distilled water (30.0 mL) to which 5.0 ml aqueous solution of CdCl$_2$ (1.0 M) was added, followed by 10.0 ml NH$_4$Cl solution (1.0 M). The mixture was made basic by adding 15 ml NH$_4$OH (1.0 M) while firmly stoppered with continued mixing. A solution (5.0 mL) of thiourea (1.0 M) was added. The mixture was heated for 60 min at 85°C with continuous stirring. The reddish solid particles of TiO$_2$/CdS were then isolated by filtration, washed with distilled water and dried in air followed by drying in a furnace at 120°C for 1 hr to evaporate excess water. The CdS uptake content in the final TiO$_2$/CdS system was calculated to be 2.7 % by mass. The TiO$_2$/CdS solid was annealed at 400°C under nitrogen for 3 hrs. Solid state electron absorption spectra and photoluminescence emission spectra have been measured. The CdS spectra were compatible with literature [4, 22-27].

Preparation of Silica/TiO$_2$/CdS

The preparation was a two-step process, starting first with Silica/TiO$_2$ preparation, then with CdS deposition. The silica/TiO$_2$ system was prepared by hydrolysis of TiCl$_3$(aq) onto sand surface. The Rutile TiO$_2$ particle preparation was conducted as described earlier [28] with some technical modifications. Natural sea shore sand particles were cleaned by acidification with conc. HCl and HNO$_3$ solutions successively. The solid was then filtered and washed several times with water before drying and sieving, with the fraction 100-50 mesh taken (particle diameter 150-300 micron.).

Sand (50.0 g) was then placed inside a 500 mL round bottomed flask. Distilled water was added to the flask together with 3.0 g solid NaOH. The mixture was stirred mechanically and thermostated at 60°C. To the mixture was added 50.0 mL of TiCl$_3$ solution (~10% by mass) dropwise. Stirring was continued for 6 h, keeping the pH above 3.0 by gradual addition of NaOH. The white color mixture was then continuously stirred for additional 8 h before leaving overnight. The aqueous layer was pipetted out to avoid loss of the ultra-fine unsupported titania particles forming at the surface. More water was added and the mixture was allowed to settle over night before pipetting. This procedure was repeated many times to completely remove sodium and chloride ions. Removal of chloride ions was ensured by testing the aqueous phase with AgNO$_3$ solutions, which showed negative test for chloride ions. The wet solid was dried by heating under continuous stirring. The dried solid was isolated and further dried at 130°C for 60 min, in an oven, before calcination at 350°C for 4 h and slow cooling. To obtain the starting Silica/TiO$_2$ system only, the resulting gross solid mixture was sieved at 50 mesh to remove unsupported TiO$_2$ fine powder. The TiO$_2$ content in the Silica/TiO$_2$ was measured gravimetrically to be 5%. Based on the used literature preparation method, the TiO$_2$ particles are of Rutile type [29]. Calcination at 350°C does not convert the Rutile into Anatase TiO$_2$ [30]. The silica/TiO$_2$ particles were then used to prepare the Silica/TiO$_2$/CdS system by CBD method describe above.

Annealing of TiO$_2$/CdS and Silica/TiO$_2$/CdS systems was performed, by heating at 400°C for 3 h, in a tube furnace under nitrogen. Both annealed and not annealed Silica/TiO$_2$/CdS systems were used in catalytic experiments. Unless otherwise stated, the annealed systems were only used. The mass percent for CdS in the final Silica/TiO$_2$/CdS solid was analyzed and found to be 0.81%. This was achieved by complete dissolution of CdS onto measured amounts of the solid in concentrated HCl solution, yielding Cd^{2+} ions, which were in turn measured by polarography as described below. For recovery/reuse experiments, Silica/TiO$_2$/CdS samples with higher CdS uptake were prepared, as discussed below.

2.2. Equipment

Methyl orange concentrations were spectro-photometrically measured on a Shimadzu UV-1601 spectrophotometer at wavelength 480 nm. To analyze dissolved Cd^{2+} ions, anodic stripping differential pulse polarography (ADPP) was conducted using a dropping mercury electrode (MDE*150)* on a PC controlled *Polarograph (POL150)*. The Hanging Mercury Drop Electrode (HMDE) method was followed. The analysis parameters were: initial potential -700 mV, final potential -400 mV, purging time 30 sec., deposition time 20 sec., scan rate 20 mV/sec., pulse height 25 mV. A Perkin-Elmer LS50 Luminescence Spectrophotometer was used to measure fluorescence spectra for CdS on TiO$_2$ particles.

Solid state absorption spectra, for the TiO$_2$/CdS particles, were also measured as thin films deposited onto quartz surfaces, on a Shimadzu UV-1601 spectrophotometer spectrophotometer. The thin film was prepared by applying a suspension of TiO$_2$/CdS in ether onto a wall of a quartz cuvette. The solvent was then allowed to evaporate leaving a uniform layer of TiO$_2$/CdS onto the wall. The TiO$_2$/CdS absorbance spectrum resembled literature counterparts [31].

UV irradiation of the catalytic reaction mixture was conducted using an Oriel 500 W Hg/Xe lamp equipped with a Model 66901 Universal Arc Lamp Housing. The lamp was working at 150 W power at fixed distance directly above the top of reaction mixture uncovered surface. Visible irradiation was conducted using either a model 45064 - 50 W Xe solar simulator lamp (Leybold Didactic Ltd.) equipped with a housing and a concentration lens, or a Luxten spot 50 W lamp. The Xe lamp has a high stability and an intense coverage of wide spectral range, from about 450 to 800 nm, with no much preference. Exact lamp specifications are described in special manuals [32-38].

2.3. Catalytic Experiments

Catalytic experiments were conducted in magnetically stirred thermostated aqueous solutions of known concentrations of contaminant and catalyst. The pH was controlled by addition of NaOH or HCl solutions. Direct irradiation (with intensities 0.0032 W/cm^2 in case of UV, wavelengths shorter than 400 nm, and 0.0212 W/cm^2 for visible) was conducted by exposing the reaction mixture to incident irradiation directly coming out from the source lamp (placed above the reaction mixture). The reaction mixture was exposed to air under continuous stirring. The reaction was conducted in a 100 mL glass beaker, jacketed with controlled temperature water bath. The reaction vessel walls were covered from outside with reflecting aluminum foil.

Known amounts of solvent and contaminant, together with added acid or base, were placed inside the reaction vessel. The catalyst was then added with continuous stirring in the dark. The reaction mixture was then allowed to stand for a few minutes before analyzing the contaminant concentration. This was to check if contaminant loss occurred by adsorption onto solid system. Reaction time was then recorded the moment direct irradiation was started. The reaction progress was followed up by syringing out small aliquots, from the reaction mixture, at different time intervals. Each aliquot was then immediately centrifuged (500 rounds/s for 5 min), and the liquid phase was pipetted out for electronic absorption analysis, at 480 nm, for methyl orange. The aqueous phase was analyzed, for dissolved Cd^{2+} ions, by polarography as described above. Detailed qualitative analysis of methyl orange degradation has not yet been conducted here, and the reaction rate was based on measuring remaining contaminant concentration with time. Turnover frequency (contaminant reacted moles per TiO_2 moles per h) and quantum yield (contaminant reacted molecules per incident photon) were also calculated and used for efficiency comparison. Complete mineralization of methyl orange into carbon dioxide, ammonia (or nitrate) and sulphate, under UV and/or visible light, can be assumed based on earlier reports [39]

Reuse experiments were conducted using recovered catalyst, by filtration, after reaction completion. The recovered catalyst system was treated like the fresh catalyst samples, as described above.

3. RESULTS AND DISCUSSIONS

The three TiO_2, TiO_2/CdS and Silica/TiO_2/CdS systems were used as photo-degradation catalysts in degradation of methyl orange, under UV and visible radiations. Comparative study between such three catalyst systems was performed in order to assess the feasibility of CdS as a future sensitizer. For this reason, the efficiency of CdS sensitizer was studied in terms of reaction rates, values of turnover frequency and values of quantum yield. The tendency of CdS to leach out from TiO_2 surface was also investigated under different conditions. The ability of Silica support to prevent Cd^{2+} ion leaching out was investigated. Silica/TiO_2/CdS catalyst recovery and reuse was also studied.

Control experiments, conducted under irradiation in the absence of catalyst systems, showed no loss of contaminant concentration. Experiments conducted using different catalyst systems in the dark showed no significant contaminant loss, which indicates that the loss under irradiation experiments is due to photo-degradation. The effects of type of catalyst, pH, temperature, catalyst concentration and contaminant concentration were all studied. Unless otherwise stated, the TiO_2/CdS and silica/TiO_2/CdS were annealed and cooled prior to catalytic use.

3.1. UV Irradiation Experiments

Despite the limitations of UV degradation of contaminants, as discussed above, a number of UV degradation experiments were conducted here for comparison purposes and to assess the sensitizing effect of CdS dye thereafter. Aqueous solutions, pre-contaminated with methyl orange, were irradiated with UV light using the three different systems: TiO_2, TiO_2/CdS and silica/TiO_2/CdS. The contaminant concentration decreased with reaction time due to photo-degradation, as shown in Figure (1). Among the three different catalyst systems, the naked TiO_2 was the most efficient one, with quantum yield value 0.00264 (reacted molecule per incident UV photon). Figure (1) shows that the catalyst efficiency varied in the order: $TiO_2 >$

TiO$_2$/CdS > silica/TiO$_2$/CdS. The estimated turnover frequency (and quantum yield) values for different systems were 0.001167 (0.002562) , 0.001105 (0.00243) and 0.000807 (0.00177), respectively. Similar turnover frequency and quantum yield values are expected in photo-degradation studies [40].

Since irradiation is in the UV region, there is no need for dyes to interfere with TiO$_2$ excitation. With its 3.2 eV band gap TiO$_2$ demands radiations of wavelength ~390 nm or shorter (UV). The naked TiO$_2$ is thus effectively excited and shows relatively high efficiency in the UV range. The inhibition effect of CdS is attributed to blocking UV radiations away from the TiO$_2$ surface. Such activity lowering by CdS is thus expected. Silica particles do additional shielding of radiations and prevent them from reaching TiO$_2$ surface. In case of silica/TiO$_2$/CdS surface, it is assumed that particles on the upper side of reaction mixture only are effectively exposed to UV light. This should lower overall efficiency of the supported catalyst. Presumably, the overall efficiency is lower when less TiO$_2$ surface is exposed to irradiation and reaction mixture. This has been evidenced by studying effect of concentration of naked TiO$_2$ on the rate of the UV degradation. Figure (2) shows that as higher nominal TiO$_2$ concentrations were used, the rate of degradation of methyl orange was only slightly higher.

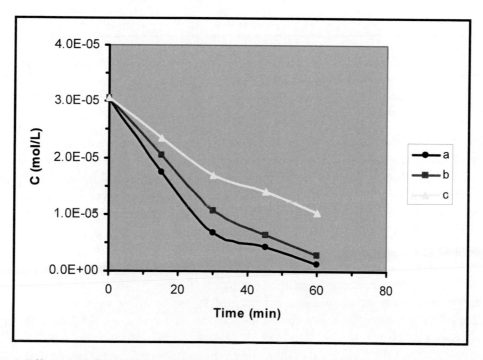

Figure 1. Effect of catalyst type on methyl orange degradation. All experiments were conducted under UV (0.0032 W/cm^2) at 25°C in a neutral 50 mL mixture using a) 0.1g naked TiO$_2$, b) 0.1 g TiO$_2$/CdS, c) 2.0 g silica/TiO$_2$/CdS (contains 0.1g TiO$_2$). Turnover frequency (and quantum yield) values were: 0.001167 and (0.002562) 0.001105 (0.00243) and 0.000807 (0.00177) respectively.

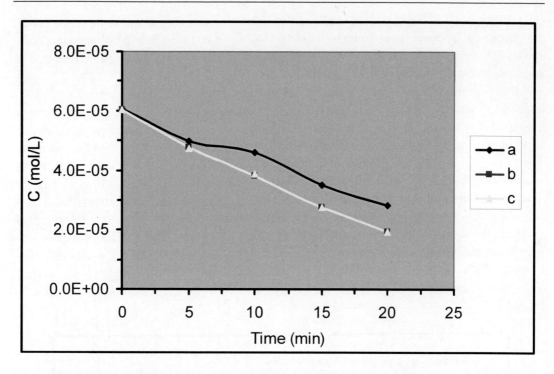

Figure 2. Effect of naked TiO_2 catalyst nominal amount on photo-degradation of methyl orange (neutral 50 ml, 20 ppm) at room temperature under UV (0.0032 W/cm^2) . Catalyst amount: a) 0.05 g b) 0.1 g c) 0.2g. Turnover frequency (and quantum yield) values are: 0.00259 (0.00284) , 0.00164 (0.003594) and 0.0008126 (0.00357) respectively.

The initial reaction rate was slightly increased with higher TiO_2 amounts. The order, of UV degradation reaction of methyl orange, was only 0.1315 with respect to naked TiO_2 catalyst. The fact that the order is less than a unity supports the idea of screening effect, since higher TiO_2 concentrations screen more irradiation. The fact that higher concentrations of TiO_2 increase the rate, and such effect is not a first order support the idea of screening effect in case of Silica/TiO_2/CdS systems. Values of turnover number per h (and quantum yield) for the naked TiO_2 UV reactions with different amounts (0.05, 0.10 and 0.20 g) were calculated as: 0.00259 (0.00284), 0.00164 (0.003594) and 0.0008126 (0.00357) respectively. Despite the increase in rate, the decrease in values of turnover frequency and quantum yield, with increasing catalyst amount, is a direct evidence of screening effect by naked TiO_2. Thus, shielding of TiO_2 surface from UV source, by support and/or dye, would further lower its overall catalyst efficiency.

Moreover, the CdS particles may also lower the mass-transfer rate, to and from TiO_2 surface. This would lower the overall catalytic efficiency of TiO_2 particles in UV degradation of methyl orange.

Effect of temperature on reaction rate was investigated. Figure (3) shows that UV degradation of methyl orange is only slightly enhanced by increasing temperature. Only a relatively low value for activation energy (E_{act} = 7.86 KJ/mol) for the reaction process, was observed. Unlike thermal catalytic reactions, where the ln(rate) is linearly related to inverse Kelvin temperature by Arrhenius equation, photo-degradation reactions are known to be insensitive to temperature [3]. Energy provided by heating is relatively small, *viz* heating to

90°C (363 K) provides only a fraction of an electron volt, which is far less than needed to excite the high band gap TiO$_2$. Moreover, at higher temperatures, contaminant molecules are more desorbed away from TiO$_2$ surface which lowers the reaction rate [38]. Higher temperatures are also responsible for removal of oxygen from the reaction mixture [19] which is necessary for contaminant oxidation.

The effect of methyl orange concentration on reaction rate was studied. The reaction rate was not affected with increasing contaminant concentration. Values of turnover frequency (and quantum yield), did not show significant dependence on contaminant concentration, Figure (4). More discussion on contaminant concentration effect is presented below.

3.2. Solar Simulator (Visible) Experiments

As stated earlier, photo-degradation experiments conducted in visible region are potentially more useful than in the UV region, since the former are abundant in solar energy. However, in order to function effectively in the visible, TiO$_2$ needs sensitization. CdS has a suitable band gap (~2.1 eV, 590 nm), and its conduction band bottom edge is suitably higher than the valence band upper edge for TiO$_2$ [18]. Despite the hazardous nature of CdS, it is still being widely described by other workers as an efficient sensitizer for TiO$_2$ in photo-degradation studies [4-6]. Thus it is worth to revisit such a sensitizer and to see if its efficiency can be enhanced by further modification. Due to its hazardous nature, CdS needs to be carefully considered as a sensitizer [6].

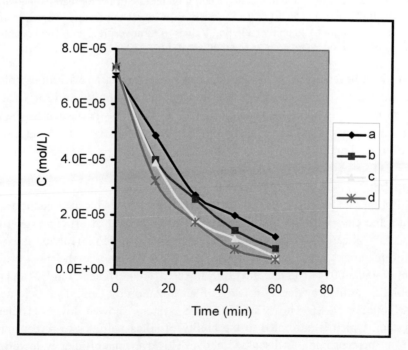

Figure 3. Effect of Temperature on photo-degradation of methyl orange (neutral 50 ml suspension, 25 ppm) using 0.1 g TiO$_2$ under UV (0.0032 W/cm^2). Reaction temperatures are: a) 25°C, b) 50°C, c) 70°C, d) 90°C.

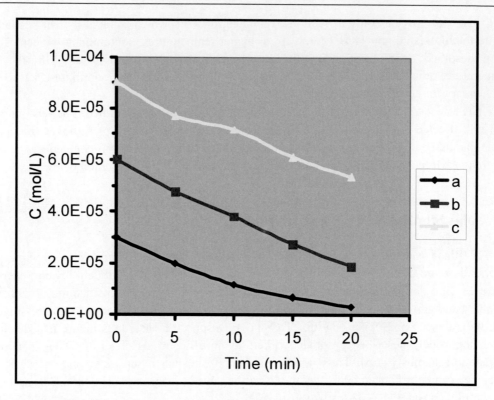

Figure 4. Effect of methyl orange contaminant concentration on rate of photo-degradation with 0.1 g TiO_2 under UV(0.0032 W/cm^2) in 50 ml neutral suspension at room temperature. Contaminant concentrations are: a) 10 ppm b) 20 ppm c) 30 ppm. Values of turnover frequency (and quantum yield) are: 0.001075 (0.00236), 0.00165 (0.00362) and 0.00147 (0.00322).

Therefore, it is necessary to study its tendency to leach out as hazardous soluble Cd^{2+} ions. Furthermore, it is necessary to asses possible techniques to prevent such possible leaching out processes by using solid support for TiO_2/CdS systems such as the abundant silica. In the visible irradiation experiments, different systems were studied as catalysts for degradation of methyl orange.

Effect of Catalyst Type on Reaction Rate

Three different catalyst systems were examined for the visible degradation of methyl orange. Unlike the case with UV region experiments, the catalyst efficiency varied in the order: TiO_2/CdS > silica/TiO_2/CdS > naked TiO_2. This was based on reaction rates, turnover frequency and quantum yield, Figure (5). The TiO_2/CdS system was the most efficient one with quantum yield 0.00094. The fact, that CdS lowered TiO_2 efficiency in UV and enhanced it in the visible, is a direct evidence in favor of sensitization by CdS. This is in accordance with earlier literature [2-8]. The Silica/TiO_2/CdS system showed lower efficiency than TiO_2/CdS system, which is attributed to the ability of silica particles to screen the catalyst sites away from incident light, as discussed above. However, the efficiency lowering due to silica support should not be considered a shortcoming, since the Silica/TiO_2/CdS systems were far easier to isolate and recover from the mixture after reaction completion, as discussed below. Moreover, the silica/TiO_2/CdS system showed higher (about two fold) efficiency than the naked TiO_2 system itself, indicating a sensitization effect.

The fact that Silica/TiO$_2$ showed lower efficiency than naked TiO$_2$ (Figure 5 a & c)) in the visible indicates a screening effect by silica. Moreover, it excludes any sensitization by the silica itself. This is another indication that sensitization is only by CdS in the Silica/TiO$_2$/CdS systems.

Despite its relatively low catalytic efficiency, TiO$_2$ still shows some activity by irradiation with the solar simulator, causing about 20% degradation of methyl orange within 60 min. This is due to availability of a minor tail, with 400 nm and shorter, within the visible light coming out of solar simulator lamps. The fact that the Quantum yield for naked TiO$_2$ in UV (0.002) is much higher than under solar simulator light (0.000095) indicates the inefficiency of TiO$_2$ in visible regions. When a cut-off filter (removing 400 nm and shorter wavelenghts) was placed between the solar simulator and the reaction mixture, the naked TiO$_2$ and Silica/TiO$_2$ showed no catalytic activities. The TiO$_2$/CdS and Silica/TiO$_2$/CdS systems functioned with the cut-off filter. This is a direct evidence in favor of CdS sensitization.

The ability of TiO$_2$ to degrade methyl orange under solar simulator radiation, due to the UV tail, may be inhibited by silica and mud particles commonly present in natural waters. The results show that silica particles screen the TiO$_2$ from the minor UV tails. This limits the use of naked TiO$_2$ on a large scale water purification. Sensitization is thus needed for water purification with TiO$_2$.

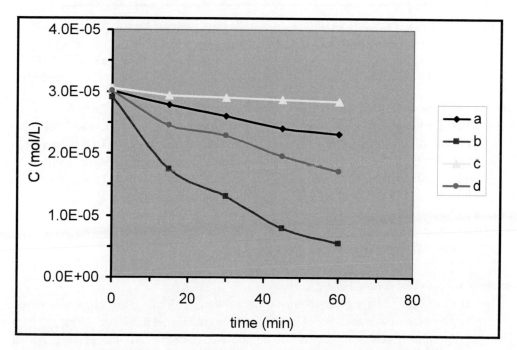

Figure 5. Effect of catalyst type on photo-degradation rate of methyl orange (neutral suspension, 50.0 ml suspension, 10.0 ppm) under solar simulator radiation (0.0212 W/cm^2) at room temperature. Nominal catalyst amounts are: a) 0.1g TiO$_2$, b) 0.1 g TiO$_2$/CdS, c) 2.0 g silica/TiO$_2$, d) 2.0g silica/TiO$_2$/CdS (contains about 0.1g TiO$_2$). Turnover frequency (and quantum yield) values are: 0.000275 (9.11649X10^{-5}), 0.000941 (31.21104X10^{-5}), 0.0000856(2.83978X10^{-5}) and 0.000519(17.22446X10^{-5}) respectively.

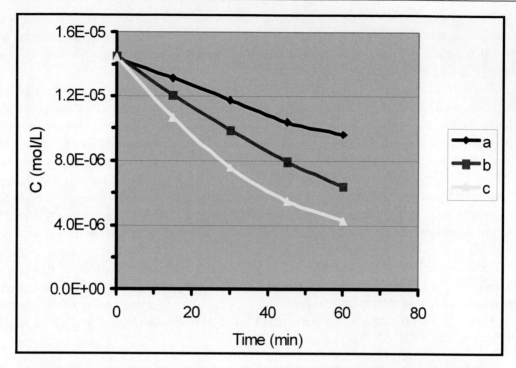

Figure 6. Effect of TiO$_2$/CdS catalyst amount on photo-degradation of methyl orange (neutral suspension, 50 ml of 5 ppm methyle orange), under solar simulator (0.0212 W/cm^2) at room temperature using TiO$_2$/CdS a) 0.05 g b) 0.1 g c) 0.15. Turnover frequency (and quantum number) values are: 0.00038 (6.284 X 10^{-5}), 0.000324 (1.074 X10^{-4}) and 0.000271(1.346 X10^{-4}) respectively.

Effect of Catalyst Concentration on Reaction Rate

As silica shows lowering of reaction rate due to screening, as discussed above, the catalyst concentration effect on reaction rate was studied for the TiO$_2$/CdS system only. Figure (6) shows the effect of TiO$_2$ nominal concentration on rate of degradation of methyl orange. Values of initial rates show a first order with respect to the TiO$_2$/CdS nominal concentration. Values of turnover frequency and quantum yield indicate that the efficiency is not lowered by increasing catalyst amount. No systematic kinetic data could be observed for Silica/TiO$_2$/CdS system, due to the screening effect of silica, as described earlier. These results are consistent with earlier reports [3].

Contaminant Concentration Effect

The effect of Methyl Orange contaminant concentration on rate of reaction was studied. The reaction rate was slower as the contaminant concentration was increased, Figure (7). This is not unexpected. Literature showed that photo-degradation is independent of the contaminant concentration, and in some cases, the rate is lowered with increased initial concentration [3]. Different explanations are proposed, all of which rely on the adsorption of contaminant molecules on the solid surface in a Langmuire Hinshelwood model. One acceptable explanation is the fact that at higher contaminant concentration, the contaminant molecules may compete with the adsorbed intermediates and inhibit degradation [3, 19]. Details of photo-degradation reaction mechanism are reported elsewhere, and are beyond the main theme of this work.

Temperature Effect on Reaction Rate

Photo-decomposition of methyl orange, in the visible range, was studied at different temperatures. Similar to UV results, the visible photo-degradation was not much affected by the reaction temperature, as shown in Figure (8). The value for activation energy was immeasurably small. The discussion to these findings follows that presented earlier for UV systems.

Effect of pH on Reaction Rate

The efficiency of Silica/TiO₂/CdS system in photo-degradation of methyl orange, with visible light, was investigated at different pH values, Figure (9). In basic solutions the reaction rates, the turnover frequency and the quantum yield values, were higher than in neutral or acidic solutions. The efficiency of naked TiO_2 is known to increase in acidic media [3]. Sulphonated TiO_2 surfaces are known to have enhanced surface acidity and to be more efficient photo-degradation catalytic activity [41-43]. The presence of OH groups may also enhance TiO_2 efficiency in photo-degradation reactions [3].

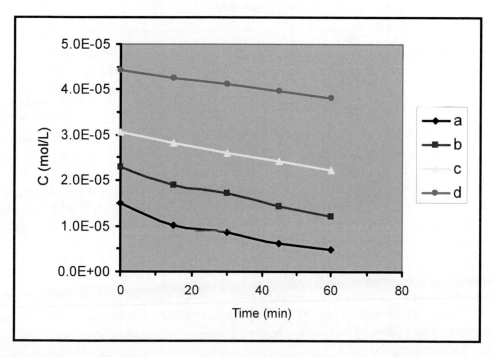

Figure 7. Effect of methyl orange concentration or rate of degradation using TiO_2/CdS (0.1 g) in 70 ml neutral suspension at room temperature under solar simulator (0.0212 W/cm²). Methyl orange concentrations are: a) 5 ppm b)7.5 ppm c) 10 ppm d) 15 ppm. Turnover frequency (and quantum yield) values are: 0.000565 (0.000134), 0.000599(0.000142), 0.000465 (0.00011) and 0.000342(0.000081), respectively.

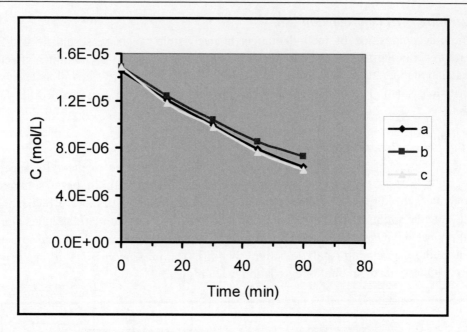

Figure 8. Temperature effect on photo-degradation rate of methyl orange (neutral suspension, 50 ml, 5 ppm) under solar simulator using TiO$_2$/CdS (0.1 g), at different temperatures: a) 30°C b) 45°C c) 60°C.

The Silica/TiO$_2$/CdS system showed higher catalytic efficiency in more basic media with higher pH value. This is possibly due to the tendency of CdS to leach out of the TiO$_2$ surface. In this study, it has been found that CdS decomposes into soluble Cd^{2+} ions under irradiation conditions. Such tendency is higher in more acidic conditions, and this is more pronounced for the Silica/TiO$_2$/CdS system, as will be shown later.

The catalytic efficiency lowering, in neutral and acidic media, is a shortcoming for Silica/TiO$_2$/CdS and TiO$_2$/CdS systems. Since natural waters commonly have pH values close to 7, the lowered efficiency of such catalytic systems limits their use for large scale natural water purification, as discussed below.

Leaching out Experiments (Catalyst Stability)

Despite the ability of CdS to successfully sensitize TiO$_2$ surfaces in photo-degradation of methyl orange, the tendency of CdS to decompose into hazardous Cd^{2+} ions has been investigated under different conditions. The ability of Silica support to prevent such tendency has also been investigated.

Concentrations of aqueous Cd^{2+} ions, leaching out during visible photo-degradation reaction experiments, have been measured with time using polarography. Effect of different parameters on tendency of Cd^{2+} ions to leach out from TiO$_2$/CdS and Silica/TiO$_2$/CdS systems, were studied. Since no noticeable effect of temperature was observed, leaching out experiments were conducted at room temperature unless otherwise stated. Figures (10) through (13) summarize the results.

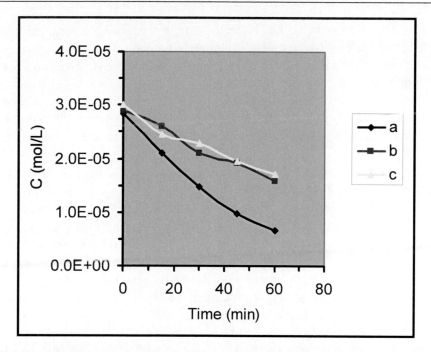

Figure 9. Effect of pH on photodegradation rate of methyl orange (50 ml, 10 ppm) using Sand/TiO$_2$/CdS (2.0 g) at room temperature. Values of pH are: a) 9.5 b) 4 c) 7.8 . Turnover frequency (and quantum yield) values are: 0.000868 (0.000288), 0.0005255 (0.000174) and 0.00051936 (0.0001723) respectively.

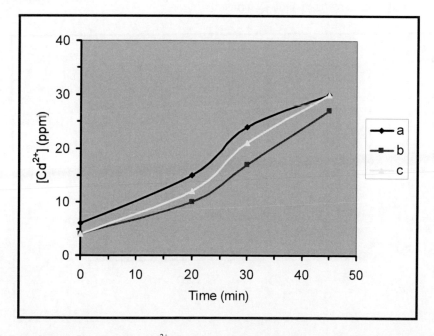

Figure 10. Reaction profiles showing Cd^{2+} ion concentrations leaching out of non-annealed TiO$_2$/CdS (0.1 g) under methyl orange degradation conditions (solar simulator 0.0212 W/cm^2, room temperature, in 70 ml suspension of 5 ppm methyl orange) at different pH values: a) 3.5 b) 9 c) neutral. Values of leaching out percentage for Cd^{2+} are: 100%, 90% and 100% respectively.

In case of TiO$_2$/CdS systems, used in methyl orange degradation experiments with visible light, the Cd^{2+} ion leaching tendency was slightly affected with solution pH value. Acidic and neutral solutions showed similar concentrations of Cd^{2+} ions, reaching about 30% decomposition within 50 min., as shown in Figures (10) and (11) with, while basic solutions exhibited lower Cd^{2+} ions. Annealing the TiO$_2$/CdS systems showed only little stabilizing effect on the TiO$_2$/CdS system, and the annealed systems showed up to 20% decomposition within 50 min. Again the acidic systems showed slightly higher Cd^{2+} concentrations than basic systems.

In case of Silica/TiO$_2$/Cds systems, Figures (12) and (13), a slightly different behavior was observed. At lower pH values, CdS showed high tendency to decompose, reaching almost 90% decomposition within 50 min. This was exhibited in both annealed and not annealed Silica/TiO$_2$/CdS systems. Under neutral conditions, significantly less decomposition was observed, reaching only up to 30% in cases of annealed and not annealed systems. In basic media, the tendency of decomposition was less than 10% within 50 min. Annealing the Silica/TiO$_2$/CdS systems did not significantly lower the leaching tendency, as shown in Figures (12) to (13). With and without annealing, the Silica/TiO$_2$/CdS systems showed some relatively high stability in basic media compared to acidic and neutral media. Despite this, Silica/TiO$_2$/CdS system should be avoided in water purification due to the hazardous nature of Cd^{2+} ions. The concentration of Cd^{2+} ions (~10 ppm) leaching out of Silica/TiO$_2$/CdS system in basic media, evidently with least leaching tendency among the used systems, is still higher than the phyto-toxic threshold limits for Cd^{2+} ions (0.005 mg/L, 5X10^{-2} ppm) in agricultural waters [44]. This imposes a limitation on applicability of the widely ongoing current studies of CdS-sensitized TiO$_2$ systems.

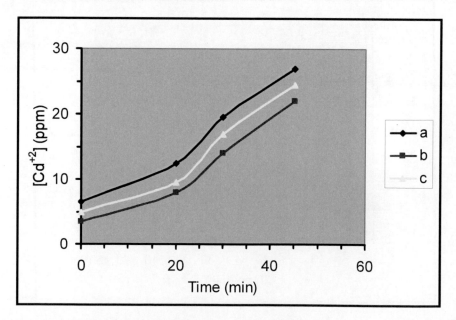

Figure 11. Reaction profiles showing Cd^{2+} ion concentrations leaching out of pre-annealed TiO$_2$/CdS (0.1 g) under methyl orange degradation conditions (solar simulator 0.0212 W/cm^2, room temperature, in 70 ml suspension of 5 ppm methyl orange) at different pH values: a) 3.5 b) 9 c) neutral. Percentage values of Cd^{2+} leaching out are: 90%, 73% and 82%, respectively.

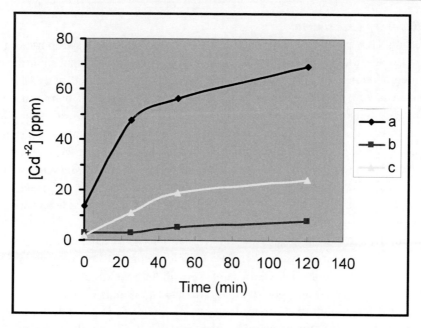

Figure 12. Reaction profiles showing Cd^{2+} ion concentrations leaching out of non-annealed Silica/TiO$_2$/CdS (1.0 g) under methyl orange degradation conditions (solar simulator 0.0212 W/cm^2, room temperature, in 70 ml suspension of 5 ppm methyl orange) at different pH values: a) 3.5, b) 9, and c) neutral. Values of leaching out percentage for Cd^{2+} are: 77%, and 9% and 24%, respectively.

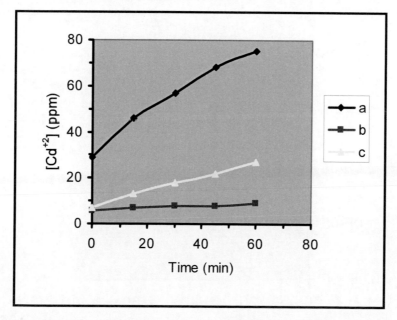

Figure 13. Reaction profiles showing Cd^{2+} ion concentrations leaching out of pre-annealed Silica/TiO$_2$/CdS (1.0 g) under methyl orange degradation conditions (solar simulator 0.0212 W/cm^2, room temperature, in 70 ml suspension of 5 ppm methyl orange) at different pH values: a) 3.5, b)9 and c)neutral. Values of leaching out percentage for Cd^{2+} are: a) 83% b) 11% c) 31%, respectively.

Catalyst Recovery Experiments

Supporting TiO_2 particles onto insoluble supports has been widely reported in literature [3, 5, 8-17, 19, 45-58]. However, only a few reports are known for supporting the combined TiO_2/Dye degradation catalysts onto insoluble supports, as discussed above [19]. TiO_2 is hydrophilic in nature, and is commonly used as fine particles with nano-scale size. This makes it difficult to isolate by simple techniques such as filtration or decantation. Supporting TiO_2/CdS onto silica is one possible solution for such technical difficulties. As stated earlier, silica is a cheap widely abundant stuff in natural lakes. Therefore, any TiO_2 systems added to natural waters will interact with silica particles. Therefore, it is necessary to study silica as support for TiO_2/CdS systems. Despite the tendency of CdS to leach out from Silica/TiO_2/CdS systems, as described above, silica support has the added value of making the catalyst system recoverable for further reuse.

Due to the tendency of CdS to leach out, Silica/TiO_2/CdS catalyst recovery and reuse study could not be performed, since complete CdS leaching out occurred within 1 h, (vide infra). Therefore, special Silica/TiO_2/CdS samples, with higher CdS uptakes (1.93% by mass) were prepared for recovery/reuse purposes. Figure (14) shows data on rates of methyl orange degradation, in neutral solutions at room temperature, under visible radiation, using a fresh, first recovery and second recovery samples of Silica/TiO_2/CdS samples. The catalyst partly looses its efficiency on reuse. This is due to the tendency of CdS to leach out of the catalyst system under photo-degradation conditions, as described above. Figure (14) shows about 30% efficiency loss, for the Silica/TiO_2/CdS system, in the third use.

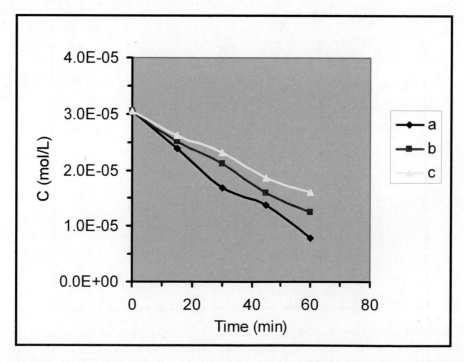

Figure 14. Effect of Silica/TiO_2/CdS (1.0 g, containing 1.9% mass CdS) catalyst recovery and reuse () on photo-degradation rate of methyl orange (neutral 50 ml suspension, 10 ppm) at room temperature. a) Fresh sample, b) 2nd use and c) 3rd use.

The results indicate that the support could not completely prevent CdS leaching out, but still retains 70% efficiency after third use. Silica surface seems to be a good candidate for future sensitized TiO$_2$ surfaces with dyes more safe than CdS. Work is underway here to utilize silica as support for TiO$_2$ surfaces sensitized with safe natural dyes, as alternatives for CdS systems.

4. CONCLUSION

CdS effectively sensitizes TiO$_2$ to visible region in degradation of organic contaminants such as methyl orange. Howevre, its tendency to degrade into soluble Cd^{2+} ions limits its applicability in water purification with solar light. The Silica support partially limited the Cd^{2+} ions leaching out only under basic condition. Moreover the support made recovery of the sensitized catalyst easier, giving relatively high efficiency reusable system. To avoid the hazardous nature of CdS system, more work is needed to completely prevent its leaching out, or better replacing it with safer dyes.

ACKNOWLEDGEMENTS

The authors wish to express their gratitude to the Palestine-French University Cooperation Scheme, for providing a mercury(Xenon) lamp and a polarograph. Support from Union of Arab Universities is acknowledged. Donation of free chemicals from Birzeit-Palestine Pharm. Co. is also acknowledged.

REFERENCES

[1]   Dai, K.; Chen, H.; Peng, T.; Ke, D.; Yi, H. "Photocatalytic degradation of methyl orange in aqueous suspension of mesoporous titania nanoparticles", *Chemosphere*. 2007, *69*, 1144-1155.

[2]   Herrmann, J.-M., Duchamp, C.; Karkmaz, M.; Hoai, B. T.; Lachheb, H.; Puzenat, E.; Guillard, C. "Environmental green chemistry as defined by photocatalysis", *J. Haz. Mater.* 2007, *145,* 624-629.

[3]   *Carp, O.; Huisman, C.L.; Reller, A. "Photoinduced reactivity* of titanium dioxide", *Progr. Solid State Chem.* 2004, *32*, 33–177.

[4]   Bessekhouad, Y. ; Robert, D.; Weber, J.V. "Bi$_2$S$_3$/TiO$_2$ and CdS/TiO$_2$ heterojunctions as an available configuration for photocatalytic degradation of organic pollutant", *J. Photochem. Photobiol. A: Chem.* 2004, *163*, 569–580; Bessekhouad, Y.; Chaoui, N.; Trzpit, M.; Ghazzal, N.; Robert, D.; Weber, J.V. "UV–vis versus visible degradation of Acid Orange II in a coupled CdS/TiO$_2$ semiconductors suspension", *J. Photochem. Photobiol. A: Chem.* 2006, *183.* 218–224; Kansal, S.K.; Singh, M.; Sud, D. "Studies on photodegradation of two commercial dyes in aqueous phase using different photocatalysts", *J. Hazard. Mater.* 2007, *141*, 581–590.

[5]   Robert, D. "Photosensitization of TiO_2 by M_xO_y and M_xS_y nanoparticles for heterogeneous photocatalysis applications", *Catal. Today.* 2007, *122*, 20–26.

[6]   Fuji, H.; Ohtaki, M.; Eguchi, K.; Arai, H.; *J. Mol. Catal. A: Chem.* 1998, *129*, 61–68; Serpone, N.; Marathamuthu, P.; Pichat, P.; Pelizzetti, E.; Hidaka, H.; *J. Photochem. Photobiol. A: Chem.* 1995, *85,* 247; Sesha, S.; Srinivasan, J. W.; Stefanakos, E. K. "Visible Light Photocatalysis via CdS/TiO_2 Nanocomposite Materials", *J. Nanomat.*, 2006, *2006*, 1-7; Reeves, P.; Ohlhausen, R.; Sloan, D.; Pamplin, K.; Scoggins, T.; Clark, C.; Hutchinson, B.; Green , D. "Photocatalytic destruction of organic dyes in aqueous TiO_2 suspensions using concentrated simulated and natural solar energy", *Sol. Ener.* 1992, *48*, 413-420.

[7]   Sakai, N.; Wang, R.; Fujishima, A.; Watanabe, T.; Hashimoto, K. "Effect of ultrasonic treatment on highly hydrophilic TiO_2 surfaces", *Langmuir,* 1998, *14*, 5918 -5920.

[8]   Greijer, H.; Karlson, L.; Lindquist, S.-E.; Hagfeldt, A. "Environmental aspects of electricity generation from a nanocrystalline dye sensitized solar cell system", *Renew. Ener.* 2001, *23*, 27-39.

[9]   Pizarro, P.; Guillard, C.; Perol, N.; Herrmann, J.-M. "Photocatalytic degradation of imazapyr in water: Comparison of activities of different supported and unsupported TiO_2-based catalysts", *Catal. Today.* 2005, *101*, 211-218.

[10]   Hidalgo, M. C.; Bahnemann, D. "Highly photoactive supported TiO_2 prepared by thermal hydrolysis of $TiOSO_4$: Optimisation of the method and comparison with other synthetic routes", *Appl. Catal.. B, Environ.*, 2005, *61*, 259-266;

[11]   Lo`pez-Mun˘oz M.J.; Grieken, R; Aguado, J.; Marugan, J. "Role of the support on the activity of -supported TiO_2 photocatalysts: Structure of the TiO_2/SBA-15 photocatalysts", *Catal. Today.* 2005, *101*, 227

[12]   Hirano, M.; Ota, K. "Preparation of photoactive anatase-type TiO_2/silica gel by direct loading anatase-type TiO_2 nanoparticles in acidic aqueous solutions by thermal hydrolysis", *J. Mater. Sci.*, 2004, *39*, 1841; and Hirano, M.; Ota, Keisuke; Inagaki, Michio; Iwata, H. "Hydrothermal Synthesis of TiO_2/SiO_2 Composite Nanoparticles and Their Photocatalytic Performances", *J. Ceram. Soc. Jpn.* 2004, *112*, 143-148.

[13]   Aguado, J; Grieken, R; Lo`pez-Mun˘oz MJ, Maruga´n, "Removal of cyanides in wastewater by supported TiO_2-based photocatalysts", *Catal. Today.* 2002, *75*, 95-102.

[14]   Najm, J.N.; Snoeyink, V.L.; Suidan, M.T.; Lee, C.H.; Richard, Z.; *J Am Water Works Ass.* 1990, *82*, 65.

[15]   Periyat, P.; Baiju, K.V.; Mukundan, P.; Pillai, P.K.; Warrier, K.G.K.; "High temperature stable mesoporous anatase TiO_2 photocatalyst achieved by addition", *Appl. Catal. A-Gen.*, *In Press, Accepted Manuscript, online (2008).*

[16]   Marugán, J.; Grieken, R.; Cassano, A. E.; Alfano, O. M. "Intrinsic kinetic modeling with explicit radiation absorption effects of the photocatalytic oxidation of cyanide with TiO_2 and -supported TiO_2 suspensions, *Appl. Catalysis B: Environm.*, In Press, Corrected Proof, Available online 3 July 2008.

[17]   Lee, J.W.; Othman, M.R.; Eom, Y.; Lee, T.G.; Kim, W.S.; Kim, J. "The effects of sonification and TiO_2 deposition on the micro-characteristics of the thermally treated SiO_2/TiO_2 spherical core–shell particles for photo-catalysis of methyl orange", *Micropor. and Mesopor. Mater.*, In Press, Corrected Proof, online 2008.

[18]   Grätzel, M.; "Photoelectrochemical cells", *Nature.* 2001, *414*, 338.

[19] Hilal, H.S.; Majjad, L.Z.; Zaatar, N.; El-Hamouz, A. "Dye-effect in TiO$_2$ catalyzed contaminant photo-degradation: Sensitization vs. charge-transfer formalism", *Solid State Sci.* 2007, *9*, 9-15.

[20] Linsbigler, A.L.; Lu, G.; Yates, J.T.; "Photocatalysis on TiO$_2$ surfaces: Principles, mechanisms, and selected results", *Chem. Rev.*, 1995, *95, 735*.

[21] Tanaka, K; Capule, M.F.V.; Hisanaga, T. "Effect of crystallinity of TiO$_2$ on its photocatalytic action", *Chem. Phys. Lett.*, 1991, *187*, 73-76.

[22] Bessekhouad, Y.; Robert, D.; Weber, J.V. "Bi$_2$S$_3$/TiO$_2$ and CdS/TiO$_2$ heterojunctions as an available configuration for photocatalytic degradation of organic pollutant", *J. Photochem. Photobiol. A: Chem.*, 2004, *163*, 569–580.

[23] Vigil-Galán, O.; Morales-Acevedo, A.; Cruz-Gandarilla, F.; Jiménez-Escamilla, M.G. "Characterization of CBD–CdS layers with different S/Cd ratios in the chemical bath and their relation with the efficiency of CdS/CdTe Solar Cells", *Thin Solid Films*, 2007, *515*, 6085–6088.

[24] Ximello-Quiebras, J.N.; Contreras-Puente, G.; Aguilar-Hernandez, J.; Santana-Rodriguez, G.; Readigos, A. A.-C. "Physical Properties of chemical bath deposited CdS thin films", *Sol. Energ. Mat. Sol. C.* 2004, *82*, 263–268.

[25] Roy, P.; Srivastava, S. K. "A new approach towards the growth of cadmium sulphide thin film by CBD method and its characterization", *Mat. Chem. Phys.* 2006, *95*, 235–241.

[26] Zinoviev, K.V.; Zelaya-Angel, O. "Influence of low temperature thermal annealing on the dark resistivity of chemical bath deposited CdS films", *Mat. Chem. Phys.*, 2001, *70*, 100–102.

[27] Chen, F.; W. Jie, "Growth and photoluminescence properties of CdS solid solution semiconductor", *Cryst. Res. Technol.* 2007, *42*, 1082 – 1086.

[28] Cassaignon, S. ; Koelsch, M.; Jolivet, J.-P. From TiCl$_3$ to TiO$_2$ nanoparticles (anatase, brookite and rutile): Thermohydrolysis and oxidation in aqueous medium; Chimie de la Matière Condensée de Paris, Université Pierre et Marie Curie-Paris6, CNRS UMR 7574, 4 place Jussieu, Paris F-75005, France; Available online 22 February 2007 ; Huanga, X. ; Pana, C.; Large-scale synthesis of single-crystalline rutile TiO$_2$ nanorods via a one-step solution route, *J. Cryst. Growth*, 2007, *306*, 117–122.

[29] Cassaignon, S. ; Koelsch , M. ; Jolivet J.-P. Soft synthesis conditions of original shaped TiO$_2$ nanoparticles from TiCl$_3$, *Chimie de la Matiere Condensee de Paris UPMC (CMCP-UPMC), 4 place jussieu, Paris 75252, France* , 2007.

[30] Finklea, H. O., in *Semiconductor Electrodes*, Finklea, H.O.; Ed.; Elsevier, Amsterdam, (1988) , pp. 45-47. See also references therein.

[31] Nath, S. S.; Chakdar, D.; Gope, G.; Avasthi, D. K. "Characterization of CdS and ZnS quantum dots prepared via a chemical method on SBR latex, http://www.azonano. com/details.asp?ArticleId=2159 , accessed Aug. 21, 2008

[32] Shoemaker, D.; Garland, C.; Nibler, J.; *Experiments in Physical Chemistry*, 5th edition, Mc Graw-Hill, Inc., N.Y., 1989.

[33] For phenol analysis, see AOAC Official Method 979.13, copyright 1998, AOAC International.

[34] The Book of Photon Tools, Oriel Instruments, Stratford, CT, USA, pp. 44, 50 (Manual).

[35] Thermo Oriel: 500 W Xenon/Mercury (Xenon) ARC Lamp Power Supply Model 68911 (Manual).

[36] Clark, B. J.; Frost, T.; and Russell, M. A. *U.V. Spectroscopy*, Chapman & Hall, London, 1993, p.p. 18-21.

[37] [http://www.uvguide.co.uk/zoolamps.htm accessed Aug. 2008.

[38] Hermann, J. M.; "Heterogeneous photocatalysis: fundamentals and applications to the removal of various types of aqueous pollutants", *Catal. Today*, 1999, *53*, 115-129.

[39] GUETTAÏ, N. ; AIT AMAR H. "Photocatalytic oxidation of methyl orange in presence of titanium dioxide in aqueous suspension. Part I: Parametric study, Desalination and the Environment. Conference, Santa Margherita Ligure , ITALIE (22/05/2005) , 2005, vol. 185, n° 1-3 (556 p.);; GE, L.; XU, M.; FANG, H.; "Photo-catalytic degradation of methyl orange and formaldehyde by $Ag/InVO_4-TiO_2$ thin films under visible-light irradiation, J. Mol. Catal. A, Chem. 2006, *258*, 68-76; Xu, Y.; Chen, Z.; "Photo-oxidation of Chlorophenols and Methyl Orange with Visible Light in the Presence of Copper Phthalocyaninesulfonate", *Chem. Lett.* 2003, *32*, 592; Xu, S.; Shangguan, W.; Yuan, J.; Shi, J.; Chen M.; "Preparations and photocatalytic degradation of methyl orange in water on magnetically separable $Bi_{12}TiO_{20}$ supported on nickel ferrite, *Sci. Technol. Adv. Mat.* 2007, *8*, 40–46; HUANG, C.; CHEN, D. H.; LI, K.; "Photocatalytic oxidation of butyraldehyde over titania in air: by-product identification and reaction pathways", *Chem. Eng. Comm.* 2003, *190*: 373-392.

[40] Hiroaki, H.; Ayako, H.; Mamoru, U.; Kunihiko, O.; Hiroshi, H. "Determination of the photo-degradation quantum yield of organic dyes in the film state", *Nippon Kagakkai Koen Yokoshu*. 2003, *83*, 747 (translated Science Links Japan, http://sciencelinks.jp/j-east/article/200321/000020032103A0641008.php); Tanaka, N.; Barashkov, N.; Heath, J.; Sisk, W. N.; "Photodegradation of polymer-dispersed perylene di-imide dyes", *Appl. Opt.* 2006, *45* (16),3846-51.

[41] Muggli, D. S.; Ding, L.; " Photocatalytic performance of sulfated TiO_2 and Degussa P-25 TiO_2 during oxidation of organics", *Appl. Catal. B: Environ.* 2001, *32*, 181

[42] Amama, P.B.; Itoh, K.; Murabayashi, M.; "Gas-phase photocatalytic degradation of trichloroethylene on pretreated TiO_2 ", *Appl. Catal. B: Environ.* 2002, *37*, 321.

[43] Lewandowski, M. M.; Ollis, D.F.; "Halide acid pretreatments of photocatalysts for oxidation of aromatic air contaminants: rate enhancement, rate inhibition, and a thermodynamic rationale", *J. Catal.* 2003, *217*, 38.

[44] Toxic chemical hazards of irrigation waste http://www.lenntech.com/irrigation/toxic-ions-hazard-of-irrigation-water.htm , accessed Aug. 13th, 2008.

[45] Bhattacharyya, A.; Kawi, S. ; Ray, M. B.; "Photocatalytic degradation of orange II by TiO_2 catalysts supported on adsorbents", Catal. today, 2004, *98*, 431-439.

[46] Jiang, Y.; Sun, Y.; Liu, H.; Zhu, F.; Yin, H. "Solar photocatalytic decolorization of C.I. Basic Blue 41 in an aqueous suspension of $TiO_2–ZnO$", *Dyes and Pigments*, 2008, *78*, 77-83.

[47] Li, Y.; Li, X.; Li, J.; Yin, J. "Photocatalytic degradation of methyl orange by TiO_2-coated activated carbon and kinetic study", *Water Res.,* 2006, *40*, 1119–1126.

[48] Matos, J.; Laine, J.; Herrmann, J. "Synergy effect in the photocatalytic degradation of phenol on a suspended mixture titania and activated carbon", *Appl. Catal. B: Environ.* 1998, *18*, 281-291.

[49] Chatterjee, D.; Dasgupta, S.; Rao, N.N. "Visible light assisted photodegradation of halocarbons on the dye modified TiO$_2$ surface using visible light", *Sol. Energ. Mat. Sol. C.* 2006, *90*, 1013-1020.

[50] Ao, C.H.; Lee, S.C.; "Combination effect of activated carbon with TiO2 for the photodegradation of binary pollutants at typical indoor air level", *J. Photochem. Photobiol. A: Chem.* 2004, *161,* 131-140.

[51] Aran˜a, J.; Don˜a-Rodrı´guez, J.M.; Rendo´n, E.T.; Garriga i Cabo, C.; Gonza´lez-Dı´az, O.; Herrera-Melia´n, J.A.; Pe´rez-Pen˜a, J.; Colo´n, G.; Navı´o, J.A.; TiO$_2$ activation by using activated carbon as a support, Part I. Surface characterisation and decantability study, *Appl. Catal. B: Environ.* 2003, *44*,161-172.

[52] Marugan, J. ; van Grieken, R.; Alfano, O. M.; Cassano A. E., "Optical and physicochemical properties of silica-supported TiO$_2$ photocatalysts", *AIChE J.*, 2006, *52*, 2832-2843.

[53] Nazir, M.; Takasaki, J.; and Kumazawa, H. "Photocatalytic degradation of gaseous ammonia and trichloroethylene over TiO$_2$ ultrafine powders deposited on activated carbon particles, *Chem. Eng. Comm.* 2003, *190,* 322-333.

[54] Kim, T. Y.; Lee, Y.-H.; Park, K.-H.; Kim, S. J.; Cho, S. Y. "A study of photocatalysis of TiO$_2$ coated onto chitosan beads and activated carbon", *Res. Chem. Intermed.*, 2005, *31*, 343–358.

[55] Liu, S.-X.; Sun, C.-L. "Preparation and performance of photocatalytic regenerationable *activated carbon* prepared via sol-gel *TiO$_2$*", J. Environ. Sc. 2006, *18*, 557-561.

[56] Jaroenworaluck, A.; Sunsaneeyametha, W.; Kosachan, N.; Stevens, R. "Characteristics of silica-coated TiO$_2$ and its UV absorption for sunscreen cosmetic applications"; Proceedings of the 11th European Conference on Applications of Surface and Interface Analysis, 25-30 September 2005, Vienna, Austria.

[57] Vohra, M. S.; Tanaka, K.; Photocatalytic degradation of aqueous pollutants using silica-modified TiO$_2$, *Water Res.* 2003, *37*, 3992-3996

[58] Chi, C.-F.; Lee, Y.-L.; Weng, H.-S. "A CdS-modified TiO$_2$ nanocrystalline photoanode for efficient hydrogen generation by visible light, Ching-Fa Chi *et al Nanotechnology.* 2008, *19,* 125704 (5pp).

In: Water Purification   ISBN 978-1-60741-599-2
Editors: N. Gertsen and L. Sønderby   © 2009 Nova Science Publishers, Inc.

Chapter 6

Photo-degradation of Methyl Orange with Direct Solar Light Using ZnO and Activated Carbon-supported ZnO

Hikmat S. Hilal[*], Ghazi Y. M. Nour and Ahed Zyoud

Department of Chemistry, An-Najah N. University, P.O. Box 7, Nablus,
West Bank, Palestine

Abstract

ZnO is a wide band gap (3.2 eV) semiconductor, with limited photo-catalytic applications to shorter wavelengths only. However, it is suitable to use in solar light photo-degradation of different contaminants, due to a number of reasons, taking into account that the reaching-in solar radiation contains only a tail in the near UV region. The high absorptivity of ZnO makes it efficient photo-catalyst under direct solar light. Moreover, it is relatively safe, abundant and non costly. In this chapter, ZnO has been investigated as a potential catalyst for photo-degradation of methyl orange (a known dye) in aqueous solutions with direct natural solar light under different conditions. The major aim was to assess the efficiency and stability of ZnO under photo-electrochemical (PEC) conditions, and to suggest techniques to enhance such features. This will shed light on the future applicability of ZnO as a candidate for economic and friendly processes in water purification.

Recovery of ZnO particles, after reaction completion, has been facilitated by supporting ZnO onto activated carbon, to yield AC/ZnO system. The AC/ZnO was used as catalyst for contaminant photo-degradation in water solutions under direct solar light.

Both catalytic systems, naked ZnO and AC/ZnO, were highly efficient in degrading both contaminants, reaching complete removal in reasonable times. The latter system showed higher efficiency. In both systems, the reaction goes faster with higher catalyst loading, until a maximum efficiency is reached at a certain concentration, after which the catalyst concentration did not show a systematic effect.

---

[*] Fax: 00-970-9-2387982; E-mail: hikmathilal@yahoo.com

In both catalytic systems, the rate of degradation reaction increases with higher contaminant concentrations until a certain limit is used. The contaminant degradation reaction was studied, using both catalysts, at different pH values. The pH value 8.0 gave the highest catalyst efficiency. The tendency of naked ZnO to degrade into soluble zinc ions, under photo-degradation experiments, was studied under different pH values. Catalyst recovery and reuse experiments were conducted on both systems. The catalytic activity of the recovered systems was only slightly lower than the fresh system in each case. The fourth time recovered catalysts showed up to 50% efficiency loss in each case, presumably due to ZnO degradation and leaching out. However, fresh and recovered catalyst systems caused complete degradation of contaminants after enough time. Temperature showed a slight effect on rate of reaction, with immeasurably small activation energy value. Details of effects of other parameters on reaction rate and catalyst efficiency are described. Using CdS as sensitizing dye failed to enhance ZnO efficiency under direct solar light. The screening effect and tendency of CdS to leach out limits its use as ZnO sensitizer. Tendency of ZnO to leach out zinc ions into solution is discussed. The naked ZnO and AC/ZnO systems are promising photo-catalysts in future water purification technologies by direct solar light.

Keywords: ZnO , CdS , Activated Carbon, methyl orange, photo-degradation, solar light, zinc oxide, activated carbon, solar, photo-degradation, methyl orange, water

1. INTRODUCTION

ZnO is a low-cost, abundant, stable, non-hazardous substrate [1-12]. Its semiconducting (SC) characteristics are based on its wide band gap (3.2 eV) which demands UV light (390 nm or shorter) for excitation. For these reasons, ZnO has been investigated as a catalyst for contaminant photo-degradation processes in water using UV or UV/Solar light.

ZnO has often been associated with another SC namely TiO_2 in PEC applications. Both systems are stable under photo-electrochemical conditions with comparable band gaps, non-hazardous nature, abundance and high oxidizing power. The TiO_2 systems are far more widely investigated than ZnO system. Strictly speaking, both systems demand UV light for PEC activity. In order to function effectively under solar light, the TiO_2 demands dye sensitizers. This adds to the cost and technical difficulties associated with dye stability and degradation concerns. On the other hand, ZnO has an edge over TiO_2. ZnO is more absorptive to UV light that is available as a small tail within the reaching solar light [4]. The perspectives of using naked ZnO in large scale PEC technology must therefore be thoroughly investigated. Dye-sensitized ZnO systems must assumably have higher efficiency in the absence of UV, than naked ZnO systems. ZnO/CdS systems (with ~0.2% CdS uptakes), are reported to have high catalytic efficiency in degrading water contaminants [1]. Due to technical and cost considerations, techniques should be developed to enhance efficiency of ZnO under solar light without the use of hazardous sensitizers, such as CdS. This stems from the fact that CdS readily degrades into hazardous Cd^{2+} ions under PEC conditions [13-19].

Supporting nano-scale ZnO particles onto insoluble supports has been reported in literature [20-24]. Silica, aerogel carbon, carbon nanotubes and carbon composites have all be described as supports, causing enhanced ZnO catalytic efficiency. Activated carbon (AC) also enhanced the efficiency of nano-scale ZnO catalysts in different contaminant photo-degradations with solar light [20-26]. Synergistic effect between the ZnO and the AC has been described to explain the observed efficiency enhancement. On the other hand the negative effects of AC have not been widely studied. AC may cause screening of light away from ZnO active sites. How screening affects the supported ZnO catalytic efficiency, needs to be thoroughly investigated.

The major theme of this study is two fold. Firstly, to assess the feasibility of using naked and ZnO/CdS catalysts in degrading a commonly studied contaminant (methyl orange) under direct solar light rather than solar simulator light. Secondly, to assess the effect of supporting ZnO, catalyst particles onto AC surfaces, in terms of efficiency and recovery perspectives. The study has been conducted keeping an eye on the potential value of the ZnO systems in technology scale perspectives.

2. EXPERIMENTAL

Chemicals

Thiourea, $CdCl_2$ and organic solvents were all purchased from either Aldrich-Sigma Co. or Frutarom Co. in pure form. Methyl Orange, with the structural formula shown below, and a molar mass 327.33, was purchased from Aldrich.

Methyl Orange

Commercial ZnO powder, Merck (Catalog no. 8849, with average particle size 230 nm,) was purchased directly from Merck). High surface area AC was purchased from Aldrich, and its surface area was measured in these laboratories [27] according to literature methods [28-30] using the acetic acid adsorption isotherms and was 850 m^2/g.

ZnS/CdS preparation: CdS was deposited onto commercial ZnO particles by Chemical Bath Deposition (CBD) techniques as described earlier [31-35]. ZnO particles were suspended in distilled water (30.0 mL) to which 5.0 ml aqueous solution of $CdCl_2$ (1.0 M) was added, followed by 10.0 ml NH_4Cl solution (1.0 M). The mixture was made basic by adding 15 ml NH_4OH (1.0 M) while firmly stoppered with continued mixing. A solution (5.0 mL) of thiourea (1.0 M) was added. The mixture was heated for 60 min at 85°C with continuous stirring. The reddish solid particles of ZnO/CdS were then isolated by filtration, washed with distilled water and dried in air followed by drying in a furnace at 120°C for 1 hr to evaporate excess water. The ZnO/CdS solid was annealed at 400°C under nitrogen for 3 hrs. Different CdS uptake values, in the final ZnO/CdS system, were obtained.

AC/ZnO preparation: by mixing and stirring ZnO with AC (different mass ratios as desired) inside water for 30 min. After that, the mixture was filtered and the dark solid taken and dried in the air before being used for catalytic purposes.

Equipment

Methyl orange concentrations were spectro-photometrically measured on a Shimadzu UV-1601 spectrophotometer at wavelength 480 nm. To analyze dissolved Cd^{2+} ions, anodic stripping differential pulse polarography (ADPP) was conducted using a dropping mercury electrode (MDE*150)* on a PC controlled *Polarograph (POL150)*. The Hanging Mercury Drop Electrode (HMDE) method was followed. The analysis parameters were: initial potential -700 mV, final potential -400 mV, purging time 30 sec., deposition time 20 sec., scan rate 20 mV/sec., pulse height 25 mV.

Direct solar light irradiation was conducted outdoors. Unless otherwise stated, experiments using direct natural solar light were conducted in the Summer months (July – September) at the sunny outskirts of the city of Nablus, Palestine. During Summer months, the skies are bright clear with no clouds. Because the radiation intensity varied from one hour to another throughout the day, experiments were conducted between 10:00 -14:00. The radiation intensity at the surface of the reaction mixture was measured using a light meter and a watt meter. The radiation intensity for the direct solar light was in the range 90000 – 120000 Lux ($0.0130 – 0.0176$ W/cm^2). During solar light irradiation, the reaction mixture temperature raised by about 3 degrees, but this caused no effect, since temperature did not show significant effect on rate of reaction.

In temperature effect study, in-doors irradiation was conducted using either a model 45064 - 50 W Xe solar simulator lamp (Leybold Didactic Ltd.) equipped with a housing and a concentration lens, or a Luxten spot 50 W lamp. The reaction mixture was thermostated at desired temperature. The Xe lamp has a high stability and an intense coverage of wide spectral range, from about 450 to 800 nm, with no much preference, and a UV tail. Exact lamp specifications are described in special manuals [30, 36-40]. The intensity at the reaction mixture surface was 0.0212 W/cm^2, mostly in the visible with a small tail in the UV.

Catalytic Experiment

Photo-catalytic experiments were conducted under direct solar radiation in a 500 mL beaker. The solvent, catalyst and methyl orange were placed inside the beaker with the desired amount of each species. The solution was then made acidic or basic as desired by adding $HCl_{(aq)}$ or $NaOH_{(aq)}$. The mixture was left in the dark for enough time to reach equilibrium. AC/ZnO systems were allowed longer waiting times. This was to ensure saturation of contaminant molecules onto surface. The AC, with high surface area, can adsorb too much of the contaminant molecules. Moreover, AC/ZnO catalyst systems demanded higher methyl orange concentrations, than naked ZnO system, due to higher adsorption ability.

The reaction mixture was then exposed to direct solar light, reaching it from the uncovered top, with no glass, mediators, mirrors or concentrators. The reaction progress was monitored the moment the reaction mixture was exposed to solar light. The progress was followed by measuring the amount of remaining methyl orange with time. This was performed by syringing out small aliquots of reaction mixture at certain times. The aliquot were then centrifuged at high speed (500 rps) for 5 min. The liquid phase was then politely syringed out and analyzed spectrophotometrically at 480 nm. Analysis for Cd^{2+} and Zn^{2+} ions was conducted using polarography as described above.

3. RESULTS AND DISCUSSION

Exposure of aqueous solutions of methyl orange, under different pH values, to direct solar light, in the presence of ZnO particles, caused complete de-colorization of the methyl orange solution in reasonably short times. Control experiments were conducted in the dark under otherwise same conditions, and no de-colorization was observed. Exposure of the methyl orange to solar light with no ZnO did not cause any noticeable de-colorization. The methyl orange solution, with ZnO inside, did not exhibit any noticeable de-colorization under direct solar light when a cut-off filter (eliminating 400 nm and shorter wavelengths). This indicates that the de-colorization is due to the UV tail available in the solar light, not the visible light itself. The de-colorization is a direct indication of photo-degradation of methyl orange contaminant. Literature indicated that methyl orange undergoes complete mineralization under PEC conditions [1-26]. Therefore, the degraded methyl orange is assumed to be completely mineralized.

The ability of ZnO to degrade methyl orange under direct solar light is the focal issue here. Effects of catalyst type, pH, temperature, catalyst nominal concentration, and contaminant concentration are all presented. Effect of CdS, as a sensitizer for ZnO, is also presented. Moreover, the effect of supporting ZnO onto AC is discussed.

3.1. Naked ZnO Catalyst System

Effect of Added CdS Dye
Naked ZnO showed high efficiency in methyl orange photo-degradation under direct solar light. The active solar light was the UV tail rather than the visible region. This was manifested by the fact that using a cut-off filter (eliminating 400 nm or shorter wavelengths) prevented degradation.

The ability of CdS to enhance efficiency of ZnO in degradation reaction, by sensitization, was assessed here. The ZnO/CdS system showed lower efficiency, under direct solar light, than the naked system. Figure 1 shows that the degradation reaction rate was slightly lowered by using small CdS uptakes. With higher CdS uptakes, the inhibition effect was more pronounced, as shown in the Figure. The same behavior was observed from values of quantum yield (moles contaminant removed per photon after 20 min) measured for different reaction runs (a) 0.0015 , (b) 0.00135 , (c) 0.00135, and (d) 0.00025.

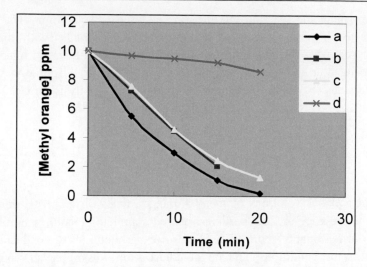

Figure 1. Effect of CdS on ZnO photo-degradation of methyl orange under direct solar light (intensity 128000 Lux, 0.0187 W/cm^2). All reactions were conducted in a neutral 250 mL contaminant solution (10 ppm) at room temperature using 0.1 g ZnO/CdS with different CdS uptakes: a) 0.0% CdS b) 0.2% CdS, c) 0.4% Cds, and d) 2% CdS.

Since no noticeable enhancement was observed, even with low CdS uptakes, the CdS system seems to be useless in the used ZnO. With higher concentrations, the CdS system caused almost complete inhibition for the degradation reaction. The inhibition is attributed to screening the radiations away from the ZnO particle surfaces. This parallels other reports, where CdS screened UV light away from SC catalyst particles and lowered the efficiency in the UV [18].

Some reports stated that ZnO/CdS (0.2% CdS by mass) enhances efficiency of ZnO in methyl orange degradation by UV [1], while other reports stated that naked ZnO showed higher efficiency in degrading contaminants [4] under solar light. To verify this matter, special experiments were conducted here. Naked ZnO and ZnO/CdS (with different CdS percent uptakes) were used in methyl orange degradation using direct solar light, with no UV tail (using a cut-off filter eliminating 400 nm and shorter UV) and with UV tail. The results indicated that CdS inhibited the reaction rate under direct solar light (with UV), as shown above in Figure 1. In case of a filter, with no allowed UV, the naked ZnO did not function, but the ZnO/CdS systems caused less than 20% degradation of methyl orange. With CdS, only very little sensitization occurred in the only-visible light, whereas in the direct solar light (with UV tail) CdS screened UV and inhibited the reaction. The inhibition (in direct solar light) is more profound than the sensitization (in only-visible light). Moreover, ZnO/CdS systems used in direct solar light showed inhibition due to CdS. When left under PEC conditions for enough time the CdS was completely degraded and washed away, and the remaining ZnO system showed enhanced efficiency. These results suggest the incompetence and limitations of CdS as sensitizer for the ZnO particles used here. Its ability to sensitize ZnO particles with smaller sizes (nano-scale) can not be ruled out, as literature suggested [1].

In CdS-sensitized TiO$_2$ particles, CdS showed high sensitization power to visible light that surpassed its screening effect, and the TiO$_2$/CdS showed higher efficiency than the naked TiO$_2$ under solar simulators [18]. In case of ZnO systems the CdS failed to behave similarly. This, together with the hazardous nature of CdS to leach out toxic Cd^{2+} ions into solution

[18], are compelling evidence against using CdS as sensitizer here. The ability of ZnO to catalyze decomposition in direct solar light puts an end to many costly and difficult sensitization-based techniques. This alone makes naked ZnO an excellent candidate for catalytic photo-degradation processes.

Effect of pH

Effect of pH on reaction rates of methyl orange degradation with direct solar light was investigated using naked ZnO catalyst. Experiments conducted at different pH values showed that the rate of the reaction was reasonably fast and not much affected with the value of pH. Figure 2 shows effect of using different solution acidities on the rate of methyl orange degradation. Because pH did not show significant effect on reaction rates, methyl orange degradation experiments were conducted using neutral solutions, unless otherwise stated.

At higher basicity (pH more than 11), a slightly different behavior occurred. The reaction demaded an induction period. It did not start efficiently at the beginning, but after a few minutes the reaction progress was more clearly noticed, Figures (2) & (3). This was presumably due to solid $Zn(OH)_2$ formation at higher pH [41]. This has something to do with the amphoteric nature of ZnO surface and its sensitivity to the high base concentrations. Such behavior should not be counted against ZnO catalyst systems, because the natural working conditions for water purification involve pH ranges closer to the pH 7. Its ability to completely degrade methyl orange in less than 30 min under pH conditions in the range 4-9, is an outstanding characteristic for the ZnO system.

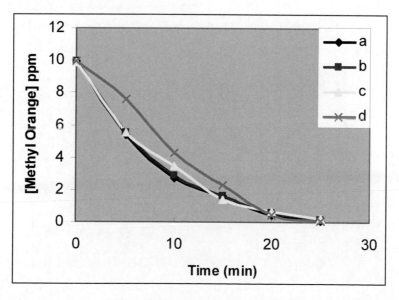

Figure 2. Effect of pH on rate of methyl orange degradation reaction. Under direct light (90000 Lux, 0.013 W cm^{-2}). All reactions were conducted at room temperature in a 250 ml solution of methyl orange (10 ppm) using 0.1 g ZnO. The pH values were: a) 4.7 b) 7 c) 9 d) 11.

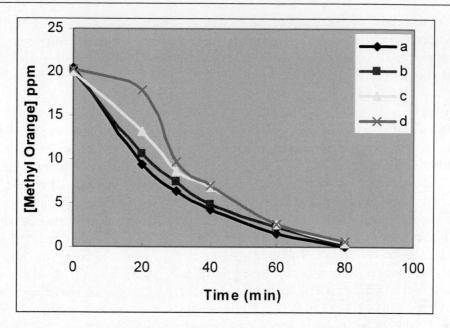

Figure 3. Effect of pH value on photodegradation under direct solar light (Radiation intensity: 127000 Lux, 0.01859 W/cm^2) of 300 ml methyl orange (20 ppm) using 0.1 g ZnO. Only basic range of pH values are: a) 8.1 b) 9.2 c) 10.4 d) 12.0.

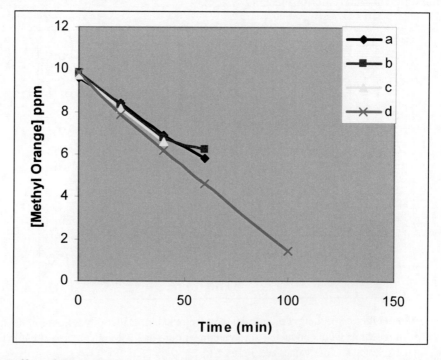

Figure 4. Eeffect of temperature on photo-degradation of neutral 300 mL solution methyl orange (10 ppm) using 0.1 g ZnO temp: a) 10°C b) 20°C c) 30°C d) 40°C. For better temperature control, radiation was conducted inside 100 mL thermostated neutral solution of methyl orange (10 ppm) with solar simulator (Radiation Intensity 0.0212 W/cm^2) and 0.05g naked ZnO.

Effect of Temperature

The effect of temperature on rate of methyl orange degradation was studied. For better control of reaction mixture a Xe lamp solar simulator was used instead of direct solar light. Figure 4 shows that the reaction rate did not increase with higher temperature, and that the value for activation energy was negligible. This is contrary to thermally induced catalytic reactions where the rate increases with higher temperature. Photochemical degradation reactions, using high band gap SC catalysts, normally have rates that are known to be independent of temperature [42-43]. What matters here is the value of the photon energy which is in the order 3.2 eV or higher. Thermal energy gained by heating at ambient temperatures (less than $100^{\circ}C$) is far less than 3.0 eV, *viz.* heat energy lies within the IR region only not UV. Moreover, higher temperatures encourage the dissolved oxygen, present in the reaction mixture, to escape. Thus the formation of OH radicals, necessary for contaminant oxidation, will be prevented by the lack of dissolved oxygen.

Effect of Catalyst Concentration

While using lower catalyst concentrations, the rate of reaction increased with increasing catalyst amount, as shown in Figure [5]. At higher concentrations the rate was independent of catalyst concentration, exhibiting approximately a pseudo first order over the used catalyst concentration range. This is a common behavior among SC catalytic systems in photo-degradation processes. This is understandable, because at higher catalysts concentrations, larger proportions if the catalyst sites are screened away from incident light, and the efficiency is expected to decrease. Turnover number values (mole reacted methyl orange per mole ZnO after 20 min) relatively decrease with higher catalyst nominal concentrations. For the lower catalyst amount system it was ~3.2 and for the higher catalyst amount it was 0.5 only. This supports the idea of radiation screening away from catalyst active sites while using higher catalyst amounts.

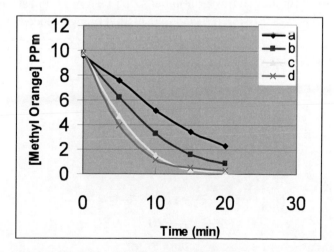

Figure 5. Effect of catalyst ZnO amount on rate of photo-degradation of methyl rang (250 ml solution, 10 ppm), at room temperature under neutral conditions using direct solar light (radiation intensity $0.01815 W/cm^{2}$). Catalyst ZnO amounts were: a) 0.05g, b) 0.1 g, c) 0.2 g, d) 0.4 g.

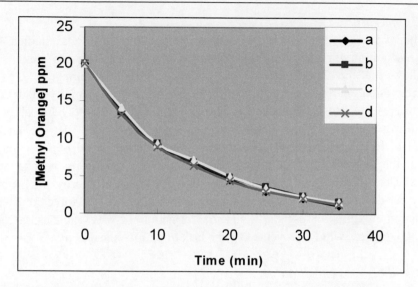

Figure 6. Effect of high catalyst amount on photo-degradation of methyl orange (300 ml solution, 20 ppm), at room temperature under neutral conditions, using direct solar light (radiation intensity 0.01625 W/cm^2). Catalyst ZnO amounts were: a) 0.6 g b) 0.8 g c) 1.0 g d) 1.2 g.

Higher catalyst concentrations exhibited no effect on rate of reaction as shown in Figure 6. The rate of reaction did not change with increasing catalyst amount. The turnover number was going lower with increased catalyst amount. This indicates that at higher catalyst amounts the relative efficiency of the catalyst decreases due to screening effect.

Effect of Contaminant Concentration

The degradation rate was higher with higher contaminant concentration, as shown in Figure [7]. Plots of ln(initial rate) vs. ln(initial conc.) showed linear relation, and the calculated reaction order was 0.7 with respect to the methyl orange concentration.

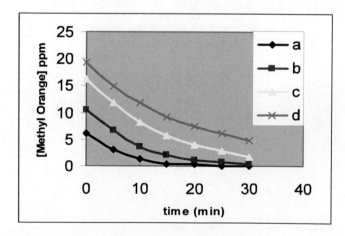

Figure 7. Effect of contaminant concentration on photo-degradation reaction using 0.1g ZnO in 300 mL neutral solution at room temperature under direct solar light (0.0149W/cm^2) . Methyl orange concentrations were: a) 5 ppm b) 10 ppm c) 15 ppm d) 20 ppm.

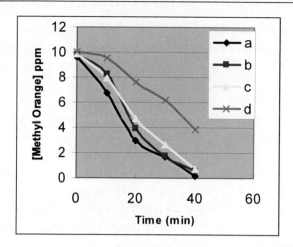

Figure 8. Photo-degradation reaction profiles for fresh and recovered ZnO catalysts. Reactions were conducted at room temperature under direct solar light (intensity 0.0144 W/cm^2) using a fresh neutral 300 mL solution of methyl orange (10 ppm) in each run. The original catalyst loading was (a) 0.1 g fresh ZnO, (b) First recovery, (c) Second recovery, and (d) Third recovery.

The results indicate that the ZnO catalyst works efficiently over a relatively wide range of contaminant concentrations, ranging from very low concentrations up to more than 20 ppm. This adds to the advantages of ZnO system.

Catalyst Reuse

Naked ZnO catalyst samples were recovered for reuse in methyl orange photo-degradation. After reaction cessation, the reaction mixture was centrifuged at high speed for up to (500 rounds/s for 5 min), after which the liquid phase was syringed out. The solid residue was washed with distilled water. The mixture was again centrifuged, and the water syringed out and the solid residue taken for catalytic reuse in fresh reactant solution. First and second recovery samples showed almost similar activity as fresh samples, albeit the little loss of ZnO during recovery. Quantum yield values for the first and second recovered systems showed quantum yields almost the same as that for the fresh sample (0.0012). The third recovery catalyst lost up to 30% of its activity compared to the fresh sample. These results are exhibited in Figure 8. Loss of efficiency on recovery and reuse should not count against the ZnO systems. Literature reports [19, 27] showed more efficiency loss in recovered SC catalysts. The technical difficulty in recovering ZnO powder out of the reaction mixture may count against ZnO system However, this difficulty is not associated only with ZnO systryems. Normally, micro-scale and nano-scale catalyst systems exhibit technical difficulty in recovery. One possible technique to facilitate recover ZnO catalyst system is using solid AC support as will be discussed below.

3.2. AC-Supported ZnO System

The AC/ZnO was a dark color solid, with 83% (or higher in some cases) ZnO and the remaining was AC. The AC/ZnO was hydrophilic in nature as it readily mixed with water with no difficulty. It was also easy to isolate from water by simple filtration. This alone makes up for any other shortcoming that could possibly be associated with using AC support. The efficiency of AC/ZnO system has been investigated using different parameters, keeping an eye on its credibility to use commercially in the future.

Effect of Catalyst Amount and Catalyst Composition

The methyl orange degradation reaction was studied using different AC/ZnO nominal amounts, as shown in Figure 9. Unlike the naked ZnO catalyst systems, the AC/ZnO system exhibited relatively high adsorption onto the AC surface which accounted to about 10 ppm of contaminant concentration on the AC surface (0.02 g). For this reason, relatively high contaminant concentrations were used in the AC/ZnO systems. The catalyst and contaminant mixture was allowed to stand for prolonged time intervals before irradiation started. This was to ensure AC surface was saturated with contaminant molecules. Any contaminant loss after that was due to degradation.

Figure 9 shows that measured values of initial contaminant concentrations did not match with the nominal used concentrations (25 ppm). Moreover, the initial rate calculations could not be accurately measured for the AC/ZnO systems. Therefore, efficiency comparisons were performed based on turnover number values only.

Complete methyl orange degradation was observed in less than 80 min reaction time, accounting for turnover number values 0.9-1.2 (moles reacted per mole ZnO after 30 min). This is a reasonably high turnover number value. It should be noted that only surface sites of ZnO system are catalytically active sites, and the majority of the ZnO active sites are inaccessible bulk type. This shows the competitiveness of the ZnO system in methyl degradation. As shown in Figure 9, where the higher AC/ZnO amounts showed the lower turnover numbers, the AC/ZnO system caused some light screening.

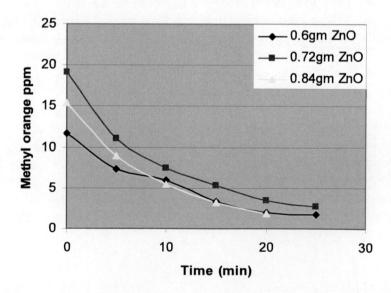

Figure 9. Effect of amount of AC/ZnO on photo-degradation rate of methyl orange (25 ppm) inside 300 mL neutral solution at room temperature under direct solar light (0.0149 W/cm^2). Catalyst amounts were: a) net ZnO 0.5 g) b) net ZnO 0.72 g), and c) net ZnO 0.84g).

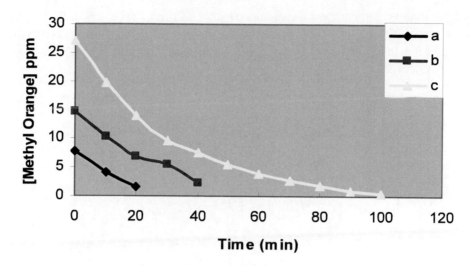

Figure 10. Effect of AC/ZnO ratio on photodegradation of neutral solutions (300 ml) of methyl orange (25 ppm) ; ratio AC:ZnO = a) 1:3 b) 1:5 d) 0:1 (pure ZnO). All reaction runs were conducted at room temperature under direct solar light (0.017 W/cm^2) with net 0.10 g ZnO.

Figure 10 shows reaction rate profiles using naked ZnO and AC/ZnO system. Two different AC/ZnO systems were used, with different AC to ZnO mass ratios, namely 0.1:0.3 and 0.1:0.5 respectively. Using relatively high irradiation intensity (0.017 W/cm^2), both AC/ZnO systems showed complete methyl orange degradation within ~30 min, compared to naked ZnO which demanded longer time. For the AC/ZnO (3:1 ratio) system, the quantum yield was 0.00413, compared to 0.00083 for the naked ZnO system. These results are in accordance with earlier reports using nano-scale SC particles [20-27]. Large size ZnO particles (230 nm), used here resemble nano-scale particles reported earlier, in the sense that their efficiency seems to increase with the AC support. As explained earlier, the AC is believed to adsorb contaminant molecules and bring them into close proximity with the ZnO active sites. This is one advantage for using the AC/ZnO system.

Effect of Contaminant Concentration

The effect of contaminant concentration on the AC/ZnO catalyzed reaction was studied. Due to methyl orange adsorption it was not possible to measure or compare initial reaction rates, as shown in Figure 11. Turnover number values measured after 70 minutes indicated that the reaction was faster with higher contaminant concentration. Turnover number values were higher for higher contaminant concentrations reaction runs. This was consistent with the naked ZnO system. The results thus indicate the ability of AC/ZnO to effectively function in the degradation reactions of relatively methyl orange high concentration, 100 ppm or higher.

Effect of Temperature

No systematic effect, for temperature on reaction rate, has been observed, for the AC/ZnO catalyzed methyl orange degradation. Figure 12 shows reaction profiles with time for different experiments conducted at different temperatures. Since methyl orange adsorbs onto the AC surface, causing inconsistency between measured contaminant concentrations and nominal counterparts, no initial rate measurement or comparison could be achieved.

Based on turnover number values calculated after 30 min, there was little rate difference between experiments conducted at different temperatures. This is consistent with results obtained using the naked ZnO system, and the discussions presented above apply to the AC/ZnO system as well.

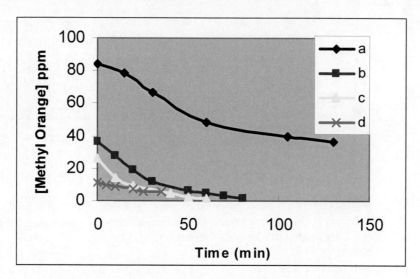

Figure 11. Effect of contaminant concentration on photo-degradation reaction of methyl orange in neutral solutions (300 mL) at room temperature under direct solar light (intensity 0.0164 W/cm^2) using AC/ZnO (0.24 g, containing 0.2 g ZnO). Contaminant concentrations were: a) 100 ppm b) 50 ppm c) 40 ppm d) 30 ppm. TN values (moles reacted per mol ZnO after 70 min) are: 7.2 , 5.4, 4.8, 3.6 respectively.

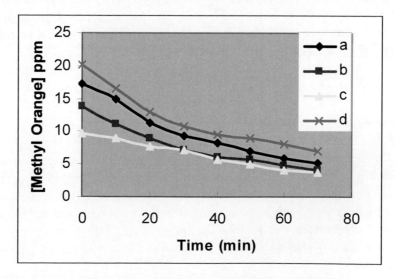

Figure 12. Effect of temperature on photo-degradation neutral 300 ml of methyl orange (25 ppm) using 0.12 g ZnO/AC (containing 0.1 g pure ZnO) under solar simulator radiation (0.0212 W/cm^2). The reaction temperature was thermostated at: a) 19 °C b) 29 °C c) 38 °C c d) 46 °C.

Effect of pH

Methyl orange degradation reaction using AC/ZnO system was investigated in solutions with different pH values. Figure 13 shows that the catalyst efficiency and turnover number did not vary with value of pH in the range 4-10. At pH 12, the reaction rate was slightly lower after 30 min. The results indicate the ability of the ZnO system to effectively function under different pH values, almost indifferently, with natural waters which have pH values closer to 7. This reflects the applicability of AC/ZnO system in future natural water purification technologies.

Catalyst Recovery and Reuse

Recovery of AC/ZnO from the methyl orange degradation reaction mixture after reaction cessation was achieved effectively and simply by filtration. This shows the importance of using the AC support. The catalyst was recovered and reused for three successive times. Compared to the fresh catalyst sample, only 20% efficiency loss occurred in the third recovery catalyst system, as shown in Figure 14. Just like the naked ZnO systems, the AC/ZnO systems also retained its efficiency for fourth reuse. This shows a promising applicability of the AC/ZnO system in the future.

It is worthwhile to compare between recovered naked ZnO systems and AC/ZnO systems in terms of efficiency loss on recovery. In case of naked ZnO systems, no efficiency loss on recovery was observed for the first and second recoveries. In case of AC/ZnO efficiency loss occurred (Figure 14) to a higher extent. This is presumably due to degradation of ZnO catalyst itself under PEC conditions, being higher in the supported system than in the naked system. Tendency of ZnO to degrade into Zn^{2+} ions is discussed below.

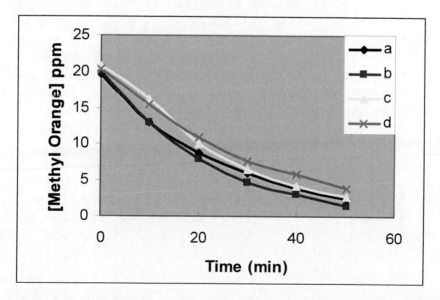

Figure 13. Effect of pH on photo-degradation of 300 ml solution of methyl orange (25 ppm) using 0.12 g ZnO/Ac (with 0.1 g net ZnO) at room temperature under direct solar light (0.0171 W/cm^2). The solution pH values were: a) 4.2 b) 8.4 c) 10 d) 12.

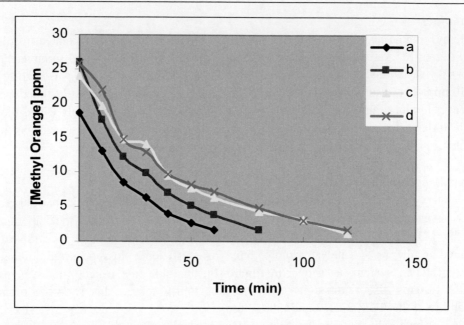

Figure 14. Reaction profiles showing photo-degradation of methyl orange (25 pm) in 300 mL neutral solutions at room temperature under direct solar light (0.01427 W/cm²) using: (a) 0.12 g fresh AC/ZnO catalyst (containing 0.1 g ZnO), (b) First Recovery, (c) Second recovey, and (d) Third recovery. Note that contaminant adsorption occurred only in fresh catalyst.

Catalyst Stability to Degradation

Under PEC conditions, the ZnO in the AC/ZnO systems underwent degradation leading to formation of aqueous Zn^{2+} ions, as confirmed by polarographic analysis of the catalytic reaction mixture with time. Such leaching tendency was more obvious in acidic conditions than in basic conditions. At pH 12, only 3.3 of the supported ZnO degraded into Zn^{2+} ions in solution. At pH 8, up to 6.3 % degradation occurred, whereas at pH 12 only 3.3 % degradation occurred. At high pH values, 12, precipitation of $Zn(OH)_2$ is possible [41] which accounts for low Zn^{2+} leaching out. Naked ZnO systems showed somewhat lower leaching out tendency. Under neutral conditions, only 3.3% of originally used naked ZnO leached out under PEC conditions. This accounts for the fact that supported ZnO lost more efficiency on recovery than the naked counterpart. However, efficiency lowering was not highly prominent and it did not undermine the major advantage of the supported system, in making recovery more feasible. Moreover, the Zn^{2+} is not considered as a highly hazardous ion [25-26]. Thus leaching out of Zn^{2+} ions should not be counted against using AC/ZnO systems. Despite this, work is underway here to develop techniques that would minimize ZnO degradation processes under PEC conditions.

4. CONCLUSION

Both naked ZnO and AC-supported ZnO systems effectively catalyzed complete photo-degradation reactions of methyl orange in aqueous solutions under direct solar light with relatively high turnover number values. The active region of the solar light was the UV tail rather than the visible region. The AC support enhanced ZnO efficiency in methyl orange degradation, while making catalyst recovery easier. Both ZnO and AC/ZnO systems were recoverable and reusable, showing the future value of the ZnO catalyst system in water purification strategies. Using CdS as sensitizer did not affect the catalyst efficiency. Its ability to leach Cd^{2+} ions is a good reason to avoid using it as sensitizer.

ACKNOWLEDGMENTS

G. N. wishes to thank I. Saadeddin, R. Ismail and M. Atatreh (all of ANU) for technical help. Support from French-Palestine University Cooperation program is acknowledged.

REFERENCES

[1]   Sakthivel, S.; Geissen, S.-U.; Bahnemann, D.W.; Murugesan, V.; Vogelpohl, A. "Enhancement of photocatalytic activity by semiconductor heterojunctions: α-Fe$_2$O$_3$, WO$_3$ and CdS deposited on ZnO", *J. Photochem. Photobiol. A: Chem.* 2002, *148*, 283–293.

[2]   Zhou, G.; Deng, J. "Preparation and photocatalytic performance of Ag/ZnO nano-composites", *Mat. Sci. Semicond. Proc.* 2007, *10*, 90-96.

[3]   Chen, C.; Wang, Z.; Ruan, S.; Zou, B.; Zhao, M.; Wu, F. "Photocatalytic degradation of C.I. Acid Orange 52 in the presence of Zn-doped TiO$_2$ prepared by a stearic acid gel method", *Dyes Pigmen.* 2008, *77*, 204-209.

[4]   Kansal, S.K.; Singh, M.; Sud, D. "Studies on photodegradation of two commercial dyes in aqueousphase using different photocatalysts", *J. Hazard. Mat.* 2007, *141*, 581–

[5]   Dabrowski, B.; Hupka, J.; Żurawska, M.; Miller, J. D. "Laboratory and pilot-plant scale photodegradation of cyanide–containing wastewaters, *Physicochem. Probl. Min. Proc.* 2005, *39*, 229-248.

[6]   Morrison, S. R.; Freund, T., "Chemical Role of Holes and Electrons in ZnO Photocatalysis", *J. Chem. Phys.*, 1967, *47*, 1543-1551.

[7]   Wang, H.; Xie, C.; Zhang, W.; Cai, S.; Yang, Z.; and Gui, Y.; "Comparison of dye degradation efficiency using ZnO powders with various size scales", *J. Hazard. Mat.* 2007, *141*, 645-652.

[8]   Yu, D.; Cai, R.; Liu, Z. ; "Studies on the photodegradation of rhodamine dyes on nanometer-sized zinc oxide", *Spectrochim. Acta A: Mol Biomol. Spectrosc.* 2004, *60*, 1617-24.

[9]   Daneshvar, N.; Aber, S.; Seyed Dorraji, M. S.; Khataee, A. R.; and Rasoulifard, M. H.; "Preparation and investigation of photocatalytic properties of ZnO nanocrystals: Effect of operational parameters and kinetic study", *Interanat. J. Cem. Biomol. Engin.* 2008; *1*, 1307-7449.

[10]   Shvalagin, V. V.; Stroyuk, A. L.; Kotenko, I. E. and Kuchmii , S. Ya.; "Photocatalytic formation of porous CdS/ZnO nanospheres and CdS nanotubes", *Theor. Exper. Chem.*, 2007, *43*, 229-234.

[11] Zhang , M.; An, T.; Hu , X.; Wang, C.; Sheng, G.; and Fu, J.; "Preparation and photocatalytic properties of a nanometer ZnO–SnO$_2$ coupled oxide", *Appl. Catal. A: Gen.*, (2004), 260, 215-222.

[12] Byrappa, K.; Subramani, A. K.; Ananda, S.; Rai, K. M. L.; Sunitha, M. H.; Basavalingu, B.; and Soga, K.; "Impregnation of ZnO onto activated carbon under hydrothermal conditions and its photocatalytic properties", *J. Mat. Sci.*, 2006, *41*, 1355-1362.

[13] Robert, D. "Photosensitization of TiO$_2$ by M$_x$O$_y$ and M$_x$S$_y$ nanoparticles for [heterogeneous photocatalysis applications", *Catal. Today.* 2007, *122*, 20–26.

[14] Fuji, H.; Ohtaki, M.; Eguchi, K.; Arai, H.; *J. Mol. Catal. A: Chem.* 1998, *129*, 61–68.

[15] Serpone, N.; Marathamuthu, P.; Pichat, P.; Pelizzetti, E.; Hidaka, H.; *J. Photochem. Photobiol. A: Chem.* 1995, *85*, 247.

[16] Sesha, S.; Srinivasan, J. W.; Stefanakos, E. K. "Visible light photocatalysis via CdS/TiO$_2$ nanocomposite materials", *J. Nanomat.*, 2006, *2006*, 1-7.

[17] Reeves, P.; Ohlhausen, R.; Sloan, D.; Pamplin, K.; Scoggins, T.; Clark, C.; Hutchinson, B.; Green , D. "Photocatalytic destruction of organic dyes in aqueous TiO$_2$ suspensions using concentrated simulated and natural solar energy", *Sol. Ener.* 1992, *48*, 413-420.

[18] Greijer, H.; Karlson, L.; Lindquist, S.-E.; Hagfeldt, A. "Environmental aspects of electricity generation from a nanocrystalline dye sensitized solar cell system", *Renew. Ener.* 2001, *23*, 27-39.

[19] Zyoud, A. H.; and Hilal, H. S.; "Silica-supported CdS-sensitized TiO$_2$ particles in photo-driven water purification: Assessment of efficiency, stability and recovery future perspectives", A chapter in this book, (in Press, 2008).

[20] Sobana, N. ; Swaminathan, M.; "Combination effect of ZnO and activated carbon for solar assisted photocatalytic degradation of Direct Blue 53", Sol. Ener. Mat. Sol. C., 2007, *91*, 727-734.

[21] Jiang, L.; and Gao, L.; "Fabrication and characterization of ZnO-coated multi-walled carbon nanotubes with enhanced photocatalytic activity", *Mat. Chem. Phys.*, 2005, *91*, 313-316

[22] Gokona, N.; Hasegawa, N.; Kaneko, H.; Aoki, H.; Tamaura, Y.; Kitamuara, M., "Photocatalytic effect of ZnO on carbon gasification with CO$_2$ for high temperature solar thermochemistry", Sol. Ener. Mater. Sol. C., 2003, *80*, 335-341.

[23] Rohe, B.; Veeman, W. S.; and Tausch, M.; "Synthesis and photocatalytic activity of silane-coated and UV-modified nanoscale zinc oxide", *Nanotechnol.* 2006, *17*, 277–282.

[24] Kalpana, D.; Omkumar, K. S.; Kumar, S. S. and Renganathan, N.G.; "A novel high power symmetric ZnO/carbon aerogel composite electrode for electrochemical supercapacitor", *Electrochim. Acta.* 2006, *52*, 1309-1315.

[25] Zinc, in http://ods.od.nih.gov/factsheets/cc/zinc.html , accessed Sept. 2008.

[26] T. Lursinsap , Zinc in Human Nutrition, http://www.sci.ru.ac.th/chem/web%20genchem%202002/research%20%20page_1/paper/ZN.pdf , accessed Sept. 2008.

[27] Hilal, H.S.; Majjad, L.Z.; Zaatar, N.; El-Hamouz, A.; "Dye-effect in TiO$_2$ catalyzed contaminant photo-degradation: Sensitization vs. charge-transfer formalism", *Solid State Sciences* 2007, *9*, 9-15.

[28] Ahnert, F.; Arafat, H.; Pinto, N.; "A study of the influence of hydrophobicity of activated carbon on the adsorption equilibrium of aromatics in nonaqueous media", *Adsorption,* 2003, *9*, 311-319.

[29] Glasstone, S.; Lewis, D.; *Elements of Physical Chemistry*, 2nd edition, Macmillan Press Ltd., New Delhi, 1983, p. 566.

[30] Shoemaker, D.; Garland, C.; Nibler, J.; *Experiments in Physical Chemistry*, 5th edition, Mc Graw-Hill, Inc., N.Y., 1989.

[31] Vigil-Galán, O.; Morales-Acevedo, A.; Cruz-Gandarilla, F.; Jiménez-Escamilla, M.G. "Characterization of CBD–CdS layers with different S/Cd ratios in the chemical bath and their relation with the efficiency of CdS/CdTe Solar Cells", *Thin Solid Films*, 2007, *515*, 6085–6088.

[32] Ximello-Quiebras, J.N.; Contreras-Puente, G.; Aguilar-Hernandez, J.; Santana-Rodriguez, G.; Readigos, A. A.-C. "Physical Properties of chemical bath deposited CdS thin films", *Sol. Energ. Mat. Sol. C.* 2004, *82*, 263–268.

[33] Roy, P.; Srivastava, S. K. "A new approach towards the growth of cadmium sulphide thin film by CBD method and its characterization", *Mat. Chem. Phys.* 2006, *95*, 235–241.

[34] Zinoviev, K.V.; Zelaya-Angel, O. "Influence of low temperature thermal annealing on the dark resistivity of chemical bath deposited CdS films", *Mat. Chem. Phys.,* 2001, *70*, 100–102.

[35] Chen, F.; Jie, W.; "Growth and photoluminescence properties of CdS solid solution semiconductor", *Cryst. Res. Technol.* 2007, *42*, 1082 – 1086.

[36] For phenol analysis, see AOAC Official Method 979.13, copyright 1998, AOAC International.

[37] The Book of Photon Tools, Oriel Instruments, Stratford, CT, USA, pp. 44, 50 (Manual).

[38] Thermo Oriel: 500 W Xenon/Mercury (Xenon) ARC Lamp Power Supply Model 68911 (Manual).

[39] Clark, B. J.; Frost, T.; and Russell, M. A. *U.V. Spectroscopy*, Chapman & Hall, London, 1993, p.p. 18-21.

[40] [http://www.uvguide.co.uk/zoolamps.htm accessed Aug. 2008.

[41] Neppolian, B.; Choi, H. C.; Sakthivel, S.; Arabindoo, B.; and Murugesan, V. "Solar/UV-induced photocatalytic degradation of three commercial textile dyes", *J. Haz. Mat.* 2002, *89*, 303-317.

[42] Hermann, J. M.; "Heterogeneous photocatalysis: fundamentals and applications to the removal of various types of aqueous pollutants", *Catal. Today*, 1999, *53*, 115-129.

[43] Robert, D. "Photosensitization of TiO_2 by M_xO_y and M_xS_y nanoparticles for heterogeneous photocatalysis applications", *Catal. Today.* 2007, *122*, 20–26.

In: Water Purification
Editors: N. Gertsen and L. Sønderby

ISBN 978-1-60741-599-2
© 2009 Nova Science Publishers, Inc.

Chapter 7

WATER PURIFICATION USING COMPRESSION AND ABSORPTION HEAT PUMPS

J. Siqueiros[*], A. Huicochea and J. A. Hernández

Centro de Investigación en Ingeniería y Ciencias Aplicadas, Universidad Autónoma del
Estado de Morelos, Av. Universidad 1001, Col. Chamilpa, C.P. 62209,
Cuernavaca, Morelos, México

ABSTRACT

A water purification process is integrated into a compression and an absorption heat pump with energy recycled. In the compressor heat pump, the heat delivered in the condenser was recycled to the heat pump evaporator and the excess heat to balance the heat pump was used to preheat the impure water supply. The advantage of this system is that it does not use waste heat to operate—because the energy recycling happens at the boiling water temperature. In the absorption heat pump also a fraction of the heat obtained by the heat pump absorber and condenser is recycled to the evaporator by the integration of the auxiliary condenser of the water purification process. This system also has the advantage that it does not use waste heat to operate because of the energy recycling. A developed thermodynamic model shows that the proposed water purification process integrated into a heat transformer is capable of increasing the original COP values more than 120% due to the energy recycling from the water purification process. Therefore, the proposed integration of the water purification process to the heat pump allows increasing the actual COP values with any working fluid. In addition, a mathematical model based in the neural network is proposed to obtain the optimum COP value considering the energy recycling of the auxiliary condenser in the water purification process.

[*] E-mail: jsiqueiros@uaem.mx; huico_chea@uaem.mx; alfredo@uaem.mx

NOMENCLATURE

Cp	heat capacity (kJ/kg °C)
COP	coefficient of performance (dimensionless)
h	specific enthalpy (kJ/kg)
F	relation of COP's
m	mass flow rate (kg/s)
Q	heat flow rate (kW)
P	absolute pressure (kPa)
T	temperature (°C)
T'	temperature increased as result of recycling heat (°C)
X	mass fraction of LiBr in solution (%)

Subscripts

AB	absorber
CO	condenser
ET	enthalpy
EV	evaporator
d	distilled water
GE	generator
HS	heat source
IN	input
N	new
OUT	outlet
S	sensible
V	vaporization
W	water
WP	purified water

Greek Symbol

η	fraction of transferred heat

INTRODUCTION

Water is an essential factor for life and for the economic activities that are developed throughout the world. Water deficiency is an important limiting factor for the development of all human activities, due to their dependency on water for agriculture, industry and domestic use.

In the present time, the accelerated growth of the world-wide population constantly demands greater potable water availability for domestic and industrial use. The population growth will put further pressure on that need in the near future. People still resort to the purification of water sources in barren and semi-arid world regions.

The selection of the water purification process depends on supply source type, mainly, because the costs of capital, operation, maintenance, as well as the pre-treatment costs must be considered.

On the other hand, in the industrial processes a great amount of power is consumed daily. Most of it is for product heat treatments in multiple processes. This amount of thermal energy mainly comes from the combustion of fuels and unwanted by-products; gases of combustion and the remaining heat at low temperature are obtained causing contamination in the atmosphere.

The rise in international fuel prices has motivated some people to implement short-term systems for energy saving in thermal processes, where the rejected energy can be reused for other product procurement. An alternative for reusability of the industrial remainder heat is the thermodynamic cycles of the heat pumps for distilled water. With this alternative can come the benefits of diminishing pollution and providing purified water.

Heat pumps have enormous potential for reducing the primary energy requirements of various processes. An efficient heat pump system is potentially attractive for use in single process plants which tend to predominate more in developing countries [1, 2].

BASIC CONCEPT

A heat pump is a device that extracts the heat of a thermal source at low temperature and transfers the heat to a heat sink at greater temperature. So that the heat pump increases the temperature of a heat source of low quality to a more useful level, it is necessary to apply it to supply a relatively good amount of energy of high quality.

There are different classifications of heat pumps [3, 4], the most mentioned are indicated in Figure 1.

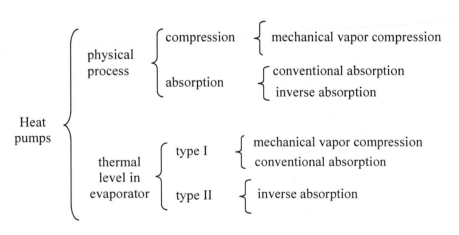

Figure 1. Heat pumps classification.

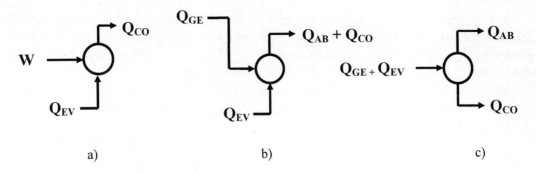

Figure 2. Thermal levels of the three main heat pumps: a) mechanical vapor compression, b) conventional absorption and c) inverse absorption.

A way to quantify energy efficiency in a heat pump is by Coefficient of Performance (COP), which is the ratio between the heat obtained and heat supplies to the components. Figure 2 shows the thermal levels between inlet and outlet heats [5].

The absorption heat pumps have some advantages over mechanical vapor compression heat pumps:

1. they use very small amounts of mechanical energy
2. they use waste heat with thermal levels between 70 and 80 °C, diminishing the environmental pollution.
3. they use working fluid, which does not pollute the environment

For every heat pump, the COP of Carnot (COP$_{Carnot}$) is ideal, due to the assumption of reversibility. Hence, it can never be achieved in practice, because there are always heat losses.

To compare heat pump systems driven by different energy sources, it is more appropriate to use the Primary Energy Ratio (PER), which is defined as,

$$PER = \frac{\text{uselful heat delivered}}{\text{primary energy input}} \qquad (1)$$

The PER is a more fundamental measure of the energy efficiency of a heat pump than the COP, because it considers the high quality energy used for obtaining useful heat [6].

It is possible, that an absorption heat pump has a lower COP than a mechanical vapor compression heat pump, although it can have a PER equal to, or even greater than, a mechanical vapor compression heat pump driven by electricity quantity used.

MECHANICAL VAPOR COMPRESSION HEAT PUMP

Due to its high yield this is the most useful heat pump. It can be used for heating as well as for cooling. The main components of this kind of heat pump are: a compressor, an evaporator, a condenser, an expansion valve and a working fluid. Schematic diagram is shown in Figure 3. The working fluid evaporates in the evaporator at T_{EV} temperature, extracting an amount of heat Q_{EV} from the heat source (waste heat), which can be in solid, liquid or gas state. The working fluid is compressed by electrical motor and gives latent heat Q_{CO} at a higher temperature T_{CO} in the condenser. Compressed working fluid is expanded through an expansion valve and returns to the evaporator. At this moment the cycle is complete.

For a mechanical vapor compression heat pump, the COP is the ratio between the heat given at the condenser Q_{CO} and the power supplied to the compressor W. The energy balance is $Q_{CO} = W + Q_{EV}$, therefore

$$COP = \frac{Q_{CO}}{W} = \frac{Q_{CO}}{Q_{CO} - Q_{EV}} \tag{2}$$

$$COP_{Carnot} = \frac{T_{CO}}{T_{CO} - T_{EV}} \tag{3}$$

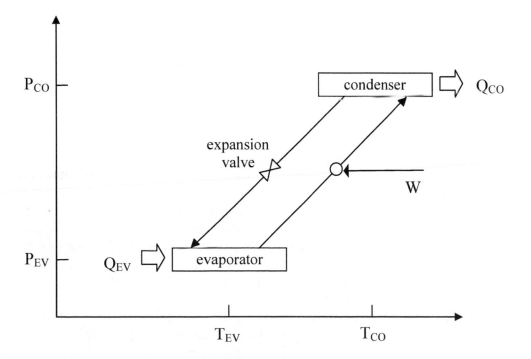

Figure 3. Schematic diagram of a mechanical vapor compression heat pump.

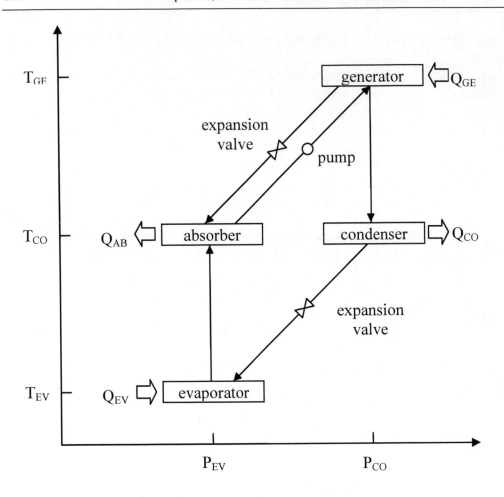

Figure 4. Schematic diagram of convectional absorption heat pump.

COVENTIONAL ABSORPTION HEAT PUMP

The schematic diagram of absorption heat pump is shown in Figure 4. High temperature energy is supplied (T_{GE}) to the generator, in which the working fluid is partially separated from the absorbent. Working fluid changes to liquid phase in the condenser (T_{CO}) and goes to the evaporator (which is at a lower pressure than the condenser) where it changes to the gas phase at a lower temperature (T_{EV}) thanks to the waste energy supplied. Steam coming from the evaporator is conduced to the absorber where it gets in contact with a working fluid-absorbent stream coming from the generator, producing heat at a temperature T_{AB} higher than T_{EV}.

For an absorption heat pump used for heating, the COP definition is the relation between total heat obtained (absorber and condenser) and heat supply in the generator.

$$COP = \frac{Q_{AB} + Q_{CO}}{Q_{GE}} \qquad (4)$$

$$COP_{Carnot} = \left(1 - \frac{T_{EV}}{T_{GE}}\right)\left(\frac{T_{CO}}{T_{CO} - T_{EV}}\right) \tag{5}$$

And for cooling, the COP is defined as the relation between heat absorbed in the evaporator and heat supply in the generator.

$$COP = \frac{Q_{EV}}{Q_{GE}} \tag{6}$$

$$COP_{Carnot} = \left(1 - \frac{T_{CO}}{T_{GE}}\right)\left(\frac{T_{CO}}{T_{CO} - T_{EV}}\right) \tag{7}$$

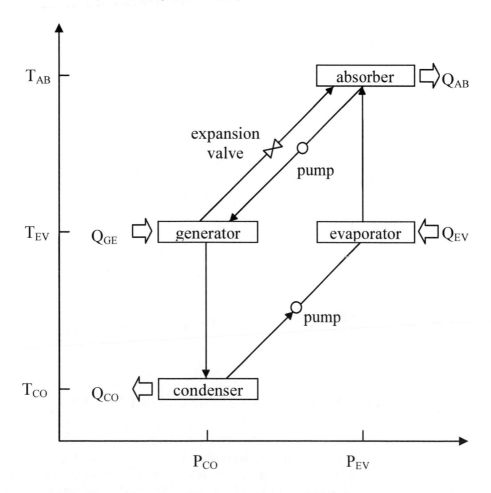

Figure 5. Schematic diagram of inverse absorption heat pump or heat transformer.

INVERSE ABSORPTION HEAT PUMP

The schematic diagram of the inverse absorption heat pump is shown in Figure 5, where medium temperature energy is supplied (T_{GE}, T_{EV}) to the generator and evaporator. In the generator a partial separation of the working fluid coming from the absorber concentrated solution in working fluid takes place. In the condenser the working fluid changes to liquid phase and then is pumped to the evaporator (which is at a higher pressure than the condenser and the generator) where it is evaporated and then conducted to the absorber, where it gets in contact with the mixture pumped from the generator, producing heat at a temperature T_{AB} higher than T_{EV}.

Also existing is an inverse absorption heat pump with an economizer. It has a basic cycle with a heat interchanger situated between a generator and an absorber. The main objective is to pre heat the working solution that goes to the absorber, using part of the energy of the working solution that comes from the absorber. This heat interchange increases the thermodynamic efficiency, due to the work solution temperature being higher. Inverse absorption heat pumps are sometimes called heat transformers or temperature boosters.

Its COP is defined as [4]:

$$COP = \frac{Q_{AB}}{Q_{GE} + Q_{EV}} \tag{8}$$

When the generator and evaporator temperatures are different, the COP_{Carnot} applied to the heat transformer is,

$$COP_{Carnot} = \frac{T_{AB}(T_{EV} - T_{CO})}{T_{GE}(T_{EV} - T_{CO}) + T_{EV}(T_{AB} - T_{GE})} \tag{9}$$

and, when the generator and evaporator temperatures are the same, COP_{Carnot} is defined as,

$$COP_{Carnot} = \frac{T_{GE} - T_{CO}}{T_{GE}} \frac{T_{AB}}{T_{AB} - T_{CO}} \tag{10}$$

WATER PURIFICATION USING HEAT PUMPS

The process of water purification integrated into heat pumps consists of a simple distillation as it is shown in Figure 6. The heat obtained in the heat pump is applied to a container "A", which contains impure water; the provided heat promotes the generation of steam at pressure and temperature conditions in container "A". The vapor goes in the container "B", where the condensation process of the vapor releases heat.

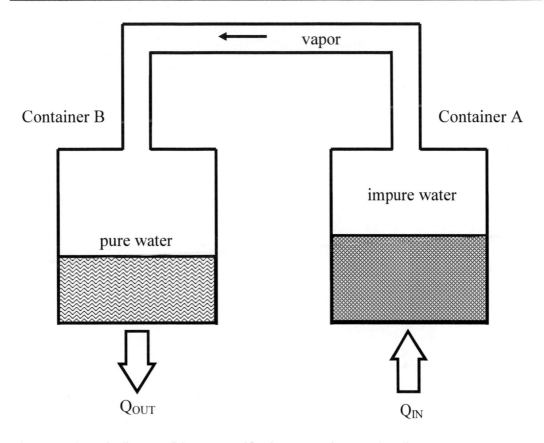

Figure 6. Schematic diagram of the water purification process integrated to a heat pump.

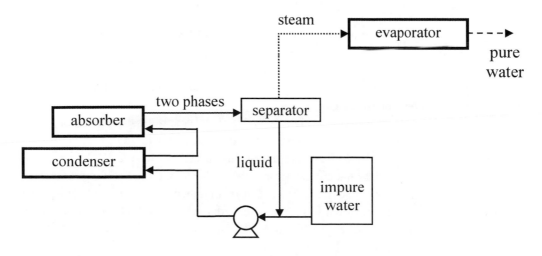

Figure 7. Detail of water purification system into a mechanical vapour compression heat pump.

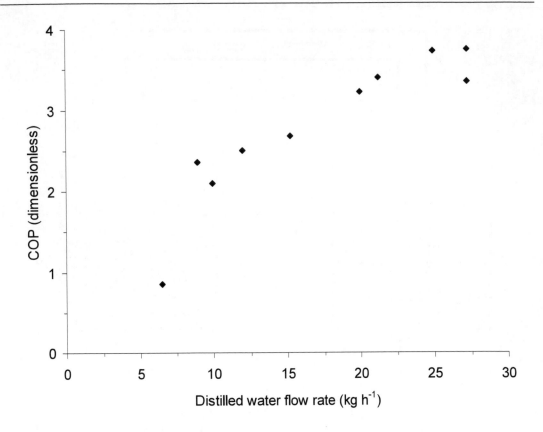

Figure 8. Coefficient of performance in function of distilled water flow rate.

Water Purification System Integrated to a Mechanical Vapour Compression Heat Pump

The impure water in a container is pumped to the condenser to retire the heat; that impure water becomes a liquid phase at low temperature and after the heat process it leaves as a mixture of water and steam at high temperature towards a separator as is shown in Figure 7. The steam leaves through the top of the separator to the evaporator to yield heat and to obtain the pure water. In this case the COP values are not modified, because COP definition is the ratio between heat delivered by the condenser and the work supplied to the compressor.

A purified water quantity of 27.25 kg/h was obtained, using a mechanical vapor compression heat pump as in shown in Figure 8. It is demonstrated, that the highest distilled water is obtained, when the highest COP exists [3]. Bromide lithium – water was used as a working fluid.

Water Purification System Integrated to an Absorption Heat Pump

Figure 9 shows details of a water purification system in an absorption heat pump, where the impure water in a container is pumped to the condenser and to the absorber by a string connection to retrieve the heat from the absorber and condenser; the impure water goes in liquid phase at low temperature and leaves a water and steam mixture at high temperature towards a separator. Steam leaves at the upper part and yields heat to an evaporate obtaining the distilled water and another part goes to the cooling system in order to eliminate the extra heat source. The COP values are not affected by energy recycling due to the fact that COP is defined by the ratio between the sum of the heat obtained in the condenser and absorber, and the heat supply to the generator.

Experimental results of a water purification system using an absorption heat pump are shown in Figure 10. It was possible to obtain distilled water between 0.348 and 4.290 kg/h to different COP´s with LiBr-water as a working fluid. The COP´s values were more than 1, due to the relation of input and output heats [3].

Water Purification System Integrated to a Heat Transformer

The impure water in a container is pumped towards the absorber and it retrieves the absorption heat, water enters in liquid phase at atmospheric pressure and it leaves like water and steam mixture at high temperature to a separator as it is indicated in Figure 11. The steam leaves the separator by the upper part yielding latent heat to a cooling system obtaining distilled water. In the water purification system using a heat transformer, part of the total heat obtained by the heat transformer is recycled, obtained an increment of COP, different from typical COP.

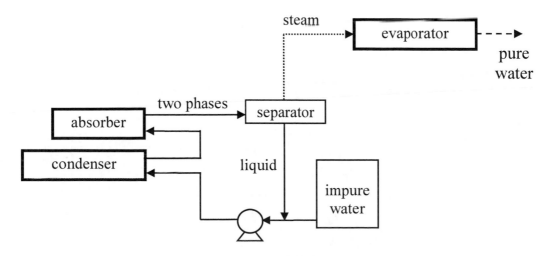

Figure 9. Detail of a water purification system in an absorption heat pump.

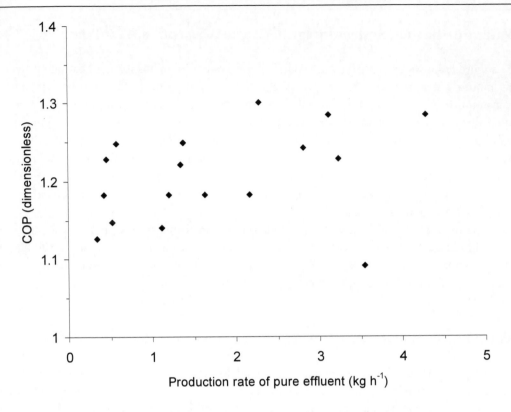

Figure 10. Performance of an absorption heat pump with a water purification system.

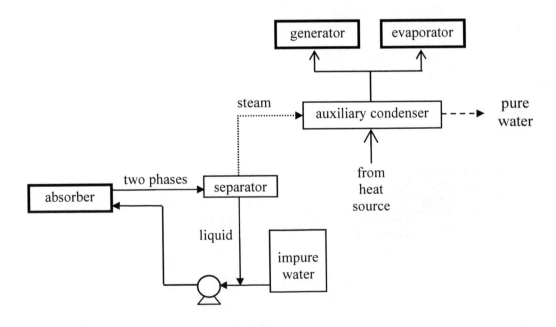

Figure 11. Detail of water purification system into a heat transformer.

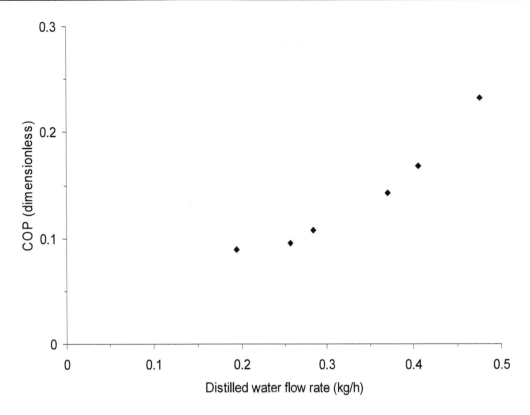

Figure 12. Behavior of COP versus distilled water flow rate.

Behavior of the coefficient of performance versus distilled water flow rate is indicated in Figure 12. The quantity of water purified is almost directly proportional to the COP. Higher COP means, major heat availability for use in a purification system. A distilled water maximum quantity of 0.476 kg/h was obtained, using a heat transformer with a 700 W design capacity and a LiBr-water solution [7].

IMPROVED ENERGY EFFICIENCY

There are two proposals to improve the heat transformer energy efficiency, by using a water purification system: a) Increasing heat source temperature and, b) diminishing the initial heat supply (without temperature increase).

a) Increasing Heat Source Temperature

The integration of a water purification system to a heat transformer allows a fraction of the amount of heat obtained by the heat transformer to be recycled, increasing the heat source temperature, then the evaporator and generator temperatures are also increased, as is shown in Figure 13. For any operating conditions, maintaining the condenser and absorber temperatures and also the heat load to the evaporator and generator, a higher value of COP is obtained when only the evaporator and generator temperatures are increased [8].

The enthalpy based coefficient of performance for water purification with a heat transformer is defined by the following equation:

$$COP_{ET} = \frac{Q_{AB}}{Q_{GE} + Q_{EV}} \tag{11}$$

The heat supplied to the evaporator and generator is provided by the heat source

$$Q_{HS} = Q_{GE} + Q_{EV} \tag{12}$$

Considering that the heat source for generator and evaporator is in liquid phase

$$Q_{HS} = m_{HS} Cp \Delta T_{HS} \tag{13}$$

The mass flow of heat source is

$$m_{HS} = \frac{Q_{HS}}{Cp \Delta T_{HS}} \tag{14}$$

Combining equations 11 and 12

$$Q_{AB} = COP_{ET} Q_{HS} \tag{15}$$

Because the temperature of the steam produced from impure water in the absorber is higher than heat source temperature, part of the absorber heat can be transferred to the heat source stream by means of an auxiliary condenser. Absorber heat is transferred to impure water as latent and sensible heat.

$$Q_{AB} = m_d (\Delta H_V + \Delta H_S) \tag{16}$$

Then part of Q_{AB} is transferred to heat source stream, and this quantity is "ηQ_{AB}". This transferred heat increases the temperature of the heat source stream by an amount, ΔT_N.

$$\eta \, Q_{AB} = m_{HS} \, Cp \, \Delta T_N \tag{17}$$

Substituting heat source mass flow rate, m_{HS}, from equation 14 and Q_{AB} from equation 15 in equation 17:

$$\eta(COP_{ET}Q_{HS}) = \left(\frac{Q_{HS}}{Cp\Delta T_{HS}}\right)Cp\Delta T_N \tag{18}$$

Considering that the Cp does not change for the heat source stream:

$$\eta COP_{ET} = \frac{\Delta T_N}{\Delta T_{HS}} \tag{19}$$

Then the increase of the temperature in the heat source stream is

$$\Delta T_N = \eta COP_{ET}\Delta T_{HS} \tag{20}$$

Where η is the ratio between latent heat and the sum of latent and sensible heat,

$$\eta = \frac{\Delta H_V}{\Delta H_V + \Delta H_S} \tag{21}$$

This equation means that transferred heat depends on impure water initial temperature and composition, as well as the boiling temperature for obtained distilled water flow (m_d).

From equation 20, the increase temperature of heat source is ΔT_N. If the same quantity Q_{HS} is transferred from the heat source to the heat transformer the outlet temperature of heat source increases ΔT_N. Then the evaporator (T'_{EV}) and generator (T'_{GE}) temperatures also increase ΔT_N, while the temperature in the absorber remains T_{AB}.

$$T'_{EV} = T_{EV} + \Delta T_N \tag{22}$$

$$T'_{GE} = T_{GE} + \Delta T_N \tag{23}$$

Therefore an increase in COP_{ET} is obtained (COP'_{ET}) at the same T_{AB} with new T'_{EV} and T'_{GE} values.

b) Diminishing the Initial Heat Supply (Without Temperature Increase)

The integration of a water purification system to a heat transformer allows that a fraction of the amount of heat obtained by the heat transformer to be recycled as is shown in Figure 14, increasing the coefficient of performance obtained from the heat transformer, by reducing the amount of initial heat supplied [9].

Applying the definition of the enthalpy based COP_{ET}

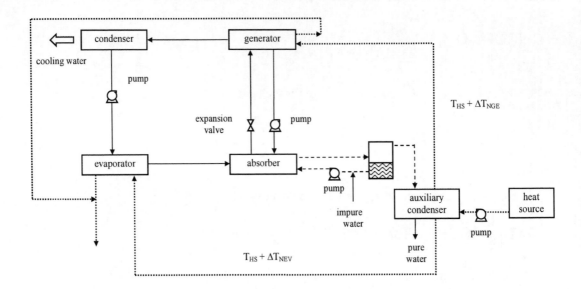

Figure13. Improved energy efficiency, increasing heat source temperature.

$$COP_{ET} = \frac{Q_{AB}}{Q_{GE} + Q_{EV}} \qquad (24)$$

and the definition of the COP_{WP} proposed for water purification with $\eta\, Q_{AB}$ recycling,

$$COP_{WP} = \frac{Q_{AB}}{Q_{GE} + Q_{EV} - \eta Q_{AB}} \qquad (25)$$

dividing all by total absorber heat, and replacing COP_{ET}, then the COP_{WP} may be defined as

$$COP_{WP} = \frac{COP_{ET}}{1 - \eta COP_{ET}} \qquad (26)$$

Now, defining the increase factor (F) for the value of the COP_{ET} as, the difference between the new (COP_{WP}) and the original COP_{ET} values divided by the original COP_{ET} value

$$F = \frac{\left[COP_{WP} - COP_{ET}\right]}{COP_{ET}} \qquad (27)$$

Simplifying and using the definition of the COP_{WP}, afterwards F is proportional to COP_{WP} value, consequently

$$F = \eta COP_{WP} \qquad (28)$$

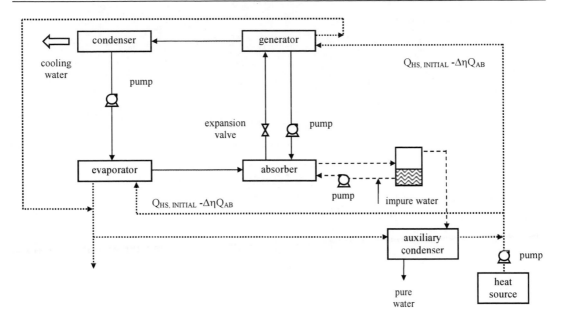

Figure 14. Improved energy efficiency, diminishing the initial heat supply (without temperature increase).

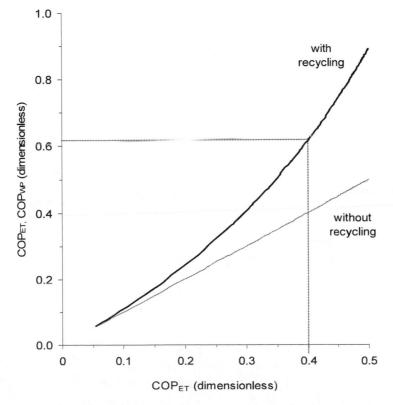

Figure 15. Behavior COP with and without heat recycling, considering $Q_{EV} = 2$ kW, $T_{CO}=32°C$, $T_{EV}=T_{GE}= 70$ to 80 °C, $T_{AB} = 100$ to 125 °C.

In all cases, η is defined as

$$\eta = \frac{\Delta H_V}{\Delta H_V + \Delta H_S} \tag{29}$$

For water purification, it has been considered an inlet temperature of 25°C for impure water and boiling temperature of 100°C at atmospheric pressure, then a constant value for η of 0.877 is obtained.

A bigger coefficient of performance COP_{WP} will be obtained from any original COP values, due to initial heat source decreases, as is described in equation 2, using a water purification system integrated into a heat transformer. The COP_{WP} increments are higher to majors COP_{ET}, as is shown in Figure 15. For COP_{ET} value equal to 0.400, the performance will increase to 0.616, representing a 54 % increment. The maximum COP_{WP} of 0.890 will be achieved with a COP_{ET} value close to 0.500 (increment of 78%), under these operating conditions.

COP PREDICTION FROM ARTIFICIAL NEURAL NETWORK

In this water purification process integrated to an absorption heat transformer, Siqueiros and Romero [8] used a thermodynamic model to simulate the COP values. Nevertheless, this model was based on a series of assumptions (i.e. heat losses and pressure drops in the tubing and the components are considered to be negligible), although these assumptions are difficult to be fulfilled in practice. Therefore, empirical models are used as an alternative to the process control, obtaining satisfactory results without considering any assumptions. However, these models have a relatively narrow validity range, but require only a limited number of simple arithmetic operations for simulation, and can be easily incorporated into control software. The progress of neurobiology has allowed researchers to build mathematical models of neurons to simulate neural behavior. Neural networks are recognized as good tools for dynamic modeling, and have been extensively studied since the publication of the perceptron identification method [10]. Interest in these models includes modeling without any assumptions about the nature of underlying mechanisms and their ability to take into account non-linearities and interactions between variables [11]. A neural model can always be identified based on the perceptron structure, with only one hidden layer, for either steady state or dynamic operations [12,13]. An outstanding feature of neural networks is their ability to learn the solution of the problem from a set of examples, and to provide a smooth and reasonable interpolation for new data. Also, in the field of process engineering, they are a good alternative to conventional empirical modeling based on polynomial and linear regressions. Recently, for absorption heat transformer processes, the application of neural networks continues to expand [14-16]. A model based from Neural Network was proposed to obtain on-line simulation of the COP in the water purification process integrated to an absorption heat pump.

NEURAL NETWORK LAYERING

The neurons are grouped into distinct layers and interconnected according to a given architecture. As in nature, the network's function is determined largely by the connections between elements (neurons). Each connection between two neurons has a weight coefficient attached to it. The standard network structure for an approximation function is the multiple-layer perceptron (or feedforward network). The feedforward network often has one or more hidden layers of sigmoid neurons followed by an output layer of linear neurons. Multiple-layers of neurons with nonlinear transfer functions allow the network to learn nonlinear and linear relationships between input and output vectors. The linear output layer lets the network produce values outside the -1 to $+1$ range [17]. For the network, the appropriate notation is used in two-layer networks [18-19].

The number of neurons in the input and output layers is given respectively by the number of input and output variables in the process under investigation. In this work, two separate feedforward are proposed. The input layer consists of 11 variables for the first model, *without* considering energy recycling and 16 variables for the second model, *with* energy recycling in the process; while the output layer contains 1 variable for each model (see Figs. 14 and 16, respectively). In this process, $LiBr+H_2O$ is used in the absorber and generator, and H_2O in the evaporator and condenser, as working fluid. The optimal number of neurons in the hidden layer(s) n_s is difficult to specify, and depends on the type and complexity of the task. This number is usually determined iteratively. Each neuron in the hidden layer has a bias b (threshold), which is added to the weighted inputs to form the neuron input n (Eq. 30). This sum, n, is the argument of the transfer function f.

$$n = Wi_{\{1,1\}}In_1 + Wi_{\{1,2\}}In_2 + \ldots\ldots + Wi_{\{1, k\}}In_k + b \tag{30}$$

The coefficients associated with the hidden layer are grouped into matrices Wi (weights) and b1 (biases). The output layer computes the weighted sum of the signals provided by the hidden layer, and the associated coefficients are grouped into matrices Wo and b2. Using the matrix notation, the network output can be given by Equation 31:

$$Out = g(Wo \times f(Wi \times In + b1) + b2) \tag{31}$$

Hidden layer neurons may use any differentiable transfer function to generate their output. In this work, a hyperbolic tangent sigmoid transfer function and a linear transfer function were used for f and g, respectively [20].

According to the transfer function used, the equation that predicts the output variable (COP) is given by the Equation 32:

$$COP = \sum_s \left\{ Wo_{(1,s)} \cdot \left[\frac{2}{1+e^{-2\cdot\left(\sum_k Wi_{(s,k)} \cdot In_k + b1_{(s,1)}\right)}} - 1 \right] \right\} + b2_{(1,1)} \tag{32}$$

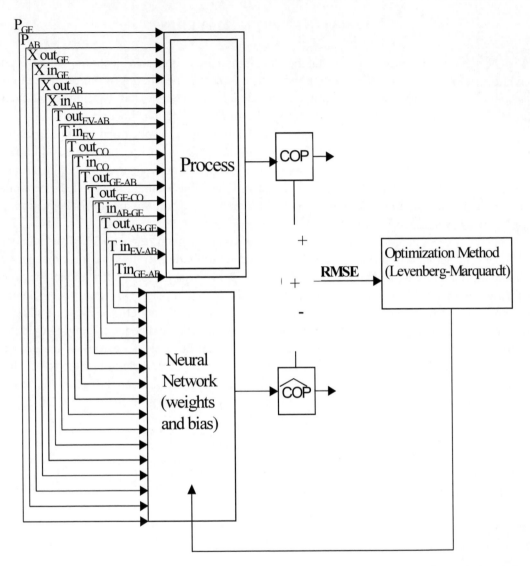

Figure 16. Recurrent network architecture for the COP values (*with* energy recycling) and the procedure used for neural network learning.

NEURAL NETWORK LEARNING

A learning (or training) algorithm is defined as a procedure that consists of adjusting the coefficients (weights and biases) of a network, to minimize an error function (usually a quadratic one) between the network outputs, for a given set of inputs, and the correct (already known) outputs. If smooth non-linearities are used, the gradient of the error function can be computed by the classical backpropagation procedure [20]. Previous learning algorithms used this gradient directly in a steepest descent optimization, but recent results have shown that second order methods are far more effective. In this work, the Levenberg-Marquardt algorithm optimization procedure - in the Matlab Neural Network Toolbox [19] - was used.

This algorithm is an approximation of Newton's method, which was designed to approach second order training speed without having to compute the Hessian matrix [20]. Despite the fact that computations involved in each iteration are more complex than in the steepest descent case, the convergence is faster, typically by a factor of 100. The root mean square error (RMSE) is calculated with the experimental values and network predictions. This calculation is used as a criterion for model adequacy (Figure 16).

EXPERIMENTAL DATA

Experimental data consisting of different COP values (*without* energy recycling and *with* energy recycling), obtained from a portable water purification process coupled to an absorption heat transformer. The experimental data set was obtained at different initial concentrations of LiBr+H$_2$O, different temperatures in AB, GE, EV, CO and different pressures in AB and GE. In addition to the experimental data of each component, the transitory and steady state were taken into account for each initial concentration used in the process. Data were obtained during four hours, two hours after start-up. These parameter changes and data acquisitions allowed us to obtain experimental information that was sufficient to develop the neural network models. A summary of the limiting conditions studied for the operation parameters used in the experimental database is presented in Tables 1 and 2. The thermodynamic properties of the LiBr+H$_2$O mixture were estimated using Alefeld correlations [7]. The input and output temperature experimental of each component (AB, GE, CO and EV) was obtained, as well as pressure of two components (AB, GE), with a temperature-pressure acquisition system (thermocouple conditioner and an Agilent equipment with commercial software). The input and output concentrations in the AB and GE were obtained by a refractometer (refraction index). Experimental data were split into learning (70% of experimental data set) and testing (30% of experimental data set) database to obtain a good representation of the situation diversity. In the first model, the network inputs (In) (see Figure 17) were 11 variables and there was 1 output (Out) for COP values *without* energy recycling. In the second model, the network inputs (see Figure 19) were 16 variables and there was 1 output for COP values *with* energy recycling. These input and output parameters were normalized (-1, +1) for calculation with Matlab [18].

MODEL FOR COP VALUES PREDICTION
WITHOUT ENERGY RECYCLING

The first neural network model which was developed (Figure 17) involved three neurons (n_s=3) in the hidden layer (36 weights (*Wi*=33; *Wo*=3) and 4 biases (b1=3 and b2=1)) to predict COP values for the water purification process coupled at an absorption heat transformer *without* energy recycling.

Table 1. Experimental operation range conditions studied to obtain the COP values *without* energy recycling

Operation parameters, °C	Limiting conditions studied
T in EV-AB	77.56-94.30
T out AB-GE	73.8-95.90
T in AB-GE	83.0-91.12
T out GE-CO	79.63-87.95
T in CO (cooling)	19.2-29.20
T out CO (cooling)	19.1-35.5
T in EV (heating)	80.79-89.50
T out EV (heating)	78.44-86.84
Operational parameters, in Hg	
P AB	43.7-49.8
P GE	31.5-50.9
Operational parameters, %	
X initial	51.18, 52.94, 53.46, 43.79, 54.39, 54.54, 55.36, 55.77

Table 2. Experimental operation range conditions studied to obtain the COP values *with* energy recycling

Operation parameters, °C	Limiting conditions studied
T in GE-AB	76.29-91.53
T in EV-AB	74.56-89.93
T out AB-GE	84.31-98.27
T in AB-GE	74.99-92.58
T out GE-CO	76.29-91.53
T out GE-AB	77.03-83.89
T in CO	40.37-65.03
T out CO	26.77-33.79
T in EV	28.52-85.33
T out EV-AB	74.56-89.93
Operational parameters, %	
X in AB	51.66-55.36
X out AB	50.75-54.36
X in GE	50.75-54.36
X out GE	53.16-56.07
Operational parameters, in Hg	
P AB	38.9-41.40
P GE	48.9-51

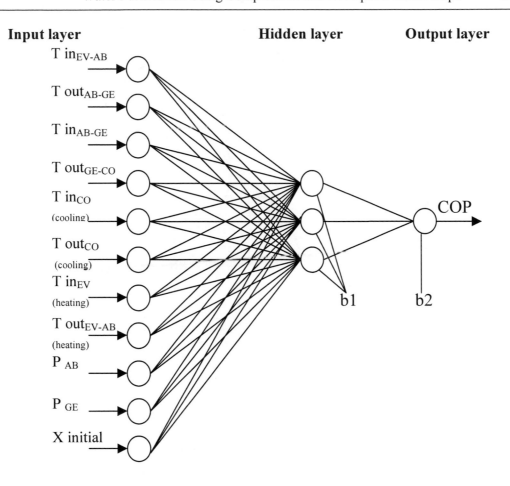

Figure 17. Model for the COP values prediction *without* energy recycling.

Learning Database for the First Model

In the learning database given by the RSME trial versus the iteration number, the algorithm was worked out for one to six neurons in the hidden layer. The obtained results (data not shown) proved that the typical learning error decreased when the number of neurons in the hidden layer increased; this is evident since the number of adjusted parameters increased. Nevertheless, one of the problems that occur during feedforward neural network training is called "over-fitting" [18, 21]. The comparison of the RSME calculated for the learning and testing database is a good criterion to optimize the number of iterations and avoid over-fitting. In this neural network the RSME showed that for four neurons in the hidden layer, the learning database value was small with respect to the testing database. Then, according to RSME results, the optimal number of neurons in the hidden layer is three.

According to the obtained parameters (*Wi, Wo,* b1, and b2) of the best fit of the first proposed model for 3 neurons in the hidden layer. These parameters are used in the proposed model to testing database.

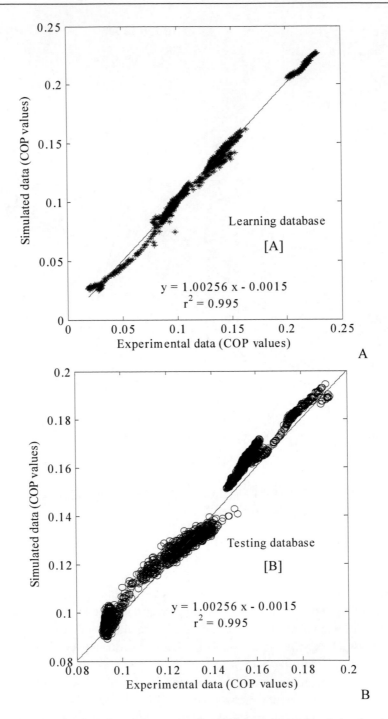

A

B

Figure 18. COP values obtained *without* energy recycling. [A] experimental and simulated data for the learning database, [B] experimental and simulated data for the testing database.

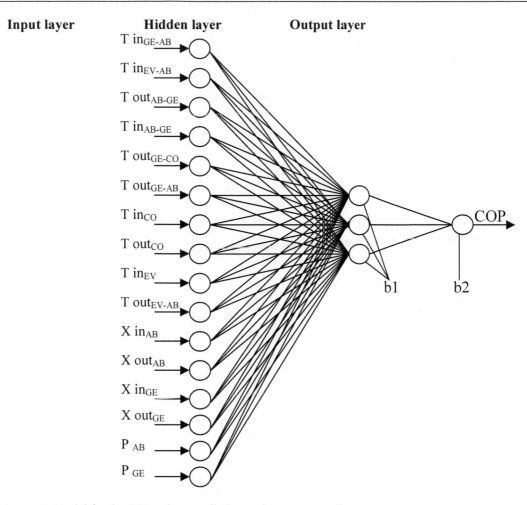

Input layer   **Hidden layer**   **Output layer**

T in$_{GE-AB}$
T in$_{EV-AB}$
T out$_{AB-GE}$
T in$_{AB-GE}$
T out$_{GE-CO}$
T out$_{GE-AB}$
T in$_{CO}$
T out$_{CO}$
T in$_{EV}$
T out$_{EV-AB}$
X in$_{AB}$
X out$_{AB}$
X in$_{GE}$
X out$_{GE}$
P$_{AB}$
P$_{GE}$

COP

b1   b2

Figure 19. Model for the COP values prediction *with* energy recycling.

Testing Database for the First Proposed Model

Figure 18B presents the COP experimental data versus COP simulated results for the testing database. It shows that the COP prediction was correlated well ($r^2>0.995$). Figure 18 shows the model's ability to predict the COP values (*without* energy recycling) at different temperatures (in AB, GE, CO and EV), pressure and initial concentration, for a given validity range. It is evident that the model was successful in predicting the experimental data of COP values. This shows the importance of the artificial neural network to simulate water purification process, coupled to an absorption heat transformer without energy recycling.

MODEL FOR COP VALUES PREDICTION WITH ENERGY RECYCLING

The second proposed neural network model (Figure 19) also involved three neurons (n_s=3) in the hidden layer (51 weights (*Wi*=48 and *Wo*=3) and 4 biases (b1=3 and b2=1)) to predict COP values for the water purification process coupled to an absorption heat transformer *with* energy recycling.

LEARNING DATABASE FOR THE SECOND MODEL

Similarly, in the learning database, the increase in the number of adjusted parameters was carried out, the typical learning error decreased because the number of neurons in the hidden layer increased. In this model, we also avoided over-fitting, considering the comparison between the RSME obtained for the learning database and that of the testing database [18]. According to the obtained parameters (Wi, Wo, b1, and b2) of the best fit of the second proposed model, for the 3 neurons in the hidden layer and iterations, as well as for learning and testing RSME. As a whole, the RSME values of the learning & testing database were similar; this accounts for a good generalization capability of the neural network. These parameters (weights and biases) are used in the second proposed model for testing database.

Testing Database for the Second Proposed Model

Figure 20A presents the experimental and simulated data of the COP values versus absorption temperature; Figure 20B depicts the COP experimental data against simulated results for testing database. It shows that the COP prediction was correct ($r^2 > 0.99$). Figure 20 shows the ability of the model to predict the COP values (*with* energy recycling) at different temperatures (in, AB, GE, CO and EV), pressure and initial concentration for a given validity range. From this correlation and statistical test, it is evident that the model was successful in predicting the experimental COP data. This shows the importance of the artificial neural network in simulating the water purification process coupled to an absorber heat transformer *with* energy recycling.

CONCLUSION

One application of heat pumps consists in obtaining purified water through a simple distillation process. A heat quantity is possible to recycle into the equipment in order to achieve both processes: distillation and a thermodynamic cycle. It has been demonstrated that an improvement in energy efficiency can be obtained when a water purification system is integrated into the heat transformer, because a fraction of the heat obtained in the absorber is recycled to the heat source. This new condition changes the initial operating conditions, in order to increase the Enthalpy Coefficient of Performance (COP_{ET}).

A neural network model is used to represent the behavior of the COP in a water purification process integrated into an absorption heat pump. The model was successfully trained with experimental database and validated with a fresh database (in the specified range of key operating conditions and for the steady and transitory state). This simulation can be carried out on-line.

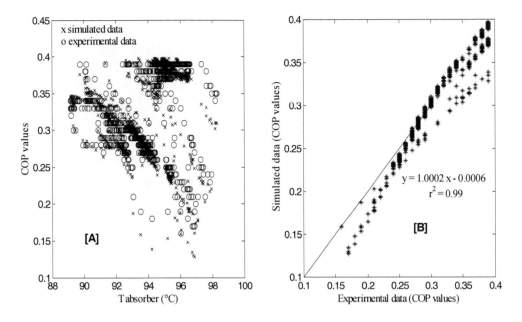

Figure 20. [A] absorber temperature data versus COP values (experimental and simulated). [B] experimental versus simulated COP values for the testing database *with* energy recycling.

Both models (*without* and *with* energy recycling) are not complex because simulation is done by simple arithmetic operations and, therefore, they can be used for on-line estimation applications for industrial processes. It is interesting to notice that the COP values obtained are different due to the energy recycling factor. We can clearly observe the high COP values (above 0.4) obtained *with* energy recycling, compared to low COP values (above 0.2) obtained *without* energy recycling.

In the case of a process with energy recycling, it was necessary to change the configuration of layering model with 5 additional input variables and consequently the number of fitted parameters to provide the operation insight was increased.

REFERENCES

[1]   Eisa MAR. Applications of heat pumps in chemical processing. *Energy Conversion and anagement* 1996; 37:360-377.

[2]   Siquieros J, Holland FA. Water desalination using heat pumps. *Energy* 2000; 25:717-729.

[3]   Holland FA, Siqueiros J, Santoyo S, Heard C L, Santoyo E R. *Water purification using heat pumps*. First Edition. E & FN Spon, London and New York; 1999.

[4]   Herold KE, Radermacher R, Kein SA. *Absorption Chillers and Heat Pumps*. First ed. Washington: CRC Press; 1996.

[5]   Holland FA. Clean water through appropriate technology transfer. *Applied Thermal Engineering* 2000; 20:863-871.

[6]   Holland FA, Watson FA, Devotta S. *Thermodynamic design data for heat pump systems*. Pergamon Press, First edition 1982.

[7]   Huicochea A, Siqueiros J, Romero R J. Portable water purification system integrated to a heat transformer. *Desalination* 2004; 165:385-91.

[8]   Siqueiros J, Romero RJ. Increase of COP for heat transformer in water purification systems, Part I – Increasing heat source temperature. *Applied Thermal Engineering* 2007; 27:1043-53.

[9]   Romero RJ, Siqueiros J, Huicochea A. Increase of COP for heat transformer in water purification system. Part II – Without increasing heat source temperature. *Applied Thermal Engineering* 2007; 27:1054-61.

[10]   Rumelhart D, Zipner D. Feature Discovering by competitive learning. *Cognitive Science* 1985; 9:75-112.

[11]   Bishop C M. Neural networks and their applications. *Rev. Sci. Instrum* 1994; 65:1803-1832.

[12]   Hornik K, Stinchcombe M, White H. Multilayer feedforward networks are universal approximations, *Neural Network* 1989; 2:359-366.

[13]   Hornik K. Some new results on neural network approximation, *Neural Network* 1993; 6:1069-1072.

[14]   Sözen A, Arcaklioglu E, Özalp M, Yücesu S. Performance parameters of an ejector-absorption heat transformer, *Appl. Energy* 2005; 80:225-339.

[15]   Sözen A, Özalp M. Solar-drive ejector-absorption cooling system, *Appl. Energy* 2005; 80:97-113.

[16]   Sözen A, Akçayol M A. Modeling (using artificial neural-networks) the performance parameters of a solar-driven ejector-absorption cycle, *Appl. Energy* 2004; 79:239-354.

[17]   Limin F. (Ed.), *Neural networks in computer intelligence*, McGraw-Hill International Series in Computer Science, 1994.

[18]   Demuth H, Beale M. (Eds.), *Neural network toolbox for Matlab-User's Guide Version 3*. Natrick, MA, The MathWorks Inc, 1998.

[19]   Rumelhart D E, Hinton G E, Williams R J. Learning internal representations by error propagation, *Parallel Data Processing* 1986; 1:318-362.

[20]   Martin T, Hagan M T, Mohammad B N. Training feedforward networks with the Marquardt Algorithm, *IEEE Transactions on Neural Networks* 1994; 6:989-993.

[21]   Hernández-Pérez JA, García-Alvarado MA, Trystram G, Heyd B. Neural networks for the heat and mass transfer prediction during drying of cassava and mango. *Innovative Food Science and Emerging Technologies* 2004; 5:57-64.

In: Water Purification
Editors: N. Gertsen and L. Sønderby

ISBN 978-1-60741-599-2
© 2009 Nova Science Publishers, Inc.

Chapter 8

WHY VAPOR PRESSURE OF A VOLATILE LIQUID DECREASES BY ADDITION OF SOLUTES: A NEW QUALITATIVE MODEL

Hikmat S. Hilal[1,] and Ali Cheknane[2]*

[1]Department of Chemistry, An-Najah N. University, PO Box 7, Nablus,
West Bank, Palestine
[2]Laboratoire d' Etudes st Developpement de Materiaux Semiconducteurs et
Dielectriques, Universite Amar Telidji de Laghouat, route de Ghardaia, BP 37G,
Laghouat 03000, Algerie

ABSTRACT

While explaining vapor pressure lowering in solutions, many general scholars mistakenly depend on rate of evaporation lowering, or on relative values of attraction forces between solvent and solute molecules. This is in spite of the correct explanations based on thermodynamics, as presented in physical chemistry references. Unfortunately, there seems to be no qualitative model available so far to explain vapor pressure lowering. The misconception is escalating, and an end needs to be made. A qualitative understandable model, to explain vapor pressure lowering, at least in solutions of nonvolatile solutes in volatile solvents, is needed. For this purpose, we propose here a *new qualitative model*, to explain vapor pressure lowering in such solutions. The new model is based on purely old thermodynamic concepts.

Keywords: Raoult's law, vapor pressure, ideal solution, entropy, rate of evaporation

* Corresponding author. Fax No.: +970-9-2387982; E-mail: hikmathilal@yahoo.com

1. INTRODUCTION AND PROBLEM DESCRIPTION

We all learn that non-volatile substances, dissolved in volatile solvents, cause significant reduction in equilibrium vapor pressure of the solvent, while keeping temperature constant. We also learn that addition of more solute, to the solution, causes more lowering in solvent vapor pressure of that solution, based on Raoult's law [1-10]. This law quantitatively works well ideal, or close-to-ideal solutions, of high dilutions, involving volatile or non-volatile solutes [7-8, 11]. Ideal solutions are characterized with zero value of ΔH_{soln}. Real solutions with ΔH_{soln} >zero tend to have negative deviations from Raoult's law. Conversely real solutions with ΔH_{soln} < zero tend to have positive deviations.

Despite the deviations from Raoult's law, ideal and real solutions exhibit solvent pressure lowering. This applies to solutions with volatile and nonvolatile solutes. To make the point easier, we avoid volatile solutes in volatile solvents, and restrict ourselves to *nonvolatile solutes in volatile solvents*. The question that concerns is: **why is the pressure, of a volatile solvent, lowered by adding a nonvolatile solute, despite the value of Δ_{soln}?**

Answers presented in several chemistry textbooks [1-18], published between 1970 and 2008, have been surveyed. Answers from chemistry students have also been surveyed. Different approaches are presented, as follows:

Some textbooks presented the facts and the equations without giving explanations [1-4, 7-9].

Rate of evaporation discussions: In a few cases, [5-6, 10] the explanation was based on *rate of evaporation*. In new and earlier editions of his well respected book, [6a-b] *Brady stated: "Figure 14.19 Effect of nonvolatile solvent on the vapor pressure of a volatile solvent. (a) Equilibrium between a pure solvent and its vapor. With a high number of solvent molecules in the liquid phase, the rate of evaporation and condensation is relatively high. (b) In the solution, some of the solvent molecules have been replaced with solute molecules. There are fewer solvent molecules available to evaporate from solution. The evaporation rate is lower. When equilibrium is reached, there are fewer molecules in the vapor. The vapor pressure of the solution is less than that of the pure solvent"*. Surprisingly, this discussion has been presented after he correctly described vapor pressure by saying that "..... *the pressure exerted by the vapor once **equilibrium has been established** is called the equilibrium vapor pressure of the liquid*" [6b, p. 379].

This explanation is not correct, and will not account for vapor pressure lowering. Vapor pressure lowering is purely a thermodynamic term, since it is considered at equilibrium only, despite the evaporation/condensation rates, which become equal at equilibrium. Solution surface area affects rate of evaporation and affects time after which equilibrium is reached, but does not affect equilibrium vapor pressure. Consider two containers with different cross-sectional areas, each containing a solution of sugar and water, having same concentration. If the two containers are kept under covers, at same temperature, then the solution with wider cross-section will evaporate faster than the other one. After equilibrium is reached, the final vapor pressures will be the same, despite the cross-sectional area. Note that the total number of surface solvent molecules in the wider container will be larger than those in the narrower container, but nonetheless the two vapor pressures are the same. Thus vapor pressure has nothing to do with time after which equilibrium is reached. Moreover, if one container has only water, and the other has sugar and water; then widening the surface area of the latter container will not prevent the equilibrium vapor pressure of the former container to be higher than that of the latter. This unequivocally means that the vapor pressure is independent of evaporation rate.

3) Intermolecular Attraction Strength: many scholars explain the phenomenon by stating that the attractive forces between the solvent and the solute are responsible for the amount of vapor pressure lowering in solutions [17, 19-20]. If this explanation is correct, then solutions with negative ΔH_{soln} values should have vapor pressure lowering, whereas solutions with positive values will have vapor pressure increase, and solutions with zero values should have same vapor pressure as pure solvent. The explanation mixes up between **vapor pressure** lowering and **deviation from Raoult's law**. The fact that concerns us here is that vapor pressure lowering occur even in solutions with weaker solvent-solute attractions. The explanation is thus incorrect.

The incorrect explanations are becoming widespread and misleading, and continue to accompany the scjolars after graduation. In 1998, Peckham [19] found most of general chemistry textbooks *giving mistaken answers, with the exception of only one textbooks* [21] *that successfully used a thermodynamic explanation.* Peckham questioned the credibility of ***rate of evaporation*** or ***attractive force*** effects on vapor pressure lowering. He also successfully suggested that the concepts of entropy and chemical potential are the true answers to the questions. Without bringing out a qualitative model to the question, Peckham suggested that Raoult's law could be presented as an experimental finding without presenting any explanations, *"no explanation at all is better that a faulty one"* as he suggested.

Despite Peckham's report, authors still continue to use the mistaken conception [6a, 14, 17], simply because he did not present a qualitative model to explain Raoult's law. Other writers recommended elimination of Raoult's law completely from introductory chemistry books. In his report "Raoult's law is a deception" Hawkes [22] stated:...*that this law has little practical application and should be omitted from introductory chemistry course..*

Neither Peckham nor Hawkes approaches help solve the problem. *Alternatively, we should put an end to the problem rather than leaving it unsolved.* A qualitative model should be suggested based on correct thermodynamic explanations.

2. THE THERMODYNAMIC BASIS BEHIND RAOULT'S LAW

Physical chemistry textbooks, [11-16, 18], successfully used the concepts of entropy and "chemical potential, μ" to explain Raoult's law and other colligative properties. Despite their success, they have not presented a simple qualitative thermodynamic model for Raoult's law. In his highly rated "Physical Chemistry" book, Atkins used the concept of chemical potential to explain Raoult's law and vapor pressure phenomena [14]. While presenting the molecular basis for Raoult's law, Atkins successfully used the concept of equilibrium rates of evaporation and condensation being equal. However, in his model of sphere representation of solutions, [14] he used illustrations to say that: "*A pictorial presentation of the molecular basis of Raoult's law: The large molecules represent solvent molecules at the surface of a solution (the uppermost line of spheres) and the small spheres represent solute molecules. The latter hinder the escape of solvent molecules into vapor, but do not hinder their return*".

This last statement is misleading, and reinforces the misconception of effect of evaporation rate on vapor pressure.

Before giving the qualitative model, it is necessary t show how thermodynamics (chemical potential, entropy and free energy) concepts explain vapor pressure lowering. Mathematical rigors are kept to a minimum.

2.1. Chemical Potential Treatment for Vapor Pressure Lowering:

For a given substance (**i**) in a mixture within a given phase, *chemical potential*, μ_i is defined as:

$$\mu_i = \left(\frac{\partial G}{\partial n_i} \right)_{T,P,nj,...} \tag{1}$$

where T , p and n_j (numbers of moles of other substances in solution) are kept constant.

Chemical potential of substance **i** is commonly referred to as *Gibbs free energy per mole* of substance **i**. In case of evaporation, the chemical potential is a *driving force for evaporation*. Another way of writing the total free energy of a given solution is [12-16]:

$$G = \mu_i n_i + \mu_j n_j + \dots \dots \tag{2}$$

From equation (2) it can be easily found that:

$$dG = \mu_i dn_i \tag{3}$$

if moles of other components are kept constant.

Let us now consider two systems of vapor pressures:

1) *The vapor pressure above a pure volatile solvent, such as water*:
If a certain number of moles of liquid water (n_w) is kept inside a container under vacuum at constant temperature, then water will evaporate. If a small number of moles of liquid water, ($n_{vap} = dn_w$), evaporate, then the free energy change in pure liquid water (dG_{lw}) is:

$$dG_{lw} = \mu_{lw}(-dn_w) = -\mu_{lw} \, dn_w \tag{4}$$

and the free energy change for the resulting water vapor in the gas phase is:

$$dG_{gw} = \mu_g \, dn_w \tag{5}$$

Thus the total free energy change in the system will be:

$$dG = dG_{lw} + dG_{vap} = (\mu_{vap} - \mu_{lw})dn_w \tag{6}$$

At the beginning, water tendency to evaporate is higher than its tendency to condense, thus $\mu_{lw} > \mu_{vap}$ and consequently dG is negatively valued at the beginning. This means that the evaporation process is spontaneous, and should continue. After equilibrium is reached, then dG = 0. Therefore,

$$\mu_{lw} = \mu_{vap} \tag{7}$$

2) *The vapor pressure above a solution*: Consider an ideal solution of water and a non-volatile solute, originally kept under vacuum. Suppose that a small number of water moles dn_w evaporates, then the free energy change due to water evaporation process in case of such a solution is:

$$dG = (\mu_{vap} - \mu^{sol}_{w})dn_w \tag{8}$$

For an ideal solution, the value of chemical potential of ith component is:

$$\mu_i = \mu_i^{*} + RT \ln X_i \tag{9}$$

Where X_i is the mole of fraction of ith component. μ_i^{*} is chemical potential of ith component in its pure form. R is Universal gas constant, and T is absolute temperature. The derivation for equation (9) is well documented [11].

Therefore, for ideal water solutions then the chemical potential for water in the aqueous liquid solution μ^{sol}_{w} is:

$$\mu^{sol}_{w} = \mu^{l}_{w} + RT \ln X_w \tag{10}$$

$$\text{Thus } dG = (\mu_{vap} - \mu^{l}_{w} - RT \ln X_w)dn_w \tag{11}$$

As the mole fraction of water in solution, X_w, is less than 1, the value of $\ln X_w$ will always be negative. Therefore, the value of dG for evaporation process in ideal water solutions, as shown in equation (11), should always be less negative than that in case of pure water systems as shown in equation (8). *This explains why the spontaneity of pure water to evaporate is higher than the spontaneity of water in solutions to evaporate, and thus lowering of vapor pressures in solutions. It explains why the driving force, for water in solution, to evaporation decreases as more non-volatile solute is added to the liquid solution (X_w is lower).*

At equilibrium, dG = 0 then resulting value for chemical potential of the vapor is:

$$\mu_{vap} = \mu^{l}_{w} + RT \ln X_w \tag{12}$$

which is lower than that shown above in equation (8) for pure water vapor at equilibrium.

The vapor phase chemical potential above aqueous solution is lower than that of vapor phase above pure water. For an ideal pure gas, which is a reasonable assumption for the vapor above the solution to be only water with no solute, then:

$$\mu_{vap} = \mu^{o}_{vap} + RT \ln (p \text{ atm} / 1 \text{ atm}) \tag{13}$$

Where μ^{o} is chemical potential of the pure gas having 1 atm pressure [11]. From equation (12), as μ for the vapor above the solution is lower than above pure water, the equilibrium vapor pressure is lower. **Therefore, the vapor pressure of a volatile liquid must be lowered by adding a non-volatile solvent, as dictated from equation (13).**

The chemical potential discussions may also be used to derive Raoult's law for aqueous solutions as follows:

If an aqueous solution attains equilibrium with the vapor phase, then:

$$\mu^{sol}_{w} = \mu_{vap} \tag{14}$$

From equations (12) and (13) then for a solution of non-volatile component in water then:

$$\mu^{l}_{w} + RT \ln X_w = \mu^{o}_{vap} + RT \ln (p/1 \text{ atm}) \tag{15}$$

For pure water only, with $X_w = 1$, and $p = p^{o}$ for pure water system, and $RT\ln X_w = 0$, thus equation (15) becomes:

$$\mu^{l}_{w} = \mu^{o}_{vap} + RT \ln (p^{o}/1 \text{ atm}) \tag{16}$$

From equations (15) and (16), then [18]:

$$RT\ln X_w = RT\ln(p/1 \text{ atm}) - RT\ln(P^{o}/1 \text{ atm}) \tag{17}$$

Which reduces to:

$$\ln X_w = \ln(P/P^{o}) \tag{18}$$

and yields Raoult's Law, for aqueous solutions of non-volatile liquids, which can be written as:

$$P = X_w P^{o} \tag{19}$$

Where P is water vapor pressure above the solution, P^{o} is water vapor pressure above pure water, and X_w is the mole fraction of water in solution.

2.2. Entropy Basis for Vapor Pressure Lowering

Vapor pressure lowering is essentially a colligative thermodynamic property that is considered at equilibrium, and will occur for both ideal and non-ideal solutions, despite the value of ΔH_{soln} and the deviations from Raoult's law. Let us see how entropy concepts explains vapor pressure lowering of a solvent by adding a nonvolatile solute:

Case 1. Consider a pure volatile liquid placed in an evacuated container. The liquid vaporizes under isothermal conditions until the equilibrium vapor pressure is attained. The molar entropy change of the system (liquid and gas phases) in this process is the sum of the molar entropy changes of the liquid phase and the vapor phase, after and before evaporation:

$$\Delta S_1 = \Delta S(liquid) + \Delta S_1(vapor) \tag{20}$$

Note that $\Delta S(liquid) = 0$, because the molar entropy of pure liquid does not change due to evaporation under isothermal conditions. Moreover

$$\Delta S_1(vapor) = S(vapor) \text{ because no vapor was present initially.} \tag{21}$$

Thus

$$\Delta S_1 = S_1(vapor) \tag{22}$$

Case 2. Now consider *an ideal solution* of the same volatile liquid containing a non-volatile solute placed in an evacuated chamber. Assume, for the time being, that the same vapor pressure (as in case 1) is attained at the same temperature. The entropy change of the system in this process is the sum of the entropy changes of the solution phase and the vapor phase.

$$\Delta S_2 = \Delta S(solution) + \Delta S_2(vapor) \tag{23}$$

With solvent evaporation $\Delta S(solution)$ becomes more and more negative, $\Delta S(solution) < 0$. This behavior would occur for *at least dilute solutions*. In a solution of nonvolatile solute, as solvent molecules evaporate, each solute molecule is left with fewer solvent molecules. This means that the solute molecule has fewer degrees of freedom, i.e. *less space to travel in and lower randomness*. Moreover, as some solvent molecules evaporate, each remaining solvent molecule is also left with less space to travel inside the total solution, and lower degrees of freedom. Thus the molar entropy change of solution $\Delta S(solution)$ by evaporation is negatively valued.

Since, as earlier assumed, same amount of vapor exists in both cases 1 and 2, $\Delta S_2(vapor) = \Delta S_1(vapor)$, because the composition of the vapor phase is identical in both cases, and because $\Delta S(solution)$ is negative, equation the becomes:

$\Delta S_2 < \Delta S_2(vapor)$
and $\Delta S_2 < S_2(vapor)$

Consequently, $\Delta S_2 < \Delta S_1$

This means that the molar entropy change in case 2 is lower than in case 1, assuming same amounts of vapor are attained. Thus the system in case 2 has less entropy than in case 1, and something must happen to attain maximum entropy to it.

Now consider the Gibb's energy change for each process.

$$\Delta G_1 = \Delta H_1 - T\Delta S_1 \tag{24}$$

$$\Delta G_2 = \Delta H_2 - T\Delta S_2 \tag{25}$$

Because $\Delta H_1 = \Delta H_2$, since $\Delta H_{solution} = 0$ for ideal solutions, (the molar enthalpy of neither the pure liquid nor the ideal solution changes in the process and the molar enthalpy changes of the two vapor phases are identical because their compositions are identical) and since

$$\Delta S_2 < \Delta S_1,$$

Then: $\Delta G_2 > \Delta G_1$.

That is, the process of evaporation from solution is less favored than from pure liquid. This means that the vapor pressure in case 2 must be less than that in case 1. In other words the spontaneity of evaporation in case 2 must me lower than in case 1.

The entropy concepts presented above are only basic and understandable by freshman students. Many general chemistry textbooks deal with entropy [8] and its treatment for colligative properties, such as osmosis [4]. Unlike physical chemistry books, general chemistry books present thermodynamics after colligative properties. There is no reason not to follow physical chemistry books and make the student understand colligative properties after learning thermodynamics. This will help solve out the problem, if a qualitative model is then provided.

3. A PROPOSED QUALITATIVE MODEL (ANALOGY TO OSMOSIS)

Osmotic pressure demonstrates the entropy concept. Consider two solutions separated by a semi-permeable membrane. The membrane allows only the solvent molecules to pass. The solvent molecules travel from a less concentrated solution (more solvent concentration) to a more concentrated solution (less solvent concentration) until a dynamic equilibrium is reached. Such a process is associated with a pressure build-up, so-called "osmotic pressure". At equilibrium, the system acquires the highest allowable degrees of freedom (highest ΔS). After that, if more solute is added to the higher concentrated solution, then more solvent will transfer until a new equilibrium is attained with higher osmotic pressure.

The vapor pressure lowering in solutions should essentially follow similar logic. Suppose that we have a pure solvent kept in equilibrium with its vapor at constant temperature. If a non-volatile solute is added to the solvent, (despite intermolecular attraction), then vapor will undergo more condensation to reach a new state of equilibrium with lower P_{vap} than the original state. This is to give more room (higher possible randomness and higher degrees of freedom, i.e. maximum allowed ΔS) for the solute molecules to swim in. The new value of P_{vap} will be lowered as the concentration of the solvent in the new solution is lowered. If more solute is added to the liquid, the solvent concentration is further lowered, and more vapor will condense which will lower the value of P_{vap}.

In this simple comparative model, the solution/vapor interface surface is emulating the semi-permeable membrane in osmosis. This is because, as formally speaking, both allow passage of solvent molecules and disallow passage of solute molecules. In case of osmosis experiments, solute molecules are disallowed to pass due to their lack of enough kinetic energy to penetrate small pore sizes and to overcome other limitations. In case of vapor pressure experiments, non-volatile solute molecules are disallowed to pass due to their lack of enough kinetic energy.

5. CONCLUSION

There is a wide spread misconception being used to explain vapor pressure lowering in solutions. Such phenomenon needs to be explained based on thermodynamic concepts rather than rate of evaporation. A qualitative model to explain such phenomenon is presented. The model is based on thermodynamic concepts, and is understandable if presented after basic thermodynamic discussions even at the freshman level. Moreover, it is strongly recommended to present other colligative properties discussions after thermodynamic chapters in general chemistry textbooks, as commonly used in physical chemistry books..

ACKNOWLEDGMENTS

The authors wish to thank R.V. (Dick) Parish (Emeritus Professor, UMIST, Manchester, UK) and K. Abdul-Hadi and Othman Hamed (both of ANU, Palestine) for critical revisions and helpful discussions.

REFERENCES

[1]   Seinko, M. J. and Plane, R. A., *Chemistry: Principles and Properties*, McGraw Hill, Inc., Tokyo, 1966, p. 175.

[2]   Slabough, W. H. and Parsons, T. D. *General Chemistry*, J. Wiley & Sons, N.Y., 1976, p. 214.

[3]   T. Moeller, J. Bailor, Jr., Kleinberg, J.; Guss, C. O.; Castellion, M. E. and Metz, C. *Chemistry with Qualitative Analysis*, 2nd Ed., Academic Press, Orlando, 1984, p. 462.

[4]   Zumdahl S. S. and Zumdahl, S. A. *Chemistry*, 6th Ed., Houghton Mifflin Co., Boston, 2003, pp. 522-524.

[5]   Russell, J. B. *General Chemistry*, McGraw Hill, Auckland, 1980, p. 322.

[6]   (a) brady, J. E. and Senese, F. *Chemistry and its Changes*, 2nd Ed., J. Wiley, N.Y., 2005, p. 620; (b) Brady, J. E. *General Chemistry: Principles and Structure*, J. Wiley, N.Y., 1990, pp. 432-433 and p. 379.

[7]   Brown, T. L.; LeMay, Jr., H. E.; Bursten, B. E. and Murphy, C. J. *Chemistry: the General Science*, 10th Ed., Pearson-Prentice Hall, N.J., 2006, p. 548.

[8]   Ebbing, D. D. and Gammon, S. D. *General Chemistry*, 8th Ed., Houghton Mufflin Co., Boston, 2005, p. 497; and 6th ed. (1999) p. 504.

[9]   Gates, J. R.; Ott, J. B. and Butler, E. A. *General Chemistry: theory and Description*, Harcourt Brace Jovanovich, N. Y., 1981, p. 193.

[10]   Allen, T. L. and Keefer, R. M. *Chemistry: Experiment and Theory*, 2nd Ed., Harper and Row, Publ., N.Y., 1982, p. 274.

[11]   Castellan, G. W. *Physical Chemistry*, 2nd Ed., Addison Wesley Publ. Co., Reading, Ma., 1971, pp. 219-20; pp. 228-232; p. 286.

[12]   Moore, W. J. *Physical Chemistry*, Longman, London, 1962, p. 122.

[13]   Bromberg, J. P. *Physical Chemistry*, Allyn and Bacon, Inc., Boston, 1980, p. 226 and 238.

[14]   Atkins, P. W. *Physical Chemistry*, 6th ed., Oxford University Press, Oxford, 1998, pp. 171-172; Atkins, P.W. and de Paula, J., *Atkin's Physical Chemistry*, 8th ed., Oxford University Press, Oxford, pp. 136-144.

[15]   http://www.tf.uni-kiel.de/matwis/amat/def_en/kap_2/advanced/t2_4_1.html (Accessed March 27th, 2007).

[16]   http://members.aol.com/profchm/raoult.html (Accessed March 27th, 2007)

[17]   http://www.chem.arizona.edu/~salzmanr/480a/480ants/mixpmqis/mixpmqis.html (Accessed March 27th, 2007).

[18]   Levine, I. N. *Physical Chemistry* , 5th Edition McGraw Hill , 1978, pp. 207-213.

[19]   Peckham, G. D. "Vapor Pressure Lowering by Non-Volatile Solutes", *J. Chem. Edu.*, 1998, *76*, 787.

[20]   Mathews, P. *Advanced Chemistry I.* Cambridge University Press, Cambridge, 1992, p. 379, as cited in ref. [19].

[21]   Chang, R. *Chemistry,* 5th ed., McGraw-Hill, N.Y. 1994, p 493, as cited in ref. [19].

[22]   Hawkes, S. J. "Raoult's law is a deception", *J. Chem. Edu.* 1995, *72*, 204.

In: Water Purification
Editors: N. Gertsen and L. Sønderby

ISBN 978-1-60741-599-2
© 2009 Nova Science Publishers, Inc.

Chapter 9

THE BIOTECHNOLOGY OF PHASED DRINKING WATER PURIFICATION IN THE CONDITIONS OF ASTRAKHAN REGION

L. T. Sukhenko[*]

Astrakhan State University, 20A Tatitschev Str., 414056 Astrakhan, Russia

ABSTRACT

The problem of quality water improvement can be solved through phased water purification by electrolytic processing and then adding plant medicinal substances of an antibacterial action. Due to a low-cost electrochemical activation of running water the salts of heavy metals and chlorine-containing compounds disintegrate. The essences of well-known medical plants, extracted in an unusual way, improve taste and hygienic properties of water.

Keywords: water pollution, electrochemical processing, anolyte, the extracts of medical plants, water purification

The problem of water purity is urgent, especially now. It is known that different adverse conditions impact the state of water biocenoses. One of such factors is the pathogen and opportunistic pathogen pollution of water bodies, caused by the inflow of waste waters from inshore inhabited areas and industrial waters, filled with organic compounds. Soil microflora, washed away by underground and surface waters, bodies of rodents who died of infections also make the water bodies polluted with pathogens [1].

So water is a main transmitting way of many diseases which can be caught not only while drinking the water of poor quality, but also while swimming [2]. When a number of

---

[*] Fax: +7(8512)251718, Mobile Phone: 89608646760. e-mail: sukhenko@list.ru.

pathogenic organisms in the water is high, the term of their persistence in a water basin expends [3].

This must change and ruin any water basin essentially, but there are some processes of self-regulation and self-purification in water bodies only if supporting factors are available. Such factors can be the preservation of the trophic chains of organic substances' uptaking, disinfecting and antibacterial agents in this water biocenosis. We know that many chemical substances of plant cellular fluids have the different antimicrobial characteristics, especially against opportunistic pathogenic organisms. As the dead vegetative and generative parts of land inshore plants (Ailanthus altissima, Robinia pseudacacia, Sophora japonica, Tilia cordata Mill., Populus nigra) fall into water, these antimicrobial extractable substances can influence the state of water basin microflora [3]. It was found that many medicinal herbs (Sophora japonica, Ailanthus altissima, Robinia pseudacacia) show their elective antibacterial properties. The higher summer temperatures and lower the humidity on the habits of such plants are, the higher the concentration of bactericide agents in cellular fluids of these plants is [4].

The methods of water purification are an old problem which is still urgent. The research findings about the activated water («life-giving water» and «dead water») which has rare extra plain water properties were published [5]. Anolyte was examined much and the findings, which confirm its antibacterial (sterilizing) properties in the hospital-infection prevention and the basin water disinfection while electrochemical treatment of water solutions, using the devices and installations of special electrotechnical characteristics, were published many times [6].

It was also determined, that many components of medical and eatable plants have an effect not only on opportunistic pathogenic organisms, including soluble microorganisms in drinking water, but also can improve the quality of drinking water [7].

The staff of the Department of Biotechnology of Astrakhan State University developed the technology of phased drinking water purification. First drinking water is purified by analyte and catholyte solutions, which were produced using the installation of the electrochemical activation of water solutions (EKHA-VIR). Then the microdoses of extractable components of medicinal herbs (blossom clusters Ailanthus altissima, containing tanning agents, saponins, bitter principles, ailanthus; blossom clusters Tilia cordata Mill - essential oils, glycosides, saponins, tanning agents; fruits of Sophora japonica - 8 types of flavonoids (rutin, cempopherode-3-sophoroside, meletin-2-3-rutinoside, genystaine-4, coral bean –bioside etc.), glycosides; the buttons of Populus nigra –salicin and popyline phenol glycosides, flavonoids (pynostrobine flavonoids and pynocembrine flavonoids), organic acids) and the essences of other plants extracted in an unusual way are added. It was found that the pathogens and opportunistic pathogens disappeared under the influence of above mentioned plant substances and the quality of the drinking water improved at nutritive and hygienic levels.

The goals of the phased purification methods are the improvement of waste water conditions and also the aftertreatment of the running water in city and rural areas to prevent the spreading of gastrointestinal infections, transmitting by water and destruction of chlorine-containing compounds.

REFERENCES

[1]   G.P. Stepanov: *"Epidemiology, Clinical Picture And Prevention Of Anthroponosal And Zoonotic Infections"*. – Astrakhan, 1977 (P.21-22).

[2]   V.I. Pokrovsky: *"Epidemiology, Clinical Picture, Diagnostics And Prevention Of Anthroponosal And Zoonotic Infections"*. – Astrakhan, 1982 (P.3).

[3]   L.T. Sukhenko, A.V. Makarova, O.S. Zamyatina: "Ecological And Biological Issues Of The Caspian Sea Basin". – Astrakhan, 2003 (P.269).

[4]   L.T. Sukhenko: *Phytopathogenic Bacteria. Phytoncidology. Allelopathy.* - Kiev, 2005 (P. 298).

[5]   V.M. Bakhir: *"The Medical Methods And Means Of Bacterial Purification And Disinfection"*. - Moscow, 1992.

[6]   S.A. Panicheva: G.A. Petrushanskaya *"The Medical Methods And Means Of Bacterial Purification And Disinfection"*. - Moscow, 1993.

[7]   L.T. Sukhenko: *" Allelopathy And Modern Biology"*. - Kiev, 2006.

ISBN 978-1-60741-599-2

Chapter 10

ADDITIONAL PURIFICATION OF WATER BODIES AND WASTE WATERS WITH CONSORCIUMS OF AQUATIC PLANTS CONTAINING MICROORGANISMS

I. N. Gogotov

Institute of Basic Biological Problems RAS, Pushchino, 142290, Russia

ABSTRACT

Phototrophic and chemotrophic microorganisms and their consortia with water plants (aquatic fern *Azolla*, water hyacinth *Eichonia cassipens*) are capable of accumulating metal ions of Ni, Pt, Ru, Cu, Cr, Pb, Zn, Si, Au and other pollutants. This ability enables using them for purification of agricultural and industrial waste water from toxic heavy metals, hydrocarbon and to obtain rare trace metals. For this purpose it is possible to use growing cultures and immobilized cells or enzyme hydrogenase. Purple bacteria (*Rhodobacter* spp., *Rhodopseudomonas* spp.) are able to accumulate Cu, Zn, Ni and Hg., showing various resistance to these metals. Green algae *Chlorella* spp. and water fern *Azolla*, duckweed Lemna spp. and water hyacinth showed higher ability for biosorption of metal ions.

Keywords: Aquatic plants, phototrophic microorganisms, hydrogenase, heavy metals, hydrocarbon, purification

Many metal ions play an important role in a metabolism of microorganisms, plants, animals and man [1,2]. They are required for synthesis of pigments, electron carriers, enzymes, vitamins and other cell components. Elevated concentrations or deficit of a number of metal ions induce a toxic effect on organisms, causing various diseases or suppressing their growth. Chemotrophic and phototrophic microorganisms, as well as their consortia with plants and lichens, accumulate efficiently the metals from an environment in the amount higher than their requirements for growth, that can be used for bioremediation of waste water and soils, and for the development of technologies of obtaining dissipated expensive metals and homeopathic drugs as well [1-4]. The amount of metals accumulated by them can be different. The absorption mechanisms also vary from simple physicochemical process of absorption and desorption, up to highly specific systems of a transport of metal ions in the cells. More than 16 metals are necessary for the growth of microorganisms and plants and their metabolism. However, biological functions of many metals and mechanism of their operation are not clear. The purpose of this presentation is the analysis of the literary and own data on efficiency of a biosorption and stability to ions of metals of the growing cultures, suspensions of native (immobilized) cells and purification of waste water from different pollutants.

1. BIOSORPTION OF METALS
BY PHOTOTROPHIC MICROORGANISMS

1.1. Purple Bacteria

The purple not sulfur bacterium *Rhodobacter capsulatus* and different species of *Rhodopseudomonas* spp absorb ions of Cu, Ni and Cr from waste water up to 80 – 90 % [1]. Purple sulfur bacterium *Ectothiorhodospira shaposhnikovii*, making a biopolymer [6], is able to absorb more then 90 % Cu, Ni and Zn at a contents 100 mg/l of metal ions in the medium (Table 1). The metal-saturated biopolymer in it reached 25 – 30 % of a dry mass. According to our data, in the presence of 214 g/l Ni^{2+} and 0,00371 g/l Cu^{2+} their absorption was much higher by a biopolymer (60 – 70 %), than by intact cells (30 – 40 %) at pH 6-7.

Metals, absorbed by the cells of purple bacteria or their biopolymers, can be isolated from them, at low pH (~ 2). The reduction of ions metals up to a metal state is possible in the presence of hydrogenase isolated from purple sulfur bacterium *Thiocapsa rosopersicina*, if the intermediate low potential carrier of electrons (ferredoxin, cytochrom C_3 or methylviologene) will be added in a the reaction mixture. Hydrogenase or microorganisms, containing it, reduced ions of Ni^{2+}, Pb^{2+}, Pd^{2+} and Ru^{3+} in these conditions. However, reduction of Cd^{2+}, Co^{2+} and Cu^{2+} ions did not take place, though there are no thermodynamic limitations for this reaction. The rate of metal ions reduction, measured by consumption of H_2, was low (~ 100 nmol/hour) due to the toxic effect of metal ions on the hydrogenase. Investigation of stability of the hydrogenase from *T. roseopersicina* to metal ions has shown, that it considerably depends on temperature and time of an incubation [8,9]. Inhibition of this enzyme in the presence of Cd^{2+} ($K_i = 0,95$) and Ni^{2+} ($K_i = 3,8$ mM) depends on pH too. In the presence of Ni^{2+} and Cd^{2+} ions inactivation of hydrogenase is reversible, whereas Cu^{2+} and Hg^{2+} ions induced an irreversible inhibition [1,8.9].

Tabl. 1. Accumulation and stability to heavy metals of prokaryotic phototropic bacteria

Microorganisms	Metals	Accumulations, %/ Stability, μM
PURPLE BACTERIA		
Rhodobacter capsulatus	**Cu, Ni, Cr**	70-80 %
Rhodopseudomonas spp.	Hg, Te, Se	40-50 %
Rhodospirillum rubrum	Hg	35 %
E. shaposhnikovii cells biopolimer	Cu, Ni, Zn Cu, Ni, Zn	40-45 % 61-99 %
T. roseopersicina	Ni, Cd, Cu	43-48 %
Rh. capsulatus	Hg	0.25 μM
R. rubrum	Hg	0.1 μM
CYANOBACTERIA		
Cyanidium caldarum	Cu Fe Zn Pb	72.0 mg/g d. w. 68.0 mg/g d. w. 3.6 mg/g d. w. 0.4 mg/g d. w.
Synechococcus sp.	Mn, Co, Zn, Ag, Sn, Cs, Hg	55-80 %
Phormidium valderianum	Cd, Co	65-70 %
Nostoc muscorum	Cd, Ti	70-80 %
Anacystis nidulans	Cd	50 μM

A great number of species of the family *Rhodospirillaceae* are investigated with respect of the stability to oxyanions of heavy metals [3]. Species of this family maintain redox-control due to reduction of tellurite (TeO_3^{2-}). For *Rh. sphaeroides* the mechanism of resistance to this group of tellurite involves membrane – bound FAD-dependent metal oxyanion reductase (MOR). As a result the intracellular precipitation of metal and significant hydrogen evolution under anaerobic photoheterotrophic conditions occur. The salts of TeO_3^{2-}

and SeO_3^{2-} (~ 1 µg/ml), added to the cultural medium, kept it from the growth of Gram-negative bacteria.

1.2. Cyanobacteria

Many cyanobacteria have transport systems of metal ions from the environment in the cells. A transport system of Ni^{2+} with K_m = 117 nM, enabling to concentrate it approximately by 2700 times is known for *Anabaena cylindrica*. Absorption of Ni^{2+} depends on membrane potential and decreases in the dark or in the presence of metabolism inhibitors [1]. Cells of *Anacystis nidulans* absorb actively Cd^{2+}. This absorption is inhibited completely by Ca^{2+} and Zn^{2+}. Low concentrations $HgCl_2$ (6 µm) influence on the energy transfer inside phycobilisomes in *Spirulina platensis* and its increased concentrations (18 µm) suppress electron transport from water to methylviologene. The absorption of metals in the cells of *Cyanidium caldarum*, grown in the light, depends on E_h: (mg/g dry weight): Cu (72, E_h 40 mV) > Fe (68, E_h 400 mV) > Zn (3,56, E_h 280 mV) > Cr (1,95, E_h 40 mV) > Pb (0,43, E_h 40 mV).

Passive accumulation of Mn, Co, Zn, Ag, Sn, Sn, Cs, Hg, Np, Pu and Am was determined for *Synechococcus* sp. With concentration factors from 0 (Cs and Np) up to ~ 10^6 in the row: Pu ~ Hg ~ Sn > Am > Zn > Co > Cs ~ Np [1]. Biomass of *Phormidium valderianum* is efficient for biosorption of toxic metals. It absorbs 65 – 70 % of Cd and Co from solution, containing 25 mM each metal ions. Immobilized into polyvinyl particles cyanobacteria absorbed Cd and Co actively. About 80 % bound metal ions were desorbed with 0,1 N HCl.

1.3. Alga

The cells of *Chlorella* and *Scenedesmus* sp. bind most actively Au^{3+}, Ag^+ and Hg^{2+} at pH 2. Irrespectively, live or dead cells are used for biosorption, it takes place their marked affinity to ions of gold [1,2]. The cells of *C. pyrenoidosa* immobilized on the column with silicagel absorb the gold in reversible way. More than 50 cycles of filling the column with tetrachloraurate solution at pH 1,5 and washing it off with HCl and acidic thiourea were replicated without loss of sorbent efficiency. The experiments performed with Au^+-tetrathionate Na and *Chlorella vulgaris* on the same column showed that both algae could be used both for efficient biosorption of gold and its removing. They may be used for extraction of gold from very diluted solutions, including ocean waters.

Relative value of biosorption to general absorption and absorbing capacity of metal ions varies greatly for different types of algae. For *Ankistrodesmus brauni* and *Ch. vulgaris* the binding of Cd with cell walls satisfies 80 % from general absorption (Table 2). *Ch. vulgaris* is the main absorbing environmental component for other metals, including uranium [1]. For living cells of *Ch. vulgaris* the efficiency of binding metals occurs as follows: UO_2^{2+} > Cu^{2+} > Zn^{2+} > Cd^{2+} > Ni^{2+} > Sr^{2+}. For cell walls of *Vaucheria* sp. the order of accumulation was: Cu^{2+} > Sr^{2+} > Zn^{2+} > Cd^{2+}[1], and for marine algae the following order was obtained: Hg^{2+} > Ag^+ > Zn^{2+} > Cd^{2+} [2].

Tabl. 2. Biosorption of Metals by Alga

Green alga	Metals	Biosorption, % d. w.
Chlorella vulgaris	Cu	< 1.55
	Au	2.10
	Hg	<0.01
Chlorella sp.	Ur	0.026-0.029
	Mn	<0.08
Chlorella regularis	Mo	0.06
	Ur	0.39
	Cd	0.023
	Mn	0.017
	Pb	0.29
Chlorella fucsa	Hg, Al, Be, Au, Cu, Pb, Ni, Zn	
Chlorella pyrenoidosa	Cd	0.11
	Pb	2.28
Chlamydomonas reinhardtii	Fe	0.5-6.6
Euglena sp.	Al	0.2-1.8
	Ba	0.0012-0.0134
	Zn	0.0049-0.0210
	Mn	0.0036-0.0414
	Ni	0.0006-0.0056
	Cu	0.0011-0.0097
	Pb	0.0031-0.0366
	Ur	0.0005-0.0013
	Th	0.0011-0.0039

Au^{3+}, Au^+ and other complexes of gold could be efficiently bound by algae biomass; concentration of gold in them could reach 10 % of dry weight. Au^{3+} was shown to reduce up to Au^+ with further elementary gold in colloid state.

Biosorption of metals can be efficient even in the presence of other cations or protons. Ca^{2+}, Mg^{2+}, Na^+, Mn^{2+}, Zn^{2+}, Co^{2+} and Ni^{2+}, but not K^{2+}, inhibit the absorption of Cd^{2+} by *Ch. vulgaris*; Cd^{2+} hinders binding of Mn^{2+}. In various algae the absorption of two-valent ions of Cd, Cu, Zn and Mn is mainly decreased at low pH. However, it depends on concentration of element. *Ch. vulgaris* was shown to be able to absorb Au^{3+}, Ag^+ and Hg^{2+} at pH 2. Density of cells can also influence on efficiency of biosorption and it is reduced with increasing biomass concentration.

2. CONSORCIUMS OF AQUATIC PLANTS CONTAINING MICROORGANISMS

Symbiotic consorciums of aquatic plants containing microorganisms is the object of great attention for protecting the environment from pollutants, obtaining renewable materials and fuel. The aquatic fern *Azolla caroliniana,* the aquatic hyacinth *Eichornia crassipens* and various duckweeds *Lemna spp.* grow at high rates on waste waters purifying them from heavy metals, hydrocarbons, aromatic and organic compounds. Their symbionts (cyanobacteria and purple bacteria) due to the availability of hydrogenase, intensively accumulate metals. Thus reducing their ions (Ni, Pt, Ru, Pd) to a metal state. Both whole microorganism cells and their plant consorciums, as well as purified enzymes in the H_2 atmosphere with intermediate electron carrier being available, are capable of it. Due to the high growth rate (1000 plants in 50 days) the aquatic hyacinth produces during a season 20 – 40 t/ha of dry biomass, which contains in a ton up to 20 – 60 kg NPK and 26 kg of protein, which is essentially rich in aminoacids. Most of the metals (~ 80 %) are accumulated by its roots, while leaves can be used as fodder of cattle, for producing paper and biogas. Duckweed is also able to extensively purify waste waters from organic compounds and heavy metals. Its biomass can also be used as a food additives in fish and cattle breeding. It contains 25 – 45 % protein, essentially rich in aminoacids, 42 % of glucose, valuable pigments and microelements. Azolla is also noted for high growth, producing up to 3 kg/m^2 of aw biomass per month, rich in protein (20 – 25 %), valuable metabolites (phitohormones, vitamins, etc.), and are capable of purifying waste waters from hydrocarbons (4,5 %), heavy metals and toxic algae. Its biomass can be used as a food additive in fish breeding, poultry farming, as well as for enriching soils with organic and other biological active compounds.

Our data showed that aqueous fern *A. caroliniana*, representing symbiotic association with cyanobacterium *Nostoc azollae*, purple (*Rhodospirillum* sp.) and chemotrophic (*Artrobacter* sp.) bacteria, as well as other consortia of aqueous plants with microorganisms (*Eichornia crassipens, Salvinia herzogii*) absorbed actively of metal ions (Table 3, 4).

Table 3. Content of metals (mg/kg) in *Chlorella*, *Azolla* and *Eichornia*

Cultures	Cu		Zn		Pb		Cd	
Chlorella sp.	2083[1]	254[2]	6771[1]	827[2]	208[1]	25[2]	104[1]	13[2]
A. caroliniana	734	86	1405	163	70	8	7,03	0,82
A. caroliniana*	1786	124	3075	214	79	5,52	19,84	1,38
E. crassipens, leaves	1216	83	1419	97	54	3,7	13,5	0,92
E. cassipens, stem	1024	102	1706	170	51	5,1	3,4	0,34
E. crassipens, leaves**	2410	151	3615	227	120	7,6	12,05	0,76
E. crassipens, stem**	1612	95	9677	569	323	19	32,3	1,9

* Growth at 1 – 20 ppm; ** Biomass in blooming; 1) Ashes; 2) Dry biomass.

Table 4. Efficiency of Ni^{2+} absorption by various sorbents

Sorbent	Content, mg/l	pH	Absorption, mg/l
Absorbite	100	6,5 – 7,3	8,1
Azolla caroliniana	1000	6,5	43,4
Eichornia crassipens	2	5,0 – 6,6	11,6
Salvinia herzogii	2	5,0 – 6,6	14,4

The data obtained testify that Azolla absorbed actively metals at their low concentration in the medium (1 – 20 ppm). The absorption of metal ions by Azolla occurred in two stages: active and passive. The metals actively absorbed during the grown of Azolla in ponds or other water reservoirs. Passive absorption takes place when the effluent passes through biomass in filter. In these conditions heavy metals are concentrated in biomass of Azolla as much as 500 – 1000 times within 2 – 7 days of the growth and the content of metals in it reaches 1 % of Cu, Cd, Zn, U, Ag and 0,3 % of Cr^{6+} and Ti. About 40 – 60 % of heavy metals were removed from the water.

Total content of metal in Azolla ashes reached 5 %. Active absorption was carried out at pH 4 – 6 with waste water of electronic and electrolysis industry. At passive absorption the filters made from dried Azolla, bound 3,2 % Ag, 4,2 % Cd, 2,7 % Ni, 2,1 % Zn and 3,9 % Cu, containing up to 3000 ppm metal. Total content of metal in Azolla ashes reached 20. Up to 99 % Cd and Ni was removed from industrial waste water, containing 1383 ppm Cd and 2026 ppm Ni. Maximal amount bound by Azolla metal reached 2,8 %. Up to 74 % Cr^{6+} was removed by Azolla from waste waters, containing it in concentration of 137 ppm. Passive processes of metal binding run at pH 3-9 and they are applied in the case of high concentrated metal wastes in electronic and metallurgic industry. *Chlorella* absorbed most actively Cu, Zn, Pb and Cd, and *A. caroliniana* -nickel (Table 3). The content of metals in plants ashes depended both on the type of plants and its separate parts (leaves, stem) and the stage of plant development.

3. INTRACELLULAR ACCUMULATION OF METALS AND THEIR TRANSPORT

Absorption of metals dependent on metabolism is the more weak process than passive absorption. It is suppressed at low temperatures, absence of energy source, by inhibitors of metabolism and uncouples of an electron – transport system [1,2]. Nature and composition of culture medium and physiological state of the cells have an influence on the rate of their absorption ability [10]. For the metals, required for the growth and metabolism of microorganisms, there are transport systems of various specificity for their accumulation from the medium. However, even unessential metals are also absorbed by such systems. So, in some bacteria and algae Cd enters through Mn – absorbing system. In cyanobacterium *Anabaena sphaeroides* Zn^{2+} and Co^{2+} suppress the active transport of K^+ and Na^+, disturbing permeability of the cell membrane. There is a dependence of resistance of this cyanobacterium on the growth phase. Its cells exhibit maximum sensitivity to Co at the initial phase of the growth, and minimum are in the end. It should be mentioned that Co ions are an order more toxic that Zn ions (0,05 mg/l). Green algae *Solenastrum* sp. is more resistant to toxic effect of Zn^{2+} (0,12 mg/l). At present the mechanisms of transport participating in accumulation of heavy metals by microorganisms are not yet understood. The amount of metal accumulated at energy – dependent transport may be much higher than at passive absorption, though there are exceptions. *Ch. vulgaris, Ch. pyrenoidosa, Spirulina platensis* and *Cyanidium caldarum* bind Ca^{2+} and Mg^{2+} weakly, and this hinders binding of heavy metals ions, that is an important advantage of these algae above ion – exchange resins at their remove from hard and waste water [10].

Energy – dependent absorption of Cr by cyanobacterium *Anabaena doliolum* exhibits two- phase behavior and dependence on concentration of chrome [4]. Unlike other metals which are mainly in cationic forms, Cr presumably exists in the form of oxyanion CrO_4^{2-} which is not blocked by anionic components of bacterial shell. There are only a few data concerning ion chrome transport in algae. The data available testify that green algae retain more chrome, as well as Al and Fe, as compared to brawn and red algae. Epiphytic algae, living on plants, exhibit high affinity to atmospheric pollutants and are capable of accumulating heavy metals from air. The content of Cr and Pb was higher in epiphytic alga *Pleurococcus* sp. in places near the highways.

Cr can enter the cells of yeast through nonspecific anion carriers, permease systems, transferring various anions, like sulfate and phosphate [4]. Some tolerant to chromate mutants of *Neurospora crassa* show strong regenerating properties at sulfate transport. The researches conducted showed that toxicity of CrO^{2-} is due to the specific antagonism to sulfate absorption, whereas the toxicity of Cr^{3+} is the result of antagonism to Fe transport.

As for transport and accumulation of metals and their forms by consortia of aqueous plants with microorganisms no reliable data are available. We (Table 3, 4) and other authors gave only the efficiency of biosorption of a number of metals by some organisms and their organs (roots, leaves, stalks) and showed the possibility of their application for purification of industry and agricultural waste waters [1 – 3, 6]. The data obtained are also important for obtaining medical drugs for treatment of microelemental diseases [7].

REFERENCES

[1]   I. N. Gogotov, N. A. Zorin, O. A. Zadvorny: Accumulation of metals by phototrophic microorganisms and their extraction. In: *Ecology and soils*. Moscow: POLTEX. 3, 1999. 238 p. (in Russian).

[2]   G. M. Gadd Accumulation of metals by microorganisms and algae. In: *Biotechnology* (H.-J.Rehm ed.). VCH: Weinheim. 6 b, 401 (1988).

[3]   M. D. Moore, S. Kaplan: *ASM News*, 60, 7 (1994).

[4]   C. Cervantes, J. Campos-Garcia, S. Devars et al.: *FEMS Microb. Rev.*, 5, 335 (2001).

[5]   Y. Sag, U. Acikel, Z. Aksu, T. Kutsal: *Process Biochem.*, 33, 273 (1998).

[6]   H. Schmeichen, H. Witting, S. Martin: *BioEngineering.*, №1, 38 (1992).

[7]   H. Sigel, A. Sigel (Eds): *Metal Ions in Biological Systems. Concepts on Metal Ion Toxicity*. Mir, Moscow, 1993. 366 p. (in Russian).

[8]   L. T. Srebryakova, N. A. Zorin, I. F. Karpilova, I. N. Gogotov: *Appl. Biochem. Microbiol.*, 33, 282 (1997).

[9]   O. A. Zadvorny, N. A. Zorin, I. N. Gogotov: *Biochemistry*, 65, 1525 (2000). (in Russian).

[10]   W. Ahlf: *Appl. Microbiol. Biotechnol.*, 238, 512 (1988).

INDEX